Energy Efficiency Technologies and Policies for Productive Sectors: Environmental, Economic, and Social – Related Implications

Energy Efficiency Technologies and Policies for Productive Sectors: Environmental, Economic, and Social – Related Implications

Editors

Chiara Martini
Claudia Toro

Basel • Beijing • Wuhan • Barcelona • Belgrade • Novi Sad • Cluj • Manchester

Editors
Chiara Martini
ENEA—Italian National
Agency for New
Technologies, Energy and
Sustainable Economic
Development
Rome
Italy

Claudia Toro
ENEA—Italian National
Agency for New
Technologies, Energy and
Sustainable Economic
Development
Rome
Italy

Editorial Office
MDPI
St. Alban-Anlage 66
4052 Basel, Switzerland

This is a reprint of articles from the Special Issue published online in the open access journal *Energies* (ISSN 1996-1073) (available at: https://www.mdpi.com/journal/energies/special_issues/EETPPS).

For citation purposes, cite each article independently as indicated on the article page online and as indicated below:

Lastname, A.A.; Lastname, B.B. Article Title. *Journal Name* **Year**, *Volume Number*, Page Range.

ISBN 978-3-7258-1327-8 (Hbk)
ISBN 978-3-7258-1328-5 (PDF)
doi.org/10.3390/books978-3-7258-1328-5

© 2024 by the authors. Articles in this book are Open Access and distributed under the Creative Commons Attribution (CC BY) license. The book as a whole is distributed by MDPI under the terms and conditions of the Creative Commons Attribution-NonCommercial-NoDerivs (CC BY-NC-ND) license.

Contents

About the Editors vii

Valeria Costantini, Valentina Morando, Christopher Olk and Luca Tausch
Fuelling the Fire: Rethinking European Policy in Times of Energy and Climate Crises
Reprinted from: *Energies* 2022, 15, 7781, doi:10.3390/en15207781 1

**Anna Barwińska Małajowicz, Miroslava Knapková, Krzysztof Szczotka,
Miriam Martinkovičová and Radosław Pyrek**
Energy Efficiency Policies in Poland and Slovakia in the Context of Individual Well-Being
Reprinted from: *Energies* 2023, 16, 116, doi:10.3390/en16010116 19

**Nicolò Golinucci, Nicolò Stevanato, Negar Namazifard, Mohammad Amin Tahavori,
Lamya Adil Sulliman Hussain, Benedetta Camilli, et al.**
Comprehensive and Integrated Impact Assessment Framework for Development Policies
Evaluation: Definition and Application to Kenyan Coffee Sector
Reprinted from: *Energies* 2022, 15, 3071, doi:10.3390/en15093071 48

Annalisa Santolamazza, Vito Introna, Vittorio Cesarotti and Fabrizio Martini
The Evolution of Energy Management Maturity in Organizations Subject to Mandatory Energy
Audits: Findings from Italy
Reprinted from: *Energies* 2023, 16, 3742, doi:10.3390/en16093742 67

Matteo Piccioni, Fabrizio Martini, Chiara Martini and Claudia Toro
Evaluation of Energy Performance Indicators and Energy Saving Opportunities for the Italian
Rubber Manufacturing Industry
Reprinted from: *Energies* 2024, 17, 1584, doi:10.3390/en17071584 94

**Valeria Costantini, Mariagrazia D'Angeli, Martina Mancini, Chiara Martini and
Elena Paglialunga**
An Econometric Analysis of the Energy-Saving Performance of the Italian Plastic
Manufacturing Sector
Reprinted from: *Energies* 2024, 17, 811, doi:10.3390/en17040811 117

Tamás Bányai
Energy Efficiency of AGV-Drone Joint In-Plant Supply of Production Lines
Reprinted from: *Energies* 2023, 16, 4109, doi:10.3390/en16104109 146

Sandra Cunha, Antonella Sarcinella, José Aguiar and Mariaenrica Frigione
Perspective on the Development of Energy Storage Technology Using Phase Change Materials
in the Construction Industry: A Review
Reprinted from: *Energies* 2023, 16, 4806, doi:10.3390/en16124806 174

**Amad Ali, Hafiz Abdul Muqeet, Tahir Khan, Asif Hussain, Muhammad Waseem and
Kamran Ali Khan Niazi**
IoT-Enabled Campus Prosumer Microgrid Energy Management, Architecture, Storage
Technologies, and Simulation Tools: A Comprehensive Study
Reprinted from: *Energies* 2023, 16, 1863, doi:10.3390/en16041863 206

Crescenzo Pepe and Silvia Maria Zanoli
Digitalization, Industry 4.0, Data, KPIs, Modelization and Forecast for Energy Production in
Hydroelectric Power Plants: A Review
Reprinted from: *Energies* 2024, 17, 941, doi:10.3390/en17040941 225

About the Editors

Chiara Martini

Chiara Martini (Ph.D. in Economics and Quantitative Methods and MSc in Environmental Economics) has worked as a researcher at the Italian National Energy Efficiency Agency—ENEA since 2012. At ENEA, her work covers energy policies, sectorial analyses of energy efficiency measures from energy audits, and the development and analysis of tools to promote energy efficiency at small- and medium-sized enterprises. Previously, she worked as a researcher at the Department of Economics of the University of Rome 'Roma Tre' and was seconded to the Organization of Economic Cooperation and Development, working in the Centre for Tax Policy and Administration, Tax and Environment Unit.

Claudia Toro

Claudia Toro (Ph.D. in Energy Engineering and MSc in Mechanical Engineering) has worked as a researcher at the Energy Efficiency in Economic Sectors technical unit of the Italian National Energy Efficiency Agency—ENEA since 2018. At ENEA, her work covers energy audits, sectorial analyses, and the development and analysis of tools to promote energy efficiency at small- and medium-sized enterprises. Previously, she worked as a researcher at the National Research Council and as a post-doc at the Department of Mechanical Engineering of the University of Rome 'Sapienza'.

Article

Fuelling the Fire: Rethinking European Policy in Times of Energy and Climate Crises

Valeria Costantini [1,*], Valentina Morando [1], Christopher Olk [1,2] and Luca Tausch [1]

1 Department of Economics, Roma Tre University, 00145 Rome, Italy
2 Department of Political Science, Free University Berlin, 14195 Berlin, Germany
* Correspondence: valeria.costantini@uniroma3.it

Abstract: The European Union's relative disregard for the economic, geopolitical and climatic concerns of its peripheral Eastern countries has contributed to making the war in Ukraine possible. Its consequences are now returning in the form of energy dependence and economic instability on the Union as a whole and the risk of economic crisis and deindustrialisation. This should prompt a re-assessment of the EU's strategy towards its eastern neighbours, particularly in the energy and climate policy field. This evaluation starts from the issue of control over cheap energy as a key material foundation of state and interstate power. On this basis, we analyse the struggle between Russia and the European core states over Ukraine in terms of the ability to extract an economic surplus through the unequal exchange of energy. The current escalation should be understood as an attempt by the Russian petrostate to preserve the economic basis of its regime, which is threatened by the prospect of a low-carbon transition in Europe. We conclude that a massive acceleration of the transition away from fossil fuels is the key to economic, geopolitical and climate stabilisation, highlighting possible policy instruments the EU could use to secure its production system and protect citizens' security.

Keywords: climate policy; core-periphery; energy crisis; European Union; Russia–Ukraine conflict; world systems analysis

1. Introduction

Russia's armed invasion of Ukraine started in February 2022, has turned into a protracted, violent war of attrition that is drawing in more and more resources from an expanding North Atlantic Treaty Organization (NATO), whose members are pouring billions into their military forces and increasingly into Ukraine's [1]. The West has also imposed heavy economic sanctions on Russia. Both the war and these sanctions have disrupted global food and energy supply chains, which creates enormous pressures throughout the global economy, in particular on countries of the global political south.

The European Union (EU) states have abruptly discovered the degree of their dependence on fossil fuel imports from Russia. In order to withhold the key source of funds from Putin's government, the EU is adopting increasing sanctions on imports of Russian fossil fuels. The geography of the EU consensus on such decisions is not homogeneous, as while Poland claims to be able to ensure oil supplies to the entire region through its ports [2], Hungary, Slovakia, the Czech Republic and Bulgaria are lobbying against an EU oil embargo for fear of cutting their ties to Russia.

Many European states are hastening to decouple from Russian energy imports, mainly by importing more fossil fuels from other petrostates and building new infrastructure for this purpose, consequently burying their climate ambitions [3,4]. The newly accelerated push for increased extraction of fossil fuels to fill the energy supply gaps accompanies the escalating effects of the climate breakdown that are already being felt by those most affected people and areas of the planet. In this way, the effects of the war are rippling not only through the EU production system but also the global, interstate system and the world economy.

As a matter of fact, the Russia–Ukraine war is impacting all economic and political branches. Therefore, a systemic view combining economic policy and political economy is the way forward to foresee potential peace solutions.

The paper aims to develop a critical political economy analysis of the current set of interrelated crises, with a focus on the role of EU energy policy. To this end, we apply the theoretical framework of world systems analysis to a set of quantitative and qualitative data comprising both historical developments and current events, following a broadly historiographical approach and mixing interpretative instruments from the disciplines of economics and political science

We propose three key arguments that amount to complementary and mutually consistent, even if not exhaustive, interpretations of the current conflict: (i) it is a geopolitical struggle between Russia and the core EU members over access to cheap energy sources and the ability to extract an economic surplus from the Eastern European peripheral members through trade; (ii) it is an attempt of the Russian petrostate to preserve the economic basis of its regime, which is threatened by the prospect of a low-carbon transition in Europe; (iii) it is a sign of the vulnerability that the EU has partly subjected itself to and partly exported to its Eastern periphery by way of adopting a half-hearted and self-centred approach to energy and climate policy. The politics of energy is thus at the core of all three perspectives on the current conflict.

Ukraine is at the centre of this conflict because it is situated simultaneously in the periphery of the EU and that of Russia, de facto the corridor for Russian energy exports to Europe. The country has been the locus of the struggle between these two power blocs for decades, not least because of its rich energy and mineral sources, as well as its fertile soils. It is an important market for both Russia and the EU and is geographically located as the transit country for the flow of several raw materials between the East and the West. However, the world-systemic position of Ukraine is not fundamentally different from that of other countries in Europe's Eastern periphery or in Central Asia. Similarly, the strategic situation in which Russia finds itself may be comparable to that of other Petrostates, such as Nigeria or Saudi Arabia [5]. Accordingly, understanding this conflict is essential to draw more general lessons for the EU's energy policy at the intersecting objectives of geopolitical stability, energy security, and the prevention of climate collapse in order to reduce the risk of additional transnational crises.

The remainder of the paper is organised as follows. Section 2 traces the political reasons behind the vulnerabilities of EU–Russia–Ukraine relations throughout history according to a world-systemic approach. Section 3 applies this world-systemic view to interpret the current crisis's dimensions and challenges. Section 4 discusses the main implication for the EU, and Section 5 concludes with some policy implications on the EU climate and energy security.

2. World Systems Analysis Framework Applied to the Russia–Ukraine Crisis

2.1. The World Systems Analysis and the Types of Vulnerability

The theoretical framework of world systems analysis is the best way to understand the global political economy and international relations as a core-periphery constellation of "combined and dependent development" [6]. The key tenet of this framework is that the industrial "core" of the world system extracts an economic surplus in the form of embodied energy, land, labour and raw materials from the extractive "periphery" through the unequal exchange while exporting environmental destabilisation [7,8]. Access to this surplus allows the polities of the core to create a degree of internal socioeconomic order and security. This occurs at the expense of the periphery, whose extractive political economy implies both lower political stability and higher vulnerability to external economic shocks. The industrial core of the world system is thus able to enhance its own socioeconomic stability by displacing instability and vulnerability to the periphery [9].

In the present context, a core region of the global capitalist world system—Western Europe—has been seeking to expand its control over the supply of energy, resources, agricultural land and cheap labour as well as over export markets in its Eastern European periphery. This process started when the European core integrated large parts of this Eastern periphery (including Poland, Hungary, the Czech Republic, Slovakia, and parts of the Balkan and the Baltic states) into the EU. These states occupy a distinct position best described as an internal periphery of the EU. Other parts of Europe's Eastern periphery (including Ukraine and Moldova) have not been integrated into the EU, putting them in an even more disadvantaged position vis-à-vis the EU core.

Between the core and the periphery lie "semi-peripheral" states, which struggle to establish a degree of independence from the core and control over a periphery of their own. Russia occupies such a semi-peripheral position in the world economy [10]. Semi-peripheral countries are to some degree dependent on resource exports to the core but are sometimes also able to challenge their domination in some areas (e.g., Brazil, South Africa, Turkey or Saudi Arabia).

The key vectors of surplus extraction in the capitalist world economy, and thus important objects of international competition and conflict, are access to and control over cheap sources of energy, land, labour, raw materials and ecological sinks [11]. States strive to structure trade to perpetuate the Ecologically Unequal Exchange (EUE) of these resources [12,13]. The EUE theory posits that the asymmetric transfer of resources from the periphery to the core and the diverging compensation lead not only to high economic growth in the core [14] but also to underdevelopment and environmental degradation in the periphery [15]. Energy is a key driving factor of a EUE [16] as revealed by Figures 1 and 2, which show the energy embodied in all net imports and exports per capita, and the trade-in value added per unit of energy embodied in exports in 2015 from the Eora26 multi-regional input–output model.

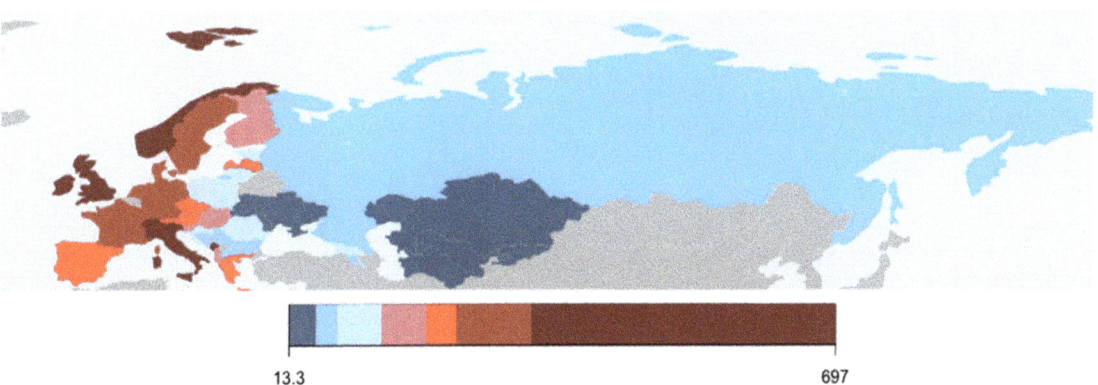

Figure 1. Trade in Value Added (USD) per unit of energy embodied in exports (TJ) (own elaboration on Eora26 [17]).

Both maps demonstrate the core status of Western Europe, the semi-peripheral status of Russia, and the existence of a post-Soviet periphery in Eastern Europe and Central Asia. This unequal distribution of surplus and the dependence of both core and (semi-) periphery on cross-border flows of energy tend to give rise to unequally distributed socioeconomic and political vulnerabilities. We trace the emergence and the distribution of three types of vulnerability: (i) the ability (or lack thereof) of a government to maintain legitimacy by pacifying social conflicts and creating consent; (ii) the economic and social instability that hinges on the access to a secure supply of energy; (iii) the exposure to direct effects of the climate and broader ecological crises.

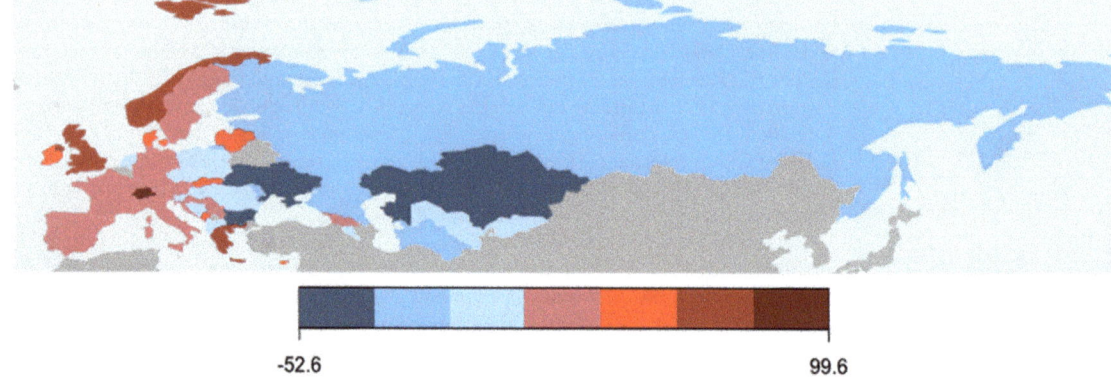

Figure 2. Net imports (negative values: exports) of embodied energy (GJ) per capita (authors' elaboration on Eora26 [17]).

The first vulnerability is typically related to the position of core states that can maintain the modicum of class compromise and popular consent required for liberal democracy when they distribute the surplus drawn from the periphery. In semi-peripheral states, the surplus is distributed among elites, which is sufficient to stabilise an authoritarian system. Lacking such a distributable surplus, peripheral states cannot follow either of these two models of regime stabilisation and typically experience a higher degree of conflict over control of the state [18].

The second type of vulnerability depends on the fact that core states use their economic and political leverage to maintain their bilateral trade agreements with semi-peripheral Petrostates, benefiting from a more secure energy supply, a more diversified energy provider base, and consequently a reduced vulnerability to external supply shocks. Petrostates, in turn, can use their advantageous position as energy-exporting countries, being vulnerable only to external demand shocks. Lacking such advantageous bargaining positions, peripheral states cannot follow either of these two models, and thus are often characterised by a concentrated mix of sources and providers and a high vulnerability to external supply shocks.

The third source of vulnerability is related to the direct effects of the climate and the broader ecological crises. Greenhouse Gas (GHG) emissions are a case of ecologically unequal exchange from which the core gains more than the periphery, while the latter is asymmetrically more vulnerable to the consequences of climate change [19]. Additionally, the overutilisation of the environmental space of the core suppresses resource use and consumption in the periphery [20] as well as increases the vulnerability to ecological impacts in the latter [21].

These types of vulnerability interpreted through the lens of the EUE framework should be complemented by a deep understanding of the historical roots of the Russia–Ukraine crisis and the current distribution of weakness in political stability to inform the EU debate on how to design the way forward for solving the international political deadlock.

2.2. When and Where Does Vulnerability Originate?

One key to the emergence of vulnerability in modern societies is their dependence on the consumption of high levels of energy, including in manufacturing and service sectors. Most core economies are not capable of internally supplying the demanded energy at sufficiently low costs; instead, depending on cheaper international imports to ensure the profitability of industries, high levels of household consumption, and continued economic growth [22]. In policy debates, this vulnerability is broadly conceptualised as a question of energy security, starting from the early 1970s after the oil crises. EUE theory adds to this debate by providing evidence on the position of semi-peripheral energy exporters as exploited economically by core countries and industries.

Russia is the world's largest exporter of fossil fuels, with most of its exports flowing to Europe. The Russian state is highly dependent on revenues from oil, gas and coal exports, representing 25% of the GDP and 45% of the federal budget [23]. Those numbers are even larger when considering the energy embodied in its industrial exports, as reported in Figure 1. This corresponds to the typical social configuration of a petrostate. Externally, such states typically depend heavily on military power, while internally, they tend towards an authoritarian model based on elite (rather than broad popular) consent of the governed [24]. Both Russia and Eastern Europe are similar in their economic positions relative to Western Europe since they represent net exporters of energy, food, and resources, with a relatively cheaper labour cost and a lower economic complexity. More importantly, Russia's share of the global domestic product has been halved over the last decade, strongly reducing any real prospects for these semi-peripheral states to achieve the core status [25,26]. At the same time, Russian economic power was historically based on controlling a periphery of its own in Eastern Europe and Central Asia. This control has been curtailed since the collapse of the USSR, including through NATO expansion to the East. Accordingly, Russia's loss in economic relevance at the world level, combined with the loss of control over the Eastern European (now semi-) periphery, has fostered the emergence of ultranationalist ideology and the neo-imperial revisionism of the Russian regime, which ultimately resulted in its invading Ukraine [27].

Ukraine is the largest country both in Europe's Eastern periphery and Russia's Western periphery. Its economy heavily depends on cereal, seed oil and iron exports. Besides fertile agricultural land, it has abundant resources such as oil, natural gas, coal and some critical minerals required for renewable energy and digital technologies, significant hydropower and biomass potential. Moreover, it exports steel and other metal products that embody large quantities of energy, much of which is originally imported from Russia. As a result, Ukraine is one of the world's ten most energy-intensive countries, with an energy intensity exceeding the OECD average [28]. Its large population, high energy consumption and its geographical position make Ukraine one of Europe's largest energy markets and the country with the most natural gas transits in the world. At the same time, its geographical position turns Ukraine into a ground of extreme competition between the EU and Russia over the ecologically unequal exchange.

Hirschman's theoretical approach thus seems to ring more clearly than ever in this instance: asymmetry confers power on the stronger power, forcing the smaller power to converge on the interests of the stronger [29]. Ukraine, in this case, as a peripheral country, found itself caught between two fires. Russia, as a semi-peripheral petrostate that has historically dominated Ukraine until recently and used its energy and economic power to include it in its own arrangements, and on the other side, the EU, a core of the world system, which with the promise of greater integration and economic support, has taken advantage of its role in the energy market, postponing and never ensuring adequate protection [18].

3. Ukraine as an In-Between Energy Node in EU-Russia Relations

3.1. The EU-Russia Energy Chain

The energy relations between Europe and the Soviet Union can be traced back to the late 1950s, including the building of pipelines to transfer oil and gas. The integration of Russia as a key energy supplier into the European semi-periphery began in the late 1990s and culminated in the opening of an energy dialogue between the two parties in 2000 [30].

In the first decade of the 21st century, the EU was hardly concerned about its dependence on Russian fossil fuels and energy security in general [31]. Despite this diffused concern, most regulations and directives in the past decades were directed towards liberalisation, trade deregulation and free market mechanism, de facto stimulating the relative convenience of investing in cheaper Russian energy sources [32].

After the enlargement process of the EU in 2004, the deeper integration of parts of the Eastern periphery into the liberalised common market increased the access for Western Europe to cheap imports of embodied labour and energy, much of which were based on Russian gas imports [33]. While many of the new post-Soviet EU members were already concerned about their high dependency on energy imports, core Western states used their bargaining power within the EU to maintain bilateral trade agreements with Russia improving their position in the hierarchy of EUE by bargaining for relatively low energy prices while ensuring preferential supply channels. The vulnerability thus remained high for the entire EU but was largely felt by the newly integrated Eastern periphery [34].

What forced the EU to amend its energy security strategy was the external threat of the gas dispute between Ukraine and Russia in 2006 and the credible risk of a gas cut-off [35]. This wave of change in 2004 also affected Ukraine, which, for the first time after a decade of a divisive but always Russia-friendly policy, experienced more open forms of conflict. The success of the "Orange Revolution" and the victory of a pro-European party in the presidential elections moved Ukraine towards the EU and NATO, competing with the economic concessions made by Russia to integrate Ukraine into its periphery.

In particular, Russia had been setting the prices for its gas and the respective transit taxes for Ukraine below European and world levels for years. After 2004, Russia began to put economic pressure on raising fuel prices for Ukraine and other peripheral countries (Estonia, Georgia, Moldova, Latvia and Lithuania). Following the refusal of Ukraine to pay these higher prices, the reaction was a cut in supplies to Ukraine, specifying, however, that the gas deliveries to the European core would not be affected. Nevertheless, the drop in delivered volumes was soon felt across the whole EU. Hungary lost up to 40% of its supplies; Austrian, Slovakian, and Romanian supplies fell by 33.3%, France's by 25–30%, Italy's by around 25%, and Poland's by 14%. The crisis did not last long, as the European outcry forced the Russian government and Gazprom to reach a provisional agreement with Ukraine. However, the text of this agreement reveals that many issues, particularly the price of gas, remained unresolved [36,37].

This was a turning point also for the EU, recognizing the still high vulnerability to external energy supply disruptions and the low effectiveness of its past energy security policies. Partly as a result, in 2014, the EU proposed the 2030 Framework on Climate and Energy, aiming at a 40% reduction of GHG emissions (from 1990 levels), a 32% share for renewables energy in the energy mix and at least 32.5% improvements in energy efficiency [38]. Within this renewed energy package, the increase in energy security was focused on suppliers' diversification and the enhancement of the interconnections and coordination among member states. Shortly after the delivery of the 2030 energy agenda, the debate upon an Energy Union emerged as the way forward for a coherent energy security strategy within the EU. It was finalised in the "Framework Strategy for a Resilient Energy Union with a Forward-Looking Climate Change Policy" [39].

The strong impact of the EU approach to climate and energy issues on global negotiations also threatens the already difficult relationships between Russia and its energy-importing partners as a side effect. The stability of the Russian economy is fundamentally based on a compromise between the state and the oligarchy that structurally depends on revenues from fossil fuels. This basis has been threatened in recent years by the decarbonisation objectives of the Paris Agreement, requiring an immediate reduction in the exploitation of fossil fuels reserves worldwide, absent feasible large-scale carbon removal technologies, and a swift phase-out of fossil fuels. At the same time, a large part of the past EU decarbonisation strategy was founded on the transition to massive use of natural gas, as supposedly less damaging to the climate than their substitutes, coal and crude oil.

3.2. Digging under the Surface: Ukraine in between

According to the evolution of EU-Russian energy relations, one element of the backdrop against which Russia invaded Ukraine is the paradoxical situation that it can no longer count on the stability of Europe's demand for its fossil fuels after the 2030s, but it can indeed rely on the dependence of Europe on its supply over the next years. As already discussed, all transition scenarios for the EU had foreseen considerable amounts of Russian gas in the energy mix. While high dependence on Russian natural gas was considered a concern for Europe regarding energy security, the continued reliance on fossil fuels remained unquestioned. Large pipeline projects such as Nord and South Streams indicated that the EU remained interested in deepening its energy relationship with Russia, as well as remained interested in the consumption of fossil fuels. However, recent developments demonstrate a break with the half-hearted approach to effectively reduce fossil fuel consumption. As the global climate justice movements put pressure on policymakers to act upon their responsibility, the EU responded with the European Green Deal and the more recent Fit for 55 package, aiming to be the first climate-neutral continent [40,41]. Within this renewed climate and energy package, the phase-out of conventional gas was supposed to bring the gross inland consumption to 22% in 2030 and 9% in 2050. Consequently, with its fossil fuels export-dependent economy and the shrinking demand for natural gas from its major trading partner in the following years, the Russian economy is on track to vanish as a portion of global GDP.

This is related to the second internal factor. Suppose Europe is serious about reducing its carbon emissions to net zero by 2050. In that case, Russia will not be among the last sources of fossil fuels to be abandoned, as the production costs of Russia's ageing oil fields are generally higher than those of West Asia. At the current price level, only a third of Russia's proven reserves will be profitable to extract [42]. The underlying physical mechanism can be expressed in terms of the Energy Return On Investment (EROI) of fossil fuels, which expresses how many units of energy are required to produce one more unit of energy [43]. Although the Russian fossil industry is attempting technological improvements, in the last years, the EROI of gas has been declining, from 1:84 in 2015 to 1:83 in 2008 to 1:74 in 2016 [44]. As the EROI has been identified as the strongest determinant of the growth of the gas industry and, indeed, of Russian GDP [45], this constitutes a true political threat to Russia's economic and political stability. Additionally, unlike other petrostates, Russia has not made serious attempts at scaling up renewable energy production as a viable solution to replace the reduction in export rents from the exploitation of exhaustible resources [46]. Provided that all transition pathways to meet the goals determined in the Paris Agreement require a swift phase-out of fossil fuel combustion, as well as net zero pledges by more than 130 countries, together responsible for around 88% of global carbon emissions, will impede Russia to replace the EU with other importers in natural gas trade. Even if demand from Asia and elsewhere may partly compensate for Europe's declining demand for fossil fuels, the combination of downward pressure on prices and the heightened uncertainty fundamentally threaten Russian fossil capital and, with it, the entire political economy of the Russian state [47].

The elements discussed below allow us to formulate a clear picture of the escalation of the conflict in Ukraine. Russian fossil capital and the associated regime, facing the spectre of a long-term decline, are currently minimizing their losses by benefiting from the short-term window of opportunity provided by the post-pandemic recovery phase. European industries' hunger for immediately available energy sources during the recovery from the 2020 production collapse, associated with a sudden rise in fossil fuels prices on the international markets, meant soaring extra profits from Russian energy exports.

Clearly, the Russian state's bargaining power in international conflicts depends on European fossil fuel consumption. Against this backdrop, it becomes clear that the Russian invasion of Ukraine itself has only been possible because the EU has, on the one hand, credibly committed to exiting fossil fuels but, on the other hand, been hesitant to transition fast enough to be independent of Russia. Consider a counterfactual scenario in which the EU had taken drastic measures to exit fossil fuels already in the 1990s. It would have drastically reduced its fossil fuel consumption and rolled out more sustainable provisioning systems and renewable energy. This would have given the European core and periphery much more bargaining power vis-a-vis Russia and withheld significant funds from the fossil fuel complex that helped stabilise and radicalise Putin's autocratic regime. In this scenario, the Russian state, starved of fossil revenues, may have adopted a rather different and probably much less aggressive model of consolidating the nation state's power.

4. The Way Forward for EU Energy Security

4.1. Energy Security: Europe Is Licking Its Wounds, Not Treating Them

In order to determine the overall degree of the EU energy security, it is essential to consider its energy mix. The energy mix is expressed as the share of gas, oil, coal, nuclear or renewables in gross available energy, the overall supply of energy for all activities on the territory of the country. Figure 3 represents the EU energy mix in 2020. It demonstrates to what extent the EU still relies on fossil fuels such as natural gas and oil but also offers a perspective on the progress toward the sustainable transition based on the share of renewables in the energy mix.

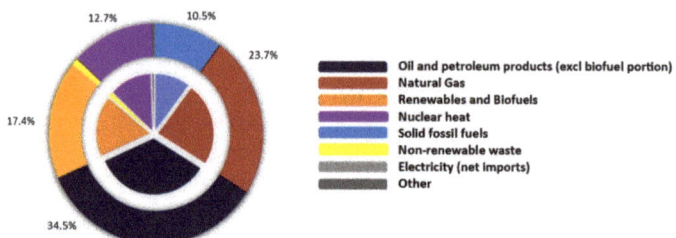

Figure 3. The EU energy mix in 2020 (authors' elaboration on [48]).

According to the statistics provided by Eurostat [48], the EU has not managed to reduce its overall dependence on natural gas and oil. While the share of renewables has increased and the share of nuclear power has decreased, natural gas and oil remain the largest components, making up almost 60% of the entire energy mix. The EU imports almost all of these fossil fuels, as only 42% of the entire gross available energy in the EU in 2020 were produced domestically, while the other 58% was due to imports, with almost 25% coming from Russia.

Furthermore, EU policy has failed to significantly increase its share of renewable energy within the energy mix. The current 17.4% share of renewables represents a partial failure to comply with the ambitious energy packages ratified in 2009 that aimed at a share of 20% renewables by 2020. Moreover, in the face of such numbers, it seems unrealistic that the EU targets of a 40% share of renewables by 2030 included in the Fit for 55 package will be accomplished, given the current path of investments. Unless radical changes occur in the upcoming years, it seems unlikely that the EU can achieve any of its targets by 2030. Lastly, these numbers perfectly fit with the large bilateral fossil fuel trade agreements between the EU core and powerful petrostates, including Russia, made to secure an energy supply at affordable prices, as discussed above. The expected benefits in terms of security of supply have been one of the major causes of delay in redirecting massive investments toward renewable sources, given the price competition of natural gas over renewables, especially in electricity production.

Related to this point, also the EU import dependency deserves attention. It highlights the extent to which a country relies upon imports in order to meet its energy needs. According to Eurostat [48], in 2019, the total energy import dependency for the EU was 60.7%, with 90% for natural gas and 97% for oil, reaching a net import dependency level among the highest in the past 30 years.

These figures represent an ambivalent position of the EU in the world system. On the one hand, as a net importer of energy, it benefits from the ecologically unequal exchange and maintains high standards of living [16]. On the other hand, the extremely high import dependency and the extremely high energy supplier concentration make the EU vulnerable to external shocks. As of 2019, the EU imports 41% of its natural gas, 27% of its crude oil and NGL and 46% of its hard coal from Russia. According to Eurostat Energy Balances, this dependency trend from Russia has even increased over the past two decades, standing in sharp contrast to the EU energy security strategy aimed at reducing its vulnerability to external shocks. Thus, the EU energy security is still characterised by a concentration of sources and suppliers, with Russian fossil fuels still being the bedrock of European prosperity. More importantly, Figure 4 shows the intra-EU division between the internal periphery and the core.

While the periphery is eager to decrease its vulnerability in terms of energy imports from Russia out of national security concerns, the core member states have been using their economic leverage to work out alternative advantageous bilateral energy trade agreements. Both for natural gas and oil imports, Europe's periphery has a much higher dependence on imports from Russia (Finland, Lithuania, Latvia, Estonia, Slovakia, Bulgaria, and the Czech Republic rely on Russian gas with at least 85% of their domestic natural gas consumption), whereas core members have a much more diversified mix of providers. In particular, core members such as Spain, Portugal, France and, more recently, Italy have a much more diversified energy import strategy than most of the peripheral EU members.

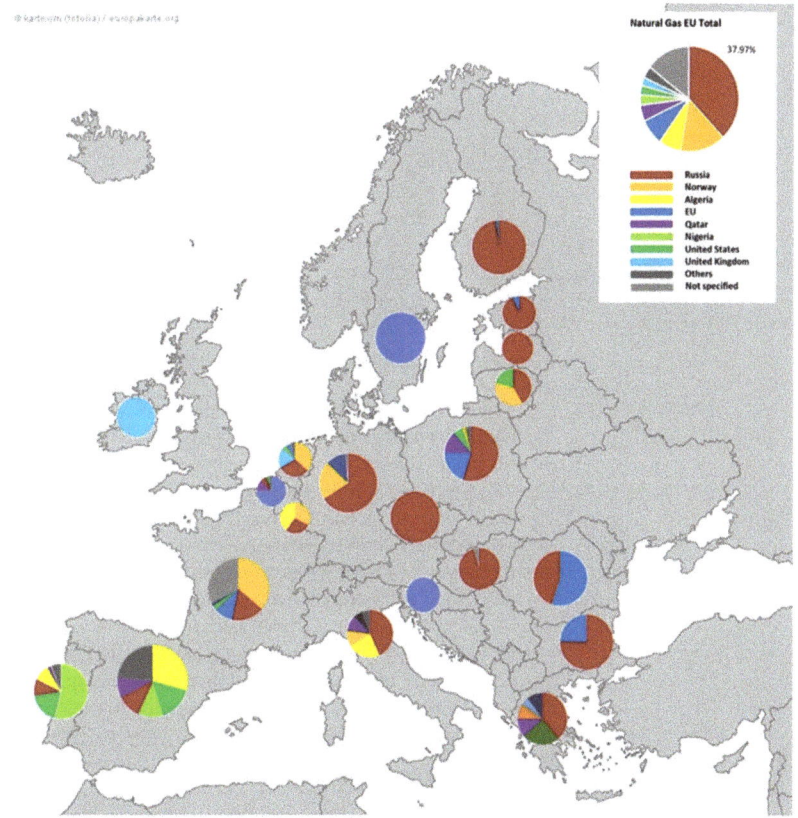

Figure 4. *Cont.*

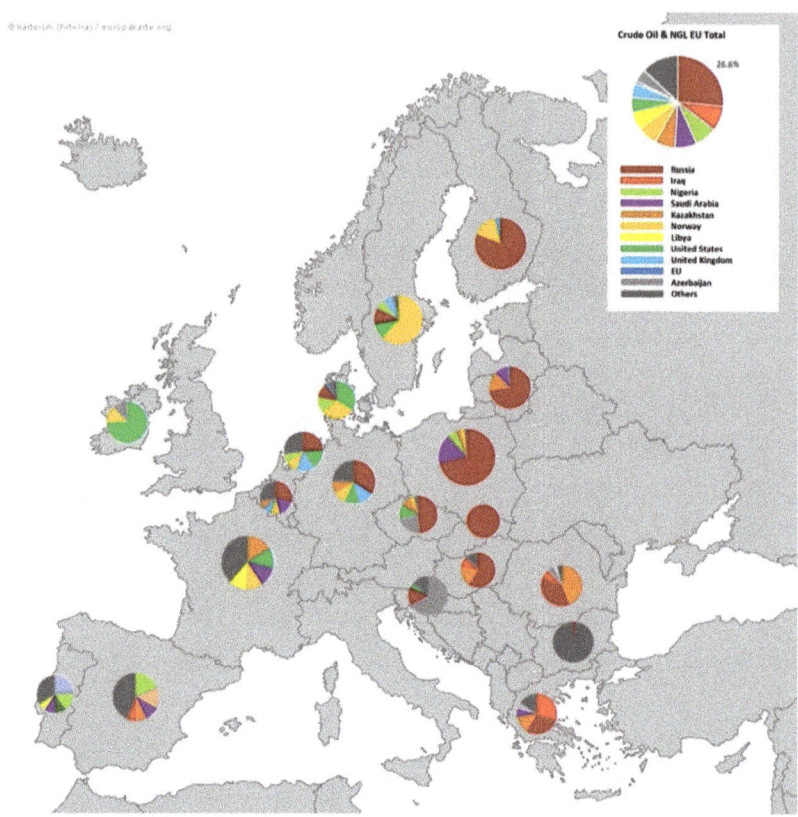

Figure 4. Share of EU import origin for natural gas and crude oil (authors' elaboration on Eurostat data).

Weak bargaining power, unreliable trade agreements, and lower fiscal capacity render the Eastern periphery relatively less capable of both preventing disruptions in the inflow of energy and dealing with their consequences. The first EU members to face an end of Russian fuel supplies over the summer of 2022 were Poland and Bulgaria [49,50]. Moreover, replacing Russian fuels and stabilizing energy prices through subsidies puts great demands on governments' balance sheets, and core members outbid Eastern EU members with larger fiscal space, particularly Germany [51]. Additionally, as a result of the ongoing war in Ukraine, food prices have increased sharply all over the world. Within the EU, people in the Eastern Periphery have been hit hardest by the sharp price increases [52]. Despite the urgency for an Energy Union strategy already discussed in 2015 [39], the reluctance of core members to effectively commit to a unified approach with respect to external relations has maintained the intra-EU division, asymmetrically felt by the Eastern periphery in terms of higher vulnerability to external energy supply and price shocks.

4.2. Geopolitics: Fuelling the War

While the dependence on energy imports from Russia renders the EU vulnerable to supply shocks, Russia is also dependent on European demand. In particular, the capacity of the Russian state to create and spend roubles on military activities indirectly depends on continued fossil fuel imports. This is not a matter of sustainability of public finance narrowly conceived; what matters for the fiscal policy space of a semi-peripheral country is, first and foremost, its balance of payments [53]. Figure 5 suggests that the largest share of Russia's constant current account surpluses over the period from 2000–2022 can be attributed to its fossil fuel exports [54].

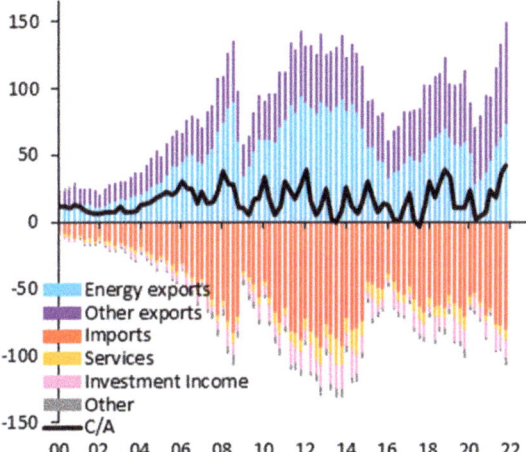

Figure 5. Composition of the Russian balance of payments (in USD billion) [54].

Oil and gas make up roughly 42% of Russian export volumes [55], and the EU accounts for 50% of those exports [56]. Furthermore, roughly 70% of exports from Russia to the EU are mineral products, with crude oil and refined petroleum accounting for roughly 55% and petroleum gas accounting for 10% [57]. Thus, it can be demonstrated that Russia's current account surplus is heavily financed by energy exports and, most importantly, by the energy imports of the EU that make up a large share of Russia's national income. Since the start of the war on February 24th, Russia has doubled its revenues from fossil fuels since it invaded Ukraine, approximately half of which come from the EU [51,58].

Continued exports of fossil fuels from Russia maintain the demand for its currency. This has stabilised the rouble's exchange rate despite massive sanctions, including on the Russian central bank, and thereby the Russian capacity to import goods necessary to make war. Only if both prices and quantity of fuels declined would the rouble crash. Combined with the current sanctions in place, this would make capital-intensive imports more expensive, and inflation would be imported, possibly threatening Russia's political economy and requiring deflationary adjustments. This would undermine both elite and popular consent (and what is left of a class compromise) on which the regime rests; a dangerous situation, especially with significant numbers of young male workers being armed. That is why the EU has indirectly financed the current war against Ukraine [59]. Only a decline in both international demand for fossil fuels and related prices would significantly hamper Russia's ability to sustain the war.

4.3. Climate: Fighting Fire with Fire

Three days after Russia's invasion, the Sixth Assessment Report of Working Group III of the Intergovernmental Panel on Climate Change (IPCC) was released. In the words of Svetilna Krakovska, the leader of the Ukrainian delegation of climate scientists to the IPCC, "almost half of the world's population is already experiencing the effects of climate change [. . .] we have one last chance to be climate resilient. But the window for action is getting narrower and narrower, and now with this war, it is closing". The report identifies 2030 as a crucial deadline. Before that year, the world must achieve a substantial reduction in emissions, and by 2050 zero net carbon emissions and a drastic decrease in all other greenhouse gases are required.

The EC's Fit for 55 package revises its target share of renewables in the energy mix for 2030 upwards from 32% (the goal set in 2019) to 40% [41]. However, unless radical changes occur in the upcoming years, it seems unlikely that the EU can achieve any of its targets by 2030. The current share of 17.4% renewables clearly represents a failure to comply with the ambitious energy packages ratified in 2009 that aimed at a share of 20% renewables by 2020. The numbers show a partial unwillingness to take climate change seriously over the last 20 years. The bilateral fossil fuel trade agreements that gave the EU core states privileged access to cheap fuels from Russia and other petrostates have played a key role in delaying the necessary transition to a more sustainable energy mix.

For the first time, the latest IPCC report explicitly mentions fossil fuels as the direct cause and economic growth as the structural driver of the climate crisis. Fossil fuels must be phased out as quickly as possible to limit the possibility of catastrophic climate collapse. Although the scientific community has established this fact for decades, the vast majority of countries have not only kept extracting and consuming fossil fuels. The war has, in fact, accelerated the extraction of fossil fuels. If the Russian state is stabilised by oil, gas and coal revenues, then, to destabilise the Russian war machine, these revenues should be cut. However, Western countries argue that it is not possible to meet the current level of energy consumption by replacing fossil fuels with renewables. As long as a change in the level of energy consumption is not part of the discussion, the only logical option then is to replace Russian fuels with fossil fuels from elsewhere.

This is precisely what the EU has been concentrating its efforts on. The REpowerEU program presented by the EC in March 2022 [60], which aims to reduce Europe's dependency on Russian gas, plans to spend EUR 195 billion to stop importing Russian fossil fuels by 2027, combining a faster rollout of renewable energy and energy savings with a switch to alternative gas suppliers and increased use of coal. Gas is supposed to be supplanted and partly replaced by hydrogen, at least partly from renewable sources. However, the goal for the share of renewable energy in the energy mix is set at 45% by 2030, which is just a 5% increase compared to the current target of the EU climate policy. Thus, the plan still envisions Europe as a massive importer of fossil fuels and focuses on identifying alternative countries that could become the basis for Europe's future supply of fossil fuels.

This approach is not conducive to stabilisation—neither of the geopolitical nor of the economic, let alone the climate crisis. The key reason is that it keeps up the overall demand for fossil fuels in the short run and locks Europe further into the fossil age, possibly for decades. This problem is aggravated as energy supplies cannot be increased in the short run without increasing emissions, while the rollout of renewables requires time (and consumes energy itself). Hence, the only way to increase supply quickly is to increase the extraction of fossil fuels.

Consequently, there are only two possibilities: either the supply of fossil fuels from West Asia, North America and elsewhere is increased, which might push down prices and thereby hurt Russia by rendering most of its reserves unprofitable to extract. However, this would imply rising emissions, which would be catastrophic for the climate. The second option is that supply does not rise but that some supply of oil and gas from West Asia and elsewhere is diverted towards Europe, and the former consumers of that gas and oil now buy from Russia. This would stabilise or even increase high fossil fuel prices, thereby benefitting Russia and other petrostates while stabilizing emissions at exorbitantly high levels. The construction of infrastructure, like pipelines and LNG terminals, would require large investments that could instead go into the transition away from fossil fuels. Ironically, the EU actually plans to fund the investments that are part of the RepowerEU program by selling more emission certificates [4].

5. Discussion

World systems analysis suggests an unequal distribution of both the economic gains and the geopolitical, economic and climatic vulnerabilities resulting from trade with Russia to the relative benefit of the EU core and the relative expense of the Eastern periphery. The complex framework of domestic and external sources of vulnerability, together with the network of international and intersectoral connections that nowadays characterises the global value chain, make clear that the route for a peaceful resolution of the war should be complemented by rethinking the energy and industrial strategy of the EU. Leaving behind the reflections on diplomatic actions, we here focus on the specific actions related to the energy system.

5.1. The Need to Exit Fossil Fuels

Over the last 20 years, the EU has failed to implement policies effectively to secure its energy supply and transition to a zero-carbon economy, partly contributing to the continued existence of semi-peripheral petrostates, rendering itself vulnerable to ecological and environmental damage impacts, asymmetrically felt by the Eastern periphery.

While ecologically unequal exchange theory and world systems analysis focuses on the economic exploitation of the periphery, the core is also vulnerable to any disruption in the inflows of cheap energy. Importantly, the resulting vulnerability is not equally distributed between the EU's core and its internal periphery. Empirical evidence demonstrates that the EU is still largely dependent on fossil fuels in its energy mix, its import dependency remains significantly high, and the concentration of Russian fossil fuels is still large. Dependence on Russian gas is asymmetrically distributed and largely felt by the Eastern periphery, which is much more vulnerable to external energy supply shocks. Additionally, current and past energy payments have and are still financing the ongoing war in Ukraine by stabilizing both Russia's federal budget and its exchange rate.

The Eastern European periphery still bears the brunt of the geopolitical risk associated with a prolonged war or even possible further Russian expansions. It also suffers much more from the current rise in energy and food costs than the European core, which has far higher levels of GDP. Their access to energy has never been carefully secured with appropriate policies. Moreover, Eastern Europe has absorbed most of the Ukrainians who were forced to flee from their homes. More than ten million refugees have migrated to Poland, Hungary, Romania, Slovakia and Moldova (as of 20 September 2022) [61].

On a broader level, the war has been impeding any serious efforts at mitigating global heating. Dealing with the climate crisis and trying to solve it must, first of all, involve peace and cooperation. Instead, a return to global power blocs seems imminent. While NATO members are collaborating on sanctions and rearmament, precisely the sanctions imposed through the dollar system have already inspired closer collaboration between BRICS states (Brazil, Russia, India, China, South Africa) on ways to circumvent Western sanctions on the Russian payment system [62]. Notably, India and China have begun to use the Rupee, the Renminbi and the Rouble as units of account for bilateral trade with Russia, thereby undermining global dollar hegemony [63]. While a transition to a more multipolar economic world order may or may not be considered desirable from a critical perspective on world systems, effective climate policy is likely to be further delayed by the formation of a bloc of authoritarian Semi-peripheral states with stakes in increased fossil fuel combustion.

This aspect also highlights a good reason to move beyond a world system based on unequal flows of energy. Without drastically reducing the use of fossil fuels, vulnerabilities are likely to persist in all three dimensions: petrostates will be trapped in a precarious model of an extractive political economy and continue to seek ways to exploit windows of opportunity for stabilizing their precarious power base. Energy-importing countries will face high and volatile energy costs, which hurts poorer countries and poorer segments of their populations the most. Their policy space for actions that foster geopolitical stability will remain constrained by their fossil fuel dependence. Lastly, and most importantly, the continued combustion of fossil fuels escalates the breakdown of the planet's climatic systems and thereby begets even more economic and geopolitical instability.

5.2. The Need to Design Interconnected Policy Actions

The different strands of the analysis presented here all lead up to one clear and straightforward policy implication: reduce the consumption of fossil fuels swiftly instead of replacing Russian fuels with imports from elsewhere. This is the key to increasing geopolitical security and economic and social stability as well as, obviously, combatting climate change.

The imperative to reduce fossil fuel use can be achieved in three ways:

1. Replace fossil fuels with renewable energy;
2. Reduce overall energy demand;
3. Accelerate the research, development and deployment of innovative technologies
4. Introduce practices that help to achieve the first two goals favouring social inclusion and equal distribution of essential need satisfiers.

Massive acceleration of the rollout of solar and wind energy and heat pumps is an inevitable step to accelerate the swift phasing out of fossil fuels. According to [64], G-7 countries, including major European core states, could save more natural gas than they import from Russia by 2025. Major reduction opportunities are found in three sectors: industry, power generation and buildings up to 18%. However, since not all fossil fuels can be replaced by sustainable energies in the short run, it is also vital that overall energy demand is reduced. This is the key to bringing energy prices down as well as reaching climate targets [65]. One way to achieve this is a significant acceleration of existing energy conservation efforts, particularly by retrofitting and insulating buildings, and also through efficiency measures in the industry.

Increasing energy efficiency is an important part of the transition but rebound effects may limit the effectiveness of efficiency measures [66]. A framework aiming for energy sufficiency should take primacy. Governments should ensure that all households have access to the essential energy services while reducing the excess energy use of high-income households and industries that do not directly contribute to social or ecological objectives [67]. Soft, market-based forms of rationing, like controlling the price for a certain quantity of gas consumed per household, could be an effective and efficient instrument towards this end. In addition, scholarship in environmental psychology has developed a range of tools to promote sufficiency in consumption behaviour [68].

The key to sufficiency is the construction of sustainable public provisioning systems. Accessible public services would allow for an improved standard of living at a low throughput of energy. In addition to offering a range of sustainable public services, the state may have to take over the fossil fuel industry and energy companies to ensure a rapid transition from fossil fuels without significant disruption to essential energy provisions. Clearly, these measures stress the incapability of the capitalist system to achieve a just and effective transition to a different and sustainable mode of production. Most importantly, they indicate the need to move beyond the current neoliberal governance hegemony and seriously consider alternative governance models, such as polycentric democratic planning, as a helpful tool to achieve such a large-scale transition.

These measures will have to be accompanied by an accelerated development and deployment of technological and social innovations. Across sectors, there are massive opportunities to scale up technologies that increase energy efficiency and sufficiency, including the key sectors of agriculture and mobility. The re-orientation of the EU's approach to innovation policy towards mission-oriented innovation should be accelerated. Even if this implies drawing resources from and accelerating the Schumpeterian creative destruction of ecologically and socially less useful sectors [69]. Importantly, the core will have to share the technologies more freely with the peripheries and semi-peripheries of the world system, as their potential must be fully used to catch up with the escalating climate crisis [70].

The issue related to creating sufficient funds to finance such actions is critical. As many of the necessary measures are unlikely to be economically profitable, at least in the short term with the current market structure, the only way to implement them is to fund them publicly. Several alternatives are available from the instruments of climate finance [71]. For example, the European Central Bank could buy government bonds to fund the necessary public investments for reducing the energy demand and increasing the supply of green technologies. In addition, the newly approved Social Climate Fund by the EU as part of the revision of the EU Emission Trading Scheme and the implementation of the Fit for 55 package might strongly redirect the investments toward a sustainable and just transition, reducing the distributive burden of rapid decarbonisation within citizens and among core and peripheral countries at least at the EU level [72]. Where increased public spending risks exacerbating inflation in the short term, policymakers can use a variety of fiscal and monetary policy measures to reduce private excess demand for non-essential activities in sustainable and equitable ways [73].

6. Conclusions

Through the lens of world systems analysis is clear that the global geopolitical economy of energy resources is still rooted in a core-periphery constellation of combined dependencies across key Western economies and semi-peripheral petrostates, with peripheral countries lagging behind in the catch-up of opportunities for sustainable and equal development pattern. The key tenet of this framework is that the industrial core of the world system extracts an economic surplus from the extractive periphery through the unequal exchange. The climate and energy security strategy adopted by the EU during the past decades was intended to escape from the dependency on heavy shares of carbon-intensive sources. However, the result of these policies has been a partial failure since the reduction in climate-related emissions has been obtained at the expense of an increased vulnerability to external shocks. The extreme concentration of natural gas imports from Russia, especially for the peripheral areas of the EU, recently revealed that the energy strategy should be updated with a radical rethinking of priorities and vulnerabilities.

The three actions—replacing fossil fuels with renewables, reducing energy demand, and sharing social and sustainable technologies globally—could run counter to ecologically unequal exchange and effectively contribute at breaking the principle of uneven and combined development. The EU should delink both from importing fossil fuels and from reaping the economic benefits of a specialisation in high-tech, high-energy modes of production.

This would flatten the hierarchy of ecologically unequal exchange within the world system, and thus possibly constitute a relative decline of European power vis-à-vis the periphery. However, the EU could simultaneously determine a new era of economic equality and geopolitical stability if it is ready to take concrete steps to reduce the vulnerabilities of the peripheral world, including its nearest Eastern partners, by drastically reducing its dependency on fossil fuels. Above all, the precondition for lasting peace and prosperity is to take the danger of climate collapse seriously and the principles of climate justice, putting into practice the complex policy instruments already available in the EU's renewed climate and energy strategy.

Author Contributions: Conceptualisation, V.C., V.M., C.O. and L.T.; methodology, V.M., C.O. and L.T.; investigation, V.M., C.O. and L.T.; data curation, V.M., C.O. and L.T.; writing—original draft preparation, V.M., C.O. and L.T.; writing—review and editing, V.C., V.M., C.O. and L.T.; supervision, V.C.; project administration, V.C.; funding acquisition, V.C. All authors have read and agreed to the published version of the manuscript.

Funding: This research was funded by the Italian Ministry of University and Research (MIUR), Scientific Research Program of National Relevance PRIN 20177J2LS9_001 project "Innovation for global challenges in a connected world: the role of local resources and socioeconomic conditions", Roma Tre CUP F88D19002210001.

Institutional Review Board Statement: Not applicable.

Informed Consent Statement: Not applicable.

Data Availability Statement: Not applicable.

Acknowledgments: Comments, discussions and suggestions from several persons enriched the current development of the research.

Conflicts of Interest: The authors declare no conflict of interest.

References

1. Tooze, A. Is Escalation in Ukraine Part of the US Strategy? Available online: https://www.theguardian.com/commentisfree/2022/may/04/us-lend-lease-act-ukraine-1941-second-world-war (accessed on 12 September 2022).
2. Liboreiro, J.; Koutsokosta, E.; Murray, S. Hungary, Slovakia, Czechia and Bulgaria Resist EU Ban on Russian Oil. Available online: https://www.euronews.com/my-europe/2022/05/09/hungary-slovakia-czech-republic-and-bulgaria-still-resisting-eu-ban-on-russian-oil (accessed on 12 September 2022).
3. Bounds, A.; Dempsey, H.; Mount, I. Europe's Push to Plug Its Energy Gaps. Available online: https://www.ft.com/content/dd4aeffe-d243-49c7-9f4e-152ee54a4f26 (accessed on 12 September 2022).
4. Bounds, A.; Fleming, S. EU Prepares to Sell More Carbon Permits to Pay for Exit from Russian Gas. Financial Times. Available online: https://www.ft.com/content/be8d95cc-273a-43b8-b6ab-e9f95685ddc7 (accessed on 12 September 2022).
5. ACLED (Armed Conflict Location & Event Data Project). Mid-term Report on 10 Conflicts to Worry About in 2022. Available online: https://acleddata.com/10-conflicts-to-worry-about-in-2022/nigeria/ (accessed on 16 October 2022).
6. Arrighi, G. *The Long Twentieth Century: Money, Power, and the Origins of Our Times*; Verso Books: London, UK, 2010.
7. Emmanuel, A. *Unequal Exchange: A Study of the Imperialism of Trade*; Monthly Review Press: New York, NY, USA, 1972.
8. Hornborg, A. The unequal exchange of time and space: Toward a non-normative ecological theory of exploitation. *J. Ecol. Anthropol.* **2003**, *7*, 4–10. [CrossRef]
9. Wallerstein, I.M. *World-Systems Analysis: An Introduction*; Duke University Press: Durham, NC, USA, 2004.
10. Dragneva, R.; Wolczuk, K. Between Dependence and Integration: Ukraine's Relations with Russia. *Eur. Stud.* **2016**, *68*, 678–698. [CrossRef]
11. Moore, J. *Capitalism in the Web of Life: Ecology and the Accumulation of Capital*; Verso Books: New York, NY, USA, 2015.
12. Ciccantell, P.S. Ecologically Unequal Exchange and Raw Materialism: The Material Foundations of the Capitalist World-Economy. In *Ecologically Unequal Exchange*; Frey, R.S., Gellert, P.K., Dahms, H.F., Eds.; Palgrave Macmillan: Cham, Switzerland, 2019; pp. 49–73. [CrossRef]
13. Jorgenson, A.K. Environment, Development, and Ecologically Unequal Exchange. *Sustainability* **2016**, *8*, 227. [CrossRef]
14. Gellert, P.K.; Frey, R.S.; Dahms, H.F. Introduction to Ecologically Unequal Exchange in Comparative Perspective. *J. World-Syst. Res.* **2017**, *23*, 226–235. [CrossRef]
15. Hornborg, A. *Global Ecology and Unequal Exchange: Fetishism in a Zero-Sum World*; Routledge: Oxfordshire, UK, 2012.
16. Dorninger, C.; Hornborg, A.; Abson, D.J.; von Wehrden, H.; Schaffartzik, A.; Giljum, S.; Engler, J.-O.; Feller, R.L.; Hubacek, K.; Wieland, H. Global patterns of ecologically unequal exchange: Implications for sustainability in the 21st century. *Ecol. Econ.* **2020**, *179*, 106824. [CrossRef]
17. The Eora26 MRIO Input-Output Model Database. Available online: https://worldmrio.com/eora26/ (accessed on 12 September 2022).
18. Desai, R.; Freeman, A.; Kagarlitsky, B. The Conflict in Ukraine and Contemporary Imperialism. *Int. Crit. Thought* **2016**, *6*, 489–512. [CrossRef]
19. Ciplet, D.; Roberts, J.T. Splintering South: Ecologically Unequal Exchange Theory in a Fragmented Global Climate. *J. World-Syst. Res.* **2017**, *23*, 372–398. [CrossRef]
20. Hornborg, A.; Martinez-Alier, J. Ecologically unequal exchange and ecological debt. *J. Politi-Ecol.* **2016**, *23*, 328–333. [CrossRef]
21. Jorgenson, A. Unequal Ecological Exchange and Environmental Degradation: A Theoretical Proposition and Cross-National Study of Deforestation, 1990–2000. *Rural Sociol.* **2006**, *71*, 685–712. [CrossRef]
22. Taylor, T.G.; Tainter, J.A. The Nexus of Population, Energy, Innovation, and Complexity. *Am. J. Econ. Sociol.* **2016**, *75*, 1005–1043. [CrossRef]
23. IEA (International Energy Agency). Gas Trade Flows—Data Product. Available online: https://www.iea.org/data-and-statistics/data-product/gas-trade-flows (accessed on 12 September 2022).
24. Meynkhard, A. long-term prospects for the development energy complex of Russia. *Int. J. Energy Econ. Policy* **2020**, *10*, 224–232. [CrossRef]
25. Kormishkina, L.A.; Kormishkin, E.D.; Koloskov, D.A. Economic growth in modern Russia: Problems and prospects in the context of neo-industrial paradigm. *J. Appl. Econ. Sci.* **2016**, *11*, 1115–1119.
26. Dabrowski, M.; Collin, A.M. Russia's Growth Problem. Bruegel Policy Contribution Issue 4, February 2019. Available online: https://www.bruegel.org/policy-brief/russias-growth-problem (accessed on 12 September 2022).
27. Way, L.A.; Casey, A. The structural sources of postcommunist regime trajectories. *Post-Soviet Aff.* **2018**, *34*, 317–332. [CrossRef]
28. IEA (International Energy Agency). Key Energy Statistics. Country Profile: Ukraine. Available online: https://www.iea.org/countries/ukraine (accessed on 12 September 2022).
29. Hirschman, A. *National Power and the Structure of Foreign Trade*; University of California Press: Berkeley, CA, USA, 1980.
30. Tichý, L. *EU-Russia Energy Relations: A Discursive Approach*; Springer International Publishing: New York, NY, USA, 2019.

31. EC (European Commission). European Energy Security Strategy, COM (2014) 330 Final. Available online: https://eur-lex.europa.eu/legal-content/EN/TXT/PDF/?uri=CELEX:52014DC0330&from=EN (accessed on 18 October 2022).
32. Belyi, A. New dimensions of energy security of the enlarging eu and their impact on relations with Russia. *J. Eur. Integr.* **2003**, *25*, 351–369. [CrossRef]
33. Kustova, I. EU–Russia Energy Relations, EU Energy Integration, and Energy Security: The State of the Art and a Roadmap for Future Research. *J. Contemp. Eur. Res.* **2015**, *11*, 288–295. [CrossRef]
34. Esakova, N. *European Energy Security: Analysing the Eu-Russia Energy Security Regime in Terms of Interdependence Theory*; Springer Science & Business Media: Heidelberg, Germany, 2013.
35. Siddi, M. EU-Russia Energy Relations: From a Liberal to a Realist Paradigm? *Russ. Polit.* **2017**, *2*, 364–381. [CrossRef]
36. Stern, J. The Russian-Ukrainian Gas Crisis of January 2006. Available online: https://efaidnbmnnnibpcajpcglclefindmkaj/http://171.67.100.116/courses/2016/ph240/lee-m1/docs/stern-jan06.pdf (accessed on 12 September 2022).
37. Stern, J.; Pirani, S.; Yafimava, K. The Russo-Ukrainian gas dispute of January 2009: A Comprehensive Assessment. Available online: https://ora.ox.ac.uk/objects/uuid:3e2ad362-0bec-478a-89c1-3974c79363b5 (accessed on 12 September 2022).
38. Dellano-Paz, F.; Fernandez, P.M.; Soares, I. Addressing 2030 EU policy framework for energy and climate: Cost, risk and energy security issues. *Energy* **2016**, *115*, 1347–1360. [CrossRef]
39. EC (European Commission). Energy Union Package—Framework Strategy for a Resilient Energy Union with a Forward-Looking Climate Change Policy. COM (2015) 080 Final. Available online: https://eur-lex.europa.eu/legal-content/EN/TXT/?uri=CELEX:52015DC0080 (accessed on 18 October 2022).
40. EC (European Commission). Communication from the Commission. The European Green Deal. COM(2019)640final, European Commission, Bruxelles. Available online: https://eur-lex.europa.eu/legal-content/EN/TXT/?uri=COM%3A2019%3A640%3AFIN (accessed on 18 October 2022).
41. EC (European Commission). Communication from the Commission. Fit for 55: Delivering the EU's 2030 Climate Target on the way to climate neutrality. COM(2021)550final, European Commission, Bruxelles. Available online: https://eur-lex.europa.eu/legal-content/EN/TXT/?uri=CELEX%3A52021DC0550 (accessed on 18 October 2022).
42. The War in Ukraine Has Reshaped the World's Fuel Markets. Available online: https://www.economist.com/interactive/briefing/2022/09/24/war-in-ukraine-has-reshaped-worlds-fuel-markets (accessed on 25 September 2022).
43. Capellán-Pérez, I.; de Castro, C.; González, L.J.M. Dynamic Energy Return on Energy Investment (EROI) and material requirements in scenarios of global transition to renewable energies. *Energy Strat. Rev.* **2019**, *26*, 100399. [CrossRef]
44. Alina, S.; Wang, Z.; Artem, D.; Maxim, R.; Svetlana, R. Is the implementation of energy savings and EROI increasing policy really effective in Russian gas companies? The case of JSC "Gazprom". *Nat. Gas Ind. B* **2019**, *6*, 639–651. [CrossRef]
45. Steblyanskaya, A.; Bi, K.; Denisov, A.; Wang, Z.; Wang, Z.; Bragina, Z. Changes in sustainable growth dynamics: The case of China and Russia gas industries. *Energy Strategy Rev.* **2021**, *33*, 100586. [CrossRef]
46. Chebotareva, G.; Strielkowski, W.; Streimikiene, D. Risk assessment in renewable energy projects: A case of Russia. *J. Clean. Prod.* **2020**, *269*. [CrossRef]
47. Makarov, I.; Chen, H.; Paltsev, S. Impacts of climate change policies worldwide on the Russian economy. *Clim. Policy* **2020**, *20*, 1242–1256. [CrossRef]
48. Eurostat. EU Energy Mix and Import Dependency: Tables and Figures. Available online: https://ec.europa.eu/eurostat/statistics-explained/index.php?title=EU_energy_mix_and_import_dependency (accessed on 12 September 2022).
49. Pokharel, S.; Thompson, M. Russia Shuts off Gas Supplies to Poland and Bulgaria. Available online: https://www.cnn.com/2022/04/26/energy/poland-russia-gas/index.html (accessed on 12 September 2022).
50. Hancock, A.; Fleming, S. German €200bn Energy Support Plan Sparks 'Animosity' within EU. Available online: https://www.ft.com/content/f52b06b9-3932-44ca-b831-777cf68c3dc8 (accessed on 30 September 2022).
51. Dempsey, P. Global Energy Suffers Invasion Effects. *Eng. Technol.* **2022**, *17*, 20–23. [CrossRef]
52. Eckardt, L.-M. KNA Katholische Nachrichten-Agentur; ZEIT Online. Gas aus Russland: Cem Özdemir erwartet Lieferengpässe bei einzelnen Lebensmitteln. Die Zeit. Available online: https://www.zeit.de/politik/2022-05/cem-oezdemir-lieferengpaesse-preise-gas-russland?utm_referrer=https%3A%2F%2Fwww.ecosia.org%2F (accessed on 12 September 2022).
53. Alami, I.; Alves, C.; Bonizzi, B.; Kaltenbrunner, A.; Koddenbrock, K.; Kvangraven, I.; Powell, J. International financial subordination: A critical research agenda. *Rev. Int. Politi-Econ.* **2022**, 1–27. [CrossRef]
54. Brooks, R. A German Embargo on Russian Energy Would Obviously Hurt Germany, but It Would Hurt Russia Much More, Hindering Putin's Ability to Wage War. Available online: https://twitter.com/RobinBrooksIIF/status/1508009547708964865?s=20&t=YZwsI2sOVKwEeZMduk62jg (accessed on 12 September 2022).
55. OEC. What Does Russia Export? Available online: https://oec.world/en/visualize/tree_map/hs92/export/rus/all/show/2020/ (accessed on 12 September 2022).
56. OEC. Where Does RUSSIA Export Mineral Products to? Available online: https://oec.world/en/visualize/tree_map/hs92/export/rus/show/5/2020/ (accessed on 12 September 2022).
57. OEC. What Does Russia Export to the EU? Available online: https://oec.world/en/visualize/tree_map/hs92/export/rus/aut.bel.bgr.hrv.cze.dnk.est.fin.fra.deu.grc.hun.irl.ita.lva.ltu.lux.nld.pol.prt.rou.svk.svn.esp.swe/show/2020/ (accessed on 12 September 2022).

58. CREA (Centre for Research on Energy and Clean Air). The Russian Fossil Tracker. Available online: https://www.russiafossiltracker.com/ (accessed on 12 September 2022).
59. Brzoska, M. Is the West Funding the Russian War in Ukraine? Available online: https://www.visionofhumanity.org/is-the-west-funding-the-russian-war-in-ukraine/ (accessed on 12 September 2022).
60. EC (European Commission). REPowerEU Plan. COM(2022) 230 Final. Available online: https://eur-lex.europa.eu/legal-content/EN/TXT/?uri=COM%3A2022%3A230%3AFIN (accessed on 18 October 2022).
61. UNHCR (United Nations Refugee Agency). Ukraine Refugee Situation. Available online: https://data.unhcr.org/en/situations/ukraine (accessed on 25 September 2022).
62. Mallapaty, S.; Padma, T.V.; Mega, E.R.; Van Noorden, R.; Masood, E. The countries maintaining research ties with Russia despite Ukraine. *Nature* **2022**, *604*, 227–228. [CrossRef]
63. Bhattacharjee, S. India-Russia Explore a Rupee-Rouble Payment Scheme to Bypass War. Available online: https://www.aljazeera.com/economy/2022/3/31/india-russia-explore-a-rupee-rouble-payment-scheme-to-bypass-war (accessed on 25 September 2022).
64. Polugodina, M.; Meissner, F.; Lettow, F.; Neuhoff, K.; Handrich, L. G7 Gas Reduction Plan. Available online: https://www.greenpeace.de/publikationen/S04021-greenpeace-studie-g7-gas-reduction-engl.pdf (accessed on 25 September 2022).
65. Tagliapetra, S.; Zachmann, G. The Only Quick-Fix to Europe's Energy Price Crisis Is Saving Energy. Bruegel Policy Brief. Available online: https://www.bruegel.org/2021/10/the-only-quick-fix-to-europes-energy-price-crisis-is-saving-energy/ (accessed on 12 September 2022).
66. Brockway, P.E.; Sorrell, S.; Semieniuk, G.; Heun, M.K.; Court, V. Energy efficiency and economy-wide rebound effects: A review of the evidence and its implications. *Renew. Sustain. Energy Rev.* **2021**, *141*, 110781. [CrossRef]
67. Becker, S.; Renn, O. Saving Energy Doesn't Have to Mean Social Imbalance—But We Will Need to Change Our Habits. Available online: https://www.iass-potsdam.de/en/blog/2022/03/saving-energy-doesnt-have-mean-social-imbalance-we-will-need-change-our-habits (accessed on 12 September 2022).
68. Freudenreich, B.; Schaltegger, S. Developing sufficiency-oriented offerings for clothing users: Business approaches to support consumption reduction. *J. Clean. Prod.* **2019**, *247*, 119589. [CrossRef]
69. Mazzucato, M. Mission-oriented innovation policies: Challenges and opportunities. *Ind. Corp. Chang.* **2018**, *27*, 803–815. [CrossRef]
70. Herman, K.S. Beyond the UNFCCC North-South divide: How newly industrializing countries collaborate to innovate in climate technologies. *J. Environ. Manag.* **2022**, *309*. [CrossRef]
71. Battiston, S.; Monasterolo, I.; Riahi, K.; van Ruijven, B.J. Accounting for finance is key for climate mitigation pathways. *Science* **2021**, *372*, 918–920. [CrossRef]
72. EC (European Commission). Proposal for a Regulation of the European Parliament and of the Council Establishing a Social Climate Fund. COM(2021) 568 Final. Available online: https://eur-lex.europa.eu/legal-content/en/TXT/?uri=CELEX%3A52021PC0568 (accessed on 18 October 2022).
73. Olk, C.; Schneider, C.; Hickel, J. How to Pay for Saving the World: Modern Monetary Theory for a Degrowth Transition. Available online: https://papers.ssrn.com/sol3/papers.cfm?abstract_id=4172005 (accessed on 18 October 2022).

Article

Energy Efficiency Policies in Poland and Slovakia in the Context of Individual Well-Being

Anna Barwińska Małajowicz [1,*], Miroslava Knapková [2], Krzysztof Szczotka [3], Miriam Martinkovičová [2] and Radosław Pyrek [1,*]

1. Department of Economics and International Economic Relations, Institute of Economics and Finance, University of Rzeszów, 35-601 Rzeszow, Poland
2. Department of Economics, Faculty of Economics, Matej Bel University, 974 01 Banská Bystrica, Slovakia
3. Department of Power Systems and Environmental Protection Facilities, Faculty of Mechanical Engineering and Robotics, AGH University of Science and Technology, 30-059 Kraków, Poland
* Correspondence: abarwinska@ur.edu.pl (A.B.M.); rpyrek@ur.edu.pl (R.P.)

Abstract: Improving energy efficiency includes a number of measures implemented as part of the greening of the energy industry, which in turn is a prerequisite for the creation of a sustainable energy industry to ensure energy and environmental security for the world. Despite the adoption of the EU directives on energy efficiency, there is still insufficient public awareness in this area in Poland and Slovakia. This is particularly surprising because improving energy efficiency not only brings national and global benefits, but also has a significant impact on the well-being of individuals and households. The main purpose of the paper is to analyze the national policies of Poland and Slovakia, which are based on the European Directive 2012/27/EU on energy efficiency, and which introduce new measures aimed not only at increasing energy efficiency, but also at increasing the well-being of households and individuals. Methods of desk research and content analysis were used. The current situation in both countries is illustrated by case studies that document the administrative process (Slovakia) and the calculation of energy savings (Poland) when using renewable energy sources in the case of family houses.

Keywords: energy efficiency; state policy; renewable energy sources; well-being; households; entrepreneurial opportunities; Poland; Slovakia

1. Introduction

The issue of energy efficiency is a matter of concern where energy has multifaceted impact on each type of economic activity of human beings [1–3]. Unreasonable energy management simultaneously causes a lot of negative effects on many fields (economy, environment, society), and therefore activities aimed at improvement of the energy efficiency have great importance in the modern world. They can bring many potential benefits—not only in terms of energy saving and preventing climate change, but also in the broadly understood well-being, especially in respect of improving human health and life and creating new jobs. On the other hand, changes in society, not only due to rising energy prices as a result of scarcity, but also due to the impact of the COVID-19 pandemic (the shift of work from the workplace to the home and childcare from educational institutions to the home environment, and thus the increasing demand for energy consumption), place more and more emphasis on individual well-being, as experienced by each individual [4]. Therefore, energy efficiency should not only be in the interest of multinational groups and entire countries, but also of individuals and households.

Improving energy efficiency includes a number of measures implemented as part of the greening of the energy industry, which in turn is a prerequisite for the creation of a sustainable energy industry to ensure energy and environmental security for the world. The European Commission pays great attention to the growth of energy efficiency. Investments in energy efficiency result in not only reducing energy costs, improvement of air quality and decrease in energy losses, but also improving human health and well-being. The main purpose of the paper is to analyze the national policies of Poland and Slovakia, which are based on the European Directive 2012/27/EU [5] on energy efficiency, and which introduce new measures aimed not only at increasing energy efficiency, but also at increasing the well-being of households and individuals. The Polish government is implementing a number of solutions to contribute to increasing building energy efficiency in Poland. Public campaigns are being conducted to reduce the knowledge gap in this area among the public. Especially at this time when energy security is so important, every citizen of Poland is beginning to pay attention to the energy efficiency of the buildings they inhabit.

The article is divided into five parts, namely literature review, a methodological part, a review of national policies in Poland and Slovakia, case studies describing selected measures to increase energy efficiency and well-being of households and individuals in Poland and Slovakia and a discussion. The intention of the paper is to show that the national policies of the European Union countries represent an important intermediate step between the European Union's intentions and their implementation, thus creating room for the use of nationally specific instruments for achieving energy efficiency and related well-being of the population. On the one hand, this study contributes to enriching the theoretical basis of research on energy efficiency, national policies and well-being; on the other hand, through case studies, it provides an analysis of the practical challenges of implementing national policies to increase energy efficiency in the context of increasing well-being. Case studies provide complex insight into family house renovation, to meet the requirements of energy efficiency increase.

2. Literature Review

It is difficult to give a unified and universal definition of energy efficiency. Definition aspects which are considered depend on compounds and contexts in which this conceptual category is analyzed. Generally, we can define energy efficiency as a ratio between an output of performance, service or good, and an input of energy. In a little more detail, following a rule of rational management (they occur in two separable variants: (1) as a rule of greatest effect or (2) as a principle of least effort [6]), one can show that energy efficiency means delivering more results and services using the same energy effects, or delivering the same results with less energy effect. Moreover, energy efficiency can be determined as measure of energy efficiency in economic activity (Directive 2012/27/EU) [5].

In the literature of the subject, the term "energy efficiency" is assigned different meanings (more or less extensive interpretation) [7–10]. There are accessible definitions for various terms and meanings, for example: "power efficiency", "energy efficiency", "reduction of energy consumption", "sustainable use of energy resources", "effective use of resources", "energy services provided per unit of energy consumed", "reverse energy consumption", "ratio of useful results to physical energy inputs", etc. [2,11–13].

Regardless of the definition form, it is clear that energy efficiency is one of the main factors of development of entrepreneurship and innovations, it helps to reduce energy losses and it is not only widely accepted, but also a highly desirable tool used to achieve sustainable development [14]. It is worth adding here that economic development does not have to be identified with increased consumption of energy resources, which indicate chosen measures of energy efficiency [8,15–17].

Nevertheless, energy efficiency does not have only an economic dimension, and it does not only mean saving of energy and financial resources, but it is also very closely linked with an individual's well-being. An encyclopedic definition of well-being (Oxford English Dictionary) [18] refers to a general reflection on human well-being. The conceptual category of well-being is equivalent and multidimensional, which results in using ambiguous terminology (the terms used are, for example: "quality of life", "social well-being", "subjective well-being", "personal well-being", "happiness", "satisfaction"). Regarding welfare, it often approaches a subjective point of view (when individuals express their opinion about their own perception of welfare). Among the most frequently asked questions there are issues from the index based on questions referring to eight different aspects of life [19]. In this way, a range of data describing the impact of different dimensions of life on well-being can be collected. The issues of well-being need to be addressed in an objective approach which in turn means using different types of indicators. Data connected with these indicators depend on many variables (sex, age, nationality, origin, socio-economic status, belonging to vulnerable groups, etc.). Situations where quantitative indicators cannot be measured are supplemented with qualitative information.

In the economy, well-being is used to assess the quality of life from a quantitative perspective. However, it should be noted that quality of life cannot be confused with standard of living conditioned only by income. In contrast, typical quality of life indicators involve employment, education, physical and mental health, living environment, social belonging, social relationships, level of independence or environmental quality [20,21].

It is also worth emphasizing that the "quality of life" is studied by social sciences and medicine, while "welfare" is an original psychological construct. There are at least three different approaches to measure the well-being construct, which examine its different aspects: assessment of life (life satisfaction), hedonistic well-being and endemic well-being [22,23].

Studies of the literature show the evolution of theoretical considerations from neoclassical welfare economics (Marshall, Pigou), focusing mainly on individual usability (subjective individual well-being) through so-called new economy well-being (Pareto, Hicks, Kaldor) (value neutrality and normative economic theory) to economy of happiness (Easterlin, Frey, Stutzer, Kahneman, Veenhoven), which uses concepts of prosperity and quality of life [23], although economy of happiness is not intended to replace measures based on income, but rather they should be supplemented by wider measures of prosperity [24].

Meanwhile, it is assumed that economic activity, production of goods and services, has a value only then when it contributes to human happiness [23].

In the context of excessive CO_2 emissions, activities aimed to improve energy efficiency may constitute significant support in the improvement of human well-being. They may, for example, help in preventing and relieving serious health problems including respiratory and cardiac diseases. Air pollution represents a major threat to human health and life. According to WHO data, it is the cause of approximately 3 million premature deaths a year [25].

Activities undertaken in order to improve energy efficiency may support not only physical but also mental health of humans. Among such activities is, for instance, creating a healthy living environment in both outdoor and indoor environments, in which the air temperature, level of humidity or noise level is appropriate or there is better air quality [26]. Under such conditions, the human mind and body work much more efficiently and effectively. On the other hand, energy poverty, lack of warmth and high energy bills have a negative impact on human health. They can cause different diseases, anxiety, stress or depression. Positive health effects of energy efficiency and better human well-being can be strengthened if activities aimed at increasing energy efficiency are included in financial support mechanisms [27] and supported by strong local community involvement [28].

The link between energy efficiency and well-being is not an entirely new topic. However, previous studies have rather addressed the issues of increasing energy intensity in order to increase individuals' well-being [29,30]. It is only in recent years that researchers' attention has been focused on the interrelationship between energy consumption, renewable resource use and well-being. Two contradictory findings are reported in the literature. Some studies show the absence of a relationship between energy intensity and well-being [31]; some studies, on the contrary, confirm that renewable energy sources can improve human well-being by improving environmental quality [32–34]. Some authors [35] also pointed, using the qualitative approach, to the nexus of health, well-being and energy consumption. In their conclusions, they pointed also to the necessity of targeted housing policies supporting energy efficiency solutions.

Although there are several studies and approaches focusing separately on the area of increasing energy efficiency through the use of RESs, as well as separately on the well-being of individuals, their interconnectedness and support by state policies are not sufficient and the interconnectedness is not sufficiently linked in the literature. For this reason, in this paper we focus specifically on those areas of energy efficiency policies in Poland and Slovakia that are closely linked to individuals, their housing and reducing the energy intensity of their housing. Reducing the energy intensity of housing represents not only savings on energy consumption (which is the expected effect), but also consequently higher housing comfort, higher satisfaction in terms of economic savings but also overall higher quality of life.

3. Materials and Methods

In this section, materials used for comparison of national policies of Poland and Slovakia, as well as methods of constructing case studies, are explained. Poland and Slovakia are both members of the European Union, entering the EU in within the "Eastern enlargement" in 2004. Both countries are also members of the Visegrád Four group, and they share a common history during which they have progressed from energy surpluses [36,37] to the current threat of energy shortages [38,39].

This article is based on the analysis of the strategic documents of Poland and Slovakia resulting from the EU Directive 2012/27/EU [5] (as the main legislative document for increasing energy efficiency in EU Member States). Table 1 summarizes legal acts and strategic documents involved in the analysis.

Table 1. Legal acts and strategic documents on increase in energy efficiency in Poland and Slovakia.

Poland	Slovakia
EU Directive 2012/27/EU [5]	
Dz.U. 2021 item 234 Act of 17 December 2020 on promoting electricity generation in offshore wind farms	Act no. 321/2014 Coll. on energy efficiency
	Integrated National Energy and Climate Plan for years 2021–2030
National Energy and Climate Plant for the years 2021–2030	Recovery and Resilience Plan
Poland's energy policy until 2040 (PEP2040)	Act No. 368/2021 Coll. on the Recovery and Resilience Support Mechanism and on the amendment and supplementation of certain acts
Strategy for Responsible Development until 2020	

The strategic documents and legislation themselves provide a framework for the application of real energy efficiency measures in the conditions of Polish and Slovak households. For this reason, the authors then focused on documenting the actual situation in Poland and Slovakia through case studies. Case studies describe ongoing (Poland) and planned (call for proposals) (Slovakia) instruments that aim to increase energy efficiency as well as to increase well-being of inhabitants and households.

The authors base their paper on two main assumptions, namely:

A1. Poland and Slovakia have adopted similar state policy and strategic documents on the basis of the EU Directive 2012/27/EU [5].

A2. Instruments for improving energy efficiency are different in the two countries.

The logic behind the research is displayed in Figure 1.

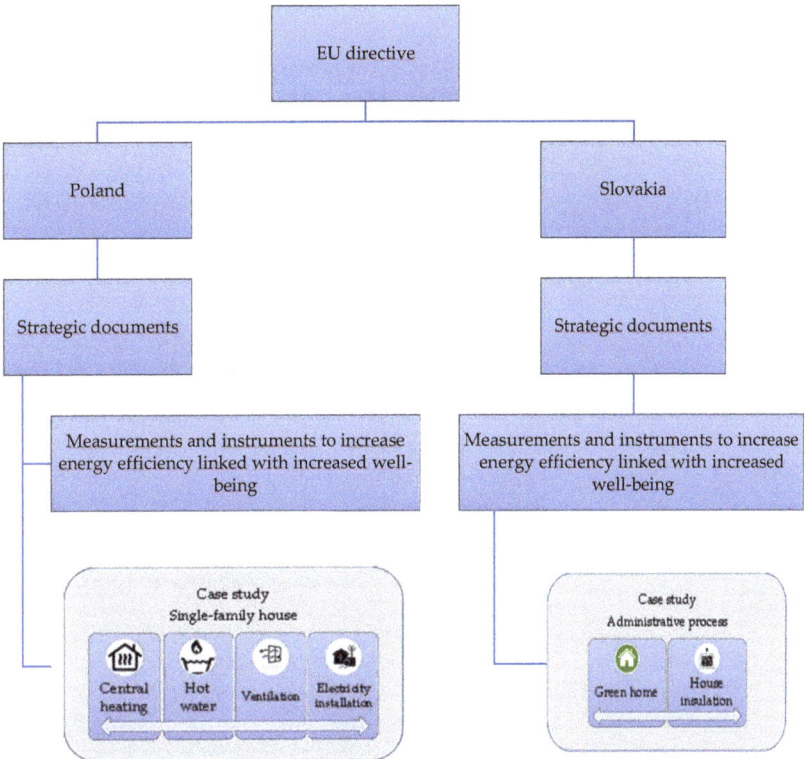

Figure 1. Research scheme.

Since the authors' intention is to illustrate that increasing energy efficiency is also linked to improving the well-being of individuals and households, they focus their analysis of the state policy and strategic documents only on those measures and instruments that are intended for households (i.e., the authors exclude those measures and instruments that are intended for the business sector and public administration from further examination).

4. Review of the State Policy in Poland and Slovakia

4.1. Review of the State Policy in Poland

Energy policy is a very wide issue and it is difficult to discuss changing trends, amendments and legal acts in a brief and accessible way. Poland as one of the European Union Member States has to adjust to several European Commission's directives regarding energy policy, reduction of CO_2 emissions and renewable energy sources. In accordance with the Regulations 2018/2019 [40], every EU Member State should take into account recommendations of the Commission concerning projects of integrated national energy and climate plans. The main goal of energy policy is energy safety while providing competitiveness of the economy, energy efficiency and reducing the influence of the energy sector on the environment with the optimal use of the country's own energy resources. Energy supply in Poland is still dominated by fossil fuels, of which the biggest part is coal, then petroleum and natural gas. Coal plays a key role in Poland's energy system and economy. A huge proportion of coal ranks Poland in second place among EU Member States in terms of the intensity of CO_2 emissions in the energy supply and in fourth place in terms of CO_2 intensity in GDP. Despite continued coal dominance, Poland has achieved significant success in the field of energy transformation. It has become one of the fastest developing PV markets in the EU due to small, distributed PV systems in residential buildings.

Moreover, Poland has a comprehensive and well-developed strategy for offshore wind energy [41] which resulted in contracts for the launch of 5.9 GW of capacity by 2030 and plans for at least 11 GW power up to 2040.

The first wind power system in Polish marine areas will start to be built in two years' time (and be finished in 2040). Polish energy policy is focused on decreasing emissions of carbon dioxide by increasing renewable energy source use and natural gas, introducing nuclear energy, greater energy demand and improvement of energy efficiency. Poland puts strong emphasis on safe energy and just transition. It will provide energy at an affordable cost in order to promote economic growth and to protect consumers. An important document determining Poland's energy and climate policy is the National Energy and Climate Plan (KPEiK) for the years 2021–2030, the Ministry of State Assets, version 4.1. from 18 December 2019 [42].

This document is required from all EU Member States and it was accepted in 2019. Another important document is Poland's Energy Policy until 2040 (PEP2040) [43], which was adopted in February 2021. In accordance with national regulations and EU directives, Poland has a various energy and climate targets. Greenhouse gas emissions from energy-intensive industrial plants and electricity production in Poland are governed by the EU Emissions Trading System.

The European Commission imposed on Poland an obligation to implement the National Energy and Climate Plan for the years 2021–2030, which was carried out by the Minister of State Assets in December 2019. This obligation is imposed on Poland by the provisions of the Regulations of the European Parliament and the Council 2018/1999 [40]. This act has been changed, the current consolidated version is: 29 July 2021 of the European Parliament and of the Council of 11 December 2018 on the Governance of the Energy Union and Climate Action, amending Regulations (EC) No 663/2009 [44] and (EC) No 715/2009 [45] of the European Parliament and of the Council, Directives 94/22/EC [46], 98/70/EC [47], 2009/31/EC [48], 2009/73/EC [49], 2010/31/EU [50], 2012/27/EU [5] of the European Parliament and of the Council and repealing Regulation (EU) No 525/2013 [51] of the European Parliament and of the Council.

The Polish energy system is one of the largest within the European Union. It is one of the top ten main macro-energy indicators. This corresponds to the potential of the Polish economy, which has seventh place in the European Union in terms of GDP (in 2018— EUR 496.4 mld at current prices), and sixth in terms of population (37.9 mln). In the category of the amount of gross primary and final energy consumption in 2018, Poland takes 6th place in the EU National Energy and Climate Plan (KPEiK) [42].

In 2018, global energy consumption was 4490,7 PJ. National gross energy consumption per person was about 116 GJ, slightly deviating from the European average which was 137.1 GJ. In 2018, direct energy consumption was 3551.8 PJ. The sector of the economy which had the largest part in direct energy consumption was industry (34.5%). The second largest sector, in terms of consumption, was transport, the share of which has increased steadily in recent years and in 2018 was 27%. In 2018, households used 23% of energy, farmers used 4.6% and other recipients 9% [52]. The National Energy and Climate Plan presents aims and objectives, policies and actions to achieve the five dimensions of the energy union: energy safety, internal energy market, energy efficiency, decarbonization and scientific research, innovation and competitiveness. The document has been prepared on the basis of national development strategies approved at the government level such as the Strategy for the sustainable development of transport by 2030, Ecological policy of the State 2030 [53], Strategy for sustainable rural development, agriculture and fisheries 2030 and regarding Poland's draft energy policy until 2040.

The National Energy and Climate plan for the years 2021–2030 sets the following climate and energy goals by 2030:

- reduction of greenhouse gas emissions in the non-ETS sectors compared to 2005's level by 7%;
- a contribution of 21–23% of RESs to gross final energy consumption considering:

- a 14% contribution of RESs in transport;
- annual increase in RES share in heating and cooling by 1.1% in an average year;
- a 23% increase in energy efficiency compared to the PRIMES forecast PRIME2007;
- reduction from 56–60% of coal's share in electricity production.

Two strategic framework documents are in the state's energy policy, consisting of: Poland's energy policy (known as PEP2040) [43] and the Strategy for Responsible Development until 2020 with prospects until 2030 (SOR) [52]. Poland's energy policy until 2040 (PEP2040) specifies how to transform energy in Poland. It includes strategic guidance on the choice of suitable technology to build a low-carbon energy system. PEP2040 contributes to implementation of the Paris Agreement signed in December 2015, taking into account the need to carry out transitions in a fair way and with solidarity. PEP2040 provides national contribution to the realization of the EU's climate and energy policy, whose dynamics have significantly increased in recent years. The policy considers the scales of the demands which are related to national economy adjustment to the EU regulatory environment related to the 2030 climate and energy goals, European Green Deal, the COVID-19 economy recovery plan and to achieve climate neutrality following national possibilities as a contribution to implementation of the Paris Agreement. A low-emission energy transformation provided for in PEP2040 will initiate wider modernization changes in the whole economy, guaranteeing the energy safety, ensuring fair sharing costs and protecting the most venerable groups in society. PEP2040 is one of the nine integrated sectoral strategies resulting from the Strategy for Responsible Development (SOR). PEP2040 is consistent with the National Energy and Climate Plan for the years 2021–2030 and it consists of a state description and terms of the energy sector. It points to three pillars [43] (just transition, zero-emission energy system and good air quality) on which eight specific objectives of PEP2040 are based, on the actions which are essential for their implementation and strategic projects.

PEP2040 foresees eight specific objectives which belong to: optimal use of own energy resources, development of manufacturing and network of electric energy, diversification of supplies and expansion of the network infrastructure of natural gas, petroleum and liquid fuels, energy market development, implementation of nuclear energy, renewable energy source development, heating and cogeneration expansion, improving the energy efficiency of the economy. The Strategy of Responsible Development (SOR) is more general in nature and it is the basis of the Polish energy policy 2040.

According to the strategy records, the main mission of the energy sector is to provide the economy, institutions and citizens constantly and optimally adapt to the needs of supplying the energy at an acceptable economic price. According to the SOR, it should be carried out using locally available materials in a rational and efficient way, having energy-efficient waste and renewable energy sources using the potential of innovation in energy generation, transmission and distribution.

At the operational level, it is planned to increase the share of stable renewable energy sources, including clusters, energy cooperatives, etc., and preserving the priority role of improving the energy efficiency of the economy including the elimination of environmentally harmful emissions. Furthermore, it is also important to develop energy storage technologies, the introduction of intelligent energy networks and develop electromobility, introducing energy-efficient and high-efficiency technologies. The strategy also covers projects to support implementation, including: power market, regional gas transport and trade center, electromobility development program, development and exploitation of geothermal potential in Poland, exploitation of the hydropower potential and also restructuring of the coal mining sector. Poland's Energy Policy is developed by the Minister for Energy on the basis of Article 12, 13–15 Act—Energy Law and in accordance with the Act of the principles of development policy and a number of entities and, in particular, the Minister for Climate and Energy and Council of Ministers are responsible for its realization.

With an ambitious funding program implemented with the support of the EU-funded project FinEERGo-Dom, buildings in Poland will become greener [54]. The PLN 100 million given to Polish citizens will finance projects that enable major renovations of buildings that will reduce greenhouse gas emissions and improve energy efficiency. In detail, matters of energy efficiency of buildings that are being designed or constructed, or reconstructed, are regulated by the Regulation on technical conditions of buildings and their location [55]. An amendment to the Regulation on the detailed scope and form of construction design expanded the obligation to conduct an analysis of the feasibility of using alternative systems in a rational way in all kinds of buildings and changed the range of it. The aim is to extend the usage of those alternatives (which include decentralized energy supply systems based on renewable energy, cogeneration, district or block heating or cooling, particularly when it is based entirely or partially on renewable energy, and heat pumps) where it makes economic, technical and environmental sense to do so. Minimum requirements include [56]:

- guarantees of the value of the EP index (kWh/(m²rok)), which consider the annual calculated demand for non-renewable primary energy for heating, ventilation, cooling and hot water,
- partitions and technical equipment of the building should meet at least the requirements of thermal insulation.

Since January 2017, the values of permissible EP ratios for newly constructed buildings and certain U-values for the building envelope have changed, in accordance with the provisions of the ordinance amendment on the technical condition ordinance to be met by buildings and their location, which came into force on 1 January 2014 [57]

The gradual introduction of regulations is aimed at bringing all participants in the construction market into compliance with current legal requirements. The solution is that all new buildings should be near-zero energy buildings.

The Ministry of Climate and Environment and National Fund for Environmental Protection and Water Management are implementing a project to help households improve household energy efficiency. The Clean Air Program [57] was launched in Poland in 2018 and will last until 2030. The program is aimed at owners and co-owners of single-family houses, or separate residential units in single-family buildings with a separate land register. Subsidies for replacement of heat sources and thermal modernization of the house include up to PLN 30, 37 or 69 thousand, and up to PLN 47 or 79 thousand for subsidies with pre-financing. The subsidy can be used to replace outdated and inefficient solid fuel as a heat sources with modern heat sources which fulfill the highest standards and carry out the necessary thermomodernization work on the building [57].

4.2. Review of the State Policy in Slovakia

The legislation amendments in energy efficiency law and other minor issues in public ownership laws entered into force on 1st February 2019 in Slovakia. The most significant among them is Act no. 321/2014 Coll. on energy efficiency. It was adopted based on the Directive 2012/27/EU [5] (this directive replaced Directive 2006/32/EC) [58]. Besides that, Slovakia adopted an Integrated National Energy and Climate Plan for years 2021–2030 [59].

The Integrated National Energy and Climate Plan for the years 2021–2030 sets the following climate and energy goals by 2030:

- reduction of greenhouse gas emissions in the non-ETS sectors compared to 2005's level by 12% (however, strategic document Envirostratégia 2030 [60] increased the plan to reducing greenhouse gas emissions up to 20% in 2030),
- to increase the contribution of RESs to gross final energy consumption to 19.2% in 2030 (an increase of 5.2% compared to 2020), considering:
- a 14% contribution of RESs in transport,
- a 19% contribution of RESs in the production of heating and cooling,
- a 27.3% contribution of RESs in electricity generation.

The war in Ukraine has intensified pressure on the energy self-sufficiency of Europe, including Slovakia. The shift away from Russian fossil fuels and the emphasis on renewable energy sources (RESs) is influencing various areas and economy sectors in Slovakia, and should be even more pronounced in the future. In 2020, the share of Russian gas in the

European Union was 38%. In Slovakia, it was significantly higher—up to 85% of Slovakia's gas came from Russia. Possible alternatives, including RESs, are therefore being considered as inevitable.

Unlike countries such as Sweden, Finland, Latvia or neighboring Austria, Slovakia is lagging far behind in the use of RESs. The share of energy that Slovakia obtained from RESs in 2020 was only 17.3%, with wood and wood chips contributing significantly. This was a relatively high share for Slovakia, as Slovakia's original target was only 14%. In comparison, Sweden, for example, had more than 60% of its energy come from renewable sources (with a target of 49%) [61].

According to data from the Statistical Office of the Slovak Republic, gross electricity production from renewable sources in Slovakia in 2020 amounted to 7279 GWh. The largest share, more than two-thirds (67.2%), was contributed by hydroelectric power plants. This was followed by wood (including wood waste and other solid waste) with a share of 15.4%. Solar photovoltaics accounted for 9.1% of the RESs in electricity generation, biogas accounted for exactly 7%. Gross heat production from RESs in 2020 was 6 518 TJ. Wood (including wood and other solid waste) accounted for the largest share, up to 82.8%. This was followed by biogas with a share of 11.1%. Energy recovery of industrial and municipal waste accounted for 3.1%, geothermal heaters for 2.8% [62].

The basic institutional and financial instrument for increasing Slovakia's energy self-sufficiency is the so-called Recovery and Resilience Plan. It is a time-bound mechanism to support recovery and resilience and it is a key instrument of the NextGenerationEU initiative [63]. It was developed and approved on the basis of the Regulation of the European Parliament and of the Council of 12 February 2021, when the Recovery and Resilience Facility 2021/241 was established. On 16 June 2021, the Government of the Slovak Republic approved Act No. 368/2021 Coll. on the Recovery and Resilience Support Mechanism and on the amendment and supplementation of certain acts.

The Recovery and Resilience Plan focuses on five key public policies in Slovakia:

- Green economy—allocation of EUR 2301 million.
- Education—allocation of EUR 892 million.
- Science, research and innovations—allocation of EUR 739 million.
- Health—allocation of EUR 1533 million.
- Efficient public administration and digitalization—allocation of EUR 1110 million.

The Slovak Republic has committed to allocate EUR 2.73 billion, 43% of the total allocation of the Recovery Plan, to the stated objectives of green transformation and climate change mitigation. The green economy area is divided into five components, namely:

- Component 1 Renewable Energy and Energy Infrastructure—allocation of EUR 232 million.
- Component 2 Renewal of Buildings—allocation of EUR 741 million.
- Component 3 Sustainable Transport—allocation of EUR 801 million.
- Component 4 Decarbonization of Industry—allocation of EUR 1368 million.
- Component 5 Adaptation to Climate Change—allocation of EUR 159 million [64].

The largest program from the Renewal of Buildings Plan will be the Single-Family Home Renewal. In total, it is planned to support the renovation of over 30,000 family houses. The subsidy per house will be a maximum of EUR 19,000 and at the same time a maximum of 60% of the total cost. It will only be possible to support projects that meet the requirements of a minimum of 30% primary energy savings. The announcement of the first call for renovation of family houses was published on 6.9.2022 and the call for project proposals was open until 17.10.2022 [65].

A restoration grant may be awarded if the project cumulatively meets the following conditions:

- the renovation must result in primary energy savings of at least 30% compared to the pre-renovation condition;

- the renovation of the house must include at least one measure from measures of group A. Group B measures are optional. Group A measures to improve the thermal performance of buildings include the following options:
  1. insulation of the building shell;
  2. insulation of the roof covering;
  3. replacement of window openings;
  4. insulation of the floor of an unheated attic (or any unheated space in the family house, under which there is a heated space);
  5. insulation of the ceiling of an unheated basement (or any unheated space in the family house above which the space is heated);
  6. insulation of the ground floor (of a heated room in the family house under which there is no other room).

Group B measures include five subgroups, namely: 1. installation of the energy source including putting it into operation (heat pump, photovoltaic panels, solar collectors, gas condensing boiler, heat recuperation, other heat source, e.g., electric boiler); 2. green roof (intensive or extensive); 3. rainwater storage tank (underground or aboveground); 4. installation of shading technology; 5. removal of asbestos.

5. Case Studies—Increasing Energy Efficiency Linked with Improvement of Well-Being

In the following, we offer a case study from Slovakia (replacement of heating equipment (measure group B) and insulation of the house (measure group A)), and a case study from Poland (impact of the use of RESs on the energy efficiency of a single-family building as well as energy savings and operating costs). Both case studies document the situation of family houses which are implementing possible measures to reduce energy costs or to increase energy efficiency. While the case study from Slovakia proposes to provide procedural aspects of the whole sustainability strategy, from the European Energy Self-Sufficiency Plan, through the National Renewal Plan, to the steps and actions that citizens have to use to put these intentions into practice in their houses, on the other hand, the case study from Poland is oriented more practically (it documents the calculation of the energy savings in the single-family house in using various kinds of RES). The basic assumption is that such changes will improve life quality (including well-being) of the population and their families, protect the environment, offer positive educational environmental role models for the younger generation and at the same time provide business and employment opportunities for companies that can be involved as contractors in the implementation of the projects.

The purpose of this section is to analyze the national policies of Poland and Slovakia, which are based on European Directive 2012/27/EU on energy efficiency and which introduce new measures aimed not only at increasing energy efficiency, but also at increasing the well-being of households and individuals. The two countries' policies are analyzed in the context indicated above. In the current phase of policy implementation in both countries, it would be expected that the results on energy savings from previously supported projects are known. However, the results of the analysis conducted show that there are information gaps in this area in both countries. They also indicated that the data refer to overall energy consumption, without being able to determine energy consumption for single-family houses (and thus being able to determine through the development trend the possible energy savings from projects supporting green households or home insulation). We consider this fact a significant shortcoming of the administrative procedure. The analysis showed that there is a lack of comprehensive information in the studied area, with which it would be possible to show how the energy-saving policies of Poland and Slovakia affect the individual well-being of the population. In both Poland and Slovakia at this moment (December 2022), there are no relevant statistical data on the actual use of RESs in family houses nor on the energy savings due to the implementation of government measures to promote the use of RESs in family houses and on the impacts of the implementation of RESs in households on the well-being of individuals.

For this reason, we decided to supplement the legislative side (national policies) with the procedure of applying for financial support for RES use in households (the currently ongoing call for RES use in households in Slovakia) and with a case study, in which we offer a technical calculation of energy savings and operating costs in the case of an exemplary house in Poland.

5.1. Slovakia—Case Study Documenting Administrative Procedure

Green households

National Green Households projects in Slovakia are prepared within the Operational Programme Environmental Quality, which has been managed by the Ministry of the Environment of the Slovak Republic since 2015. The project is part of Priority Axis 4 which is aimed at promoting an energy-efficient, low-carbon economy in all sectors. The purpose of the projects is to provide support for the installation of small-scale renewable energy sources (RESs), which will reduce the use of fossil fuels. The support is given for electricity generation installations, namely photovoltaic panels, and heat generation installations, which are solar collectors, biomass boilers and heat pumps. The goal of the project is to increase the involvement of RES use in households and the related reduction of greenhouse gas emissions.

The implementer of the national project is the Slovak Innovation and Energy Agency (SIEA). Thanks to the European and state support provided through the national Green Households project, since 2015, 18,502 devices for the use of renewable energy sources have been installed in Slovak households. Among them, there were 3673 photovoltaic panels, 6974 solar collectors, 2613 biomass boilers and 5242 heat pumps. In the pilot Green Households project, EUR 45 million was made available. Although the latest claims were received and paid in 2018, official data on the actual energy savings achieved for 2015–2018 are not yet known.

The continuation of the pilot project, which ended in 2018, is the national Green Households II project, under which households outside the Bratislava Self-Governing Region can benefit from European support for home renovation between 2019 and 2023. Green Households II is the second phase of support aimed at the use of so-called small-scale renewable energy sources in family and apartment buildings.

The rules for supporting the installation of small-scale RES installations are defined in the General Conditions for Supporting the Use of Renewable Energy Sources in Households. Support for the installation of small-scale RES installations shall be implemented through the issue of vouchers, in time-limited or otherwise limited rounds. Specific conditions are also defined for each round. The specific conditions specify, in particular, the type of small RES installations currently supported, the eligible territory or the number of individual subsidies. A small installation for electricity generation is an installation with a capacity of up to 10 kW. In the case of heat production, a small installation is an installation covering the energy needs of a building used by natural persons for residential purposes.

The support for the installation of small equipment for the use of RESs includes the provision of a financial contribution for the installation of small equipment to produce electricity or heat from RESs, which are:

1. small installations to produce electricity with an output of up to 10 kW:
    (a) photovoltaic panels (electricity production),
    (b) wind turbines (electricity production) (it is not yet possible to issue vouchers for these installations),
2. heat production installations that cover the energy needs of a family or apartment building:
    (a) solar collectors (heat production),
    (b) biomass boilers (heat production),
    (c) heat pumps (heat production).

The specific type of production facility for which a contribution may be granted must be listed in the approved list of eligible facilities maintained by the SIEA.

The following persons are entitled to install photovoltaic panels, wind turbines, solar collectors, biomass boilers and heat pumps in a family house:
1. a natural person who is the owner of the family house,
2. natural persons who are joint owners of the family house,
3. a natural person or natural persons who are co-owners of the family house and who are entitled to make decisions on the management of the common property according to the majority of their shares.

All categories of owners and co-owners are also subject to additional rules strictly regulating additional requirements concerning ownership, use or size of the heating area of the house.

In the second half of 2022, the Slovak Innovation and Energy Agency responded to an exceptional situation, as the interest of households in allowances for heat pumps, biomass boilers, solar collectors and photovoltaic panels is growing exponentially, but the delivery times of some technologies are becoming longer, and the installation capacity of contractors may also be limited. The current call for applications is therefore open from 13 June 2022. Thanks to an additional EUR 30 million increase in the project budget, households were able to benefit from support for all four installations—heat pumps, biomass boilers, solar collectors and photovoltaic panels—in the second half of 2022.

The vouchers are valid for 90 days, but when they are active and issued is now decided by households in agreement with contractors. This is based on when the equipment is ready to be installed. This is a fundamental change that responds to contractors' limited personnel, technological and time capacity.

The list of eligible contractors will be continuously updated by the Slovak Innovation and Energy Agency. Contractors can apply for the list at any time during the duration of the Green Households projects. All applicants who meet the conditions and commit to provide installations in accordance with the contract will be included in the list. SIEA publishes the list of contractors who have an effective voucher reimbursement contract.

For contractors who have agreed to provide their contact details to the public, telephone numbers and e-mail addresses are also published in the list. Additional contacts will be added upon receipt of consent at the next update of the list. As of 20 September 2022, there are 1386 eligible contractors on the list throughout the country.

House insulation

The development of interest in the contribution for insulation and reconstruction of older family houses shows that the Slovaks are more and more often engaged in the renovation of their homes. The contribution for the house insulation started to be granted in Slovakia in 2016; since then, a total of eight calls for projects and applications for support have been announced. In the first five calls from 2016–2019, the interest of the residents was so low that even within 30 days of the opening of the call, the reserved limit of 500 applications could not be met. In response, the Ministry of Transport and Construction of the Slovak Republic significantly reduced the number of applications accepted to 150 (in the 6th call, April 2020); in the last two calls (autumn 2020 and May 2021), the number of applications accepted was adjusted to 200. However, residents' interest in the insulation allowance has increased significantly since the start of the COVID-19 pandemic: in April 2020, the limit of 150 applications was reached within two days, in autumn 2020 it was reached in 24 hours, and in May 2021, the limit of 150 applications was reached within 17 minutes.

The year 2022 brought significant changes in the field of insulation of family houses. As of 8 March 2022, the contribution for the insulation of family houses has been suspended (eight calls since 2016), with the intention that contributions for the insulation of houses should be provided in the future from the SR Recovery and Resilience Plan project. Under the Recovery and Resilience Plan, the insulation of houses falls under the subprogram "Green Recovery". Owners of older family homes can therefore count on a significantly strengthened program to support home renovation and insulation. The Green Recovery Program should be one of the biggest measures to kick-start the economy after the COVID-19 pandemic. The funding in the program, which will flow to Slovakia under the Recovery and Resilience Plan, should be enough to renovate 30 thousand houses. The chances of receiving a grant for insulation and other home renovation should thus be significantly higher than before, when the state only granted a grant to 405 projects since 2016 (Table 2).

Table 2. Supported house insulation projects (2016–2021).

Round No.	Date of the Call for Proposals	No. of Applications	No. of Supported Projects
1.	16 March 2016–12 April 2016	230	39
2.	9 June 2016–6 July 2016	90	14
3.	1 August 2017–31 October 2017	246	34
4.	1 April 2018–31 August 2018	284	56
5.	17 June 2019–30 September 2019	317	63
6.	1 April 2020–30 June 2020	150	95
7.	2 September 2020–30 October 2020	150	80
8.	18 May 2021–30 June 2021	200	24 to date

According to the latest information [66], the call for applications from the Green Recovery Program opened in autumn 2022. The Ministry of the Environment has determined that the condition for obtaining the financial support will be the achievement of 30% savings of primary energy resources through a comprehensive renovation of the house. This will primarily involve insulating the façade and roof of the house, replacing windows and doors, upgrading heating and using renewable energy sources, as well as installing external shading and green roofs. Preliminarily, the maximum contribution per house is EUR 16,600. However, it is possible that the state may decide to support a larger number of houses, which will reduce the contribution somewhat. The Ministry of the Environment is considering contributing up to 50% of eligible costs.

Citizens can apply for support through the website of the Slovak Environmental Agency [67] based in the capital Bratislava. There are also ten regional offices (in 10 cities) where citizens can obtain help with their applications. The Ministry of the Environment announced that by the end of 2022 it would like to approve financial support for the first four thousand family houses.

The essential feature of the Green Recovery Program is that it will insist on the comprehensive renovation of houses. The two basic requirements for project funding are that any renovation must result in maximum energy savings for operation and reduce air emissions from heating. Grants for only partial works, such as replacing windows or roofs, would not achieve the necessary effect and it is positive that the state has ruled them out in advance.

Based on the call for proposals of 6 September 2022 [65], the eligible applicant must fulfill the following requirements:

- he/she is a citizen of an EU country,
- he/she resides permanently in a house (for the renovation of which he/she is requesting funding from the mechanism),
- he/she has a legal capacity,
- he/she is either the owner or the owner in common or the co-owner of the house,
- he/she has not been convicted of certain types of criminal offences,
- he/she has no tax debts for which the aggregate amount exceeds EUR 170, i.e., he/she is not on the list of tax debtors,
- he/she has no social security debts of more than EUR 100,
- he/she is not on the list of debtors for health insurance,
- he/she is not the subject of an execution under the Execution Code,
- he/she is the holder of the bank account mentioned in the application.

5.2. Polish Case Study of the Impact of the Use of Renewable Energy Sources on the Energy Efficiency of a Single-Family Building As Well As Energy Savings and Operating Costs

The requirements for energy consumption that must be met by new and modernized buildings are related to legal requirements and global trends. We strive for large energy savings through thermal modernization of buildings, greater use of renewable energy sources (RESs), electrification of heating and continuous improvement of the energy efficiency of the economy, including construction. Specific requirements are imposed on investors by the Act on RESs, energy performance and energy efficiency. These requirements are also included in the Regulation on technical conditions to be met by buildings and their location, which from 2021 set expectations for almost zero-energy buildings [68–71].

The obligation to produce energy performance certificates for buildings sold or rented, introduced in 2009 by the Act on the energy performance of buildings, is another contribution to improving the energy efficiency of buildings. This act also introduced the term "nearly zero-energy consumption building" (nZEB) and this is the standard applicable to all new buildings from 1 January 2021 [72,73].

Achieving the zero-emission building standard now, mandatory for all buildings erected after 2029, is not difficult and brings measurable operational benefits. The requirements and recommendations for new buildings included in the draft revision of the EPBD directive concern not only the lack of emissions at the location of the building, but also [68–70]:

- higher smart readiness indicator for buildings (SRI), i.e., the practical implementation of the idea of intelligent buildings ensuring high comfort, safety and controlled energy consumption,
- high share of renewable energy in the balance of energy consumed by the building,
- a higher level of self-consumption and autarky of energy from own (prosumer) photovoltaic systems with electricity storage,
- a healthy indoor climate thanks to effective and controlled ventilation,
- low energy costs for heating and hot water,
- the possibility of cooling rooms thanks to surface installations,
- better loans and higher creditworthiness thanks to low operating costs,
- higher value of the building on the market.

The draft revision of the EPBD directive on the energy performance of buildings from December 2021 provides that, by 2030 at the latest, new buildings will have to be erected as zero-emission buildings (ZEBs; this concept has replaced the nZEB standard, i.e., nearly zero-energy buildings), and in the public sector by 2027. Such buildings are to show low energy demand and use local renewable energy sources, they are to be devoid of fossil fuel combustion sources and they are also to have a low impact on global warming, related to CO_2-equivalent emissions throughout the life cycle. Additional conditions are: for residential buildings, EP < 65 kWh/(m^{2}year), no CO_2 emissions from heating devices used in the building and production of energy from renewable energy—e.g., a heat pump with the equivalent annual production of electricity from PV—realized in the building or as part of a local energy cooperative [71,74,75].

The analyzed single-family building

For the analysis of the impact of the use of RESs on the energy performance, a standard single-family building that meets the Technical Conditions WT2021, currently in force in Poland, was adopted. Taking into account the current trends in single-family housing, the selected construction project can be erected on a plot with minimum dimensions of 19.5 m × 17.5 m. The building has a simple and compact body, two stories with an attic, a gable roof and no basement.

The building contains a total of 15 rooms on the ground floor and attic with a total usable area of 154.89 m^2.

The building model was mapped in the SANKOM Audytor OZC 7.0 Pro software (Figure 2), which enables numerical simulation and energy performance calculations for a specific building model. Defining the model was a very important step during the construction of the model. The calculations were made on the basis of the PN-EN 12831: 2006 standard [75–80].

 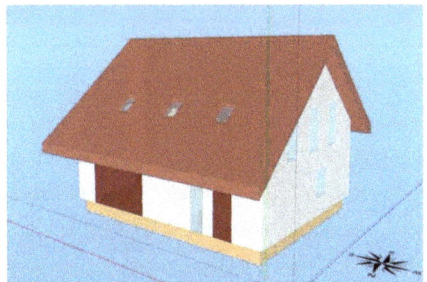

Figure 2. Analyzed building—view of the model of the analyzed building made in Audytor OZC (own elaboration).

A key element before evaluating the results and selecting the optimal solution for our building is to understand the meaning and method of determining the three key indicators in classifying the energy efficiency of a facility as a structure and the installations it contains. All buildings intended for human use, including public utility, industrial and residential, can be clearly characterized by calculating individual energy demand coefficients: primary EP, utility EU and final EK [72,73,81,82].

Moving on to a further analysis of the influence of the use of renewable energy sources on the energy performance of a building, several assumptions should be made that define the subsequent thinking. It is assumed that in terms of external partitions, such as external walls, floor on the ground, roof and ceilings with windows and external doors, the building meets all the requirements contained in the technical conditions of WT2021 [75,78,79,83,84]. In order to provide an overview spectrum and to show the differences and the impact of renewable energy sources on the energy performance of a building, several variants are set up (Table 3) that provide heating of the building surface and domestic hot water, the spectrum of which starts with energy from fossil fuels, through a partial share of renewable energy sources, up to the full share of RESs in the energy mix.

Table 3. List of variants of the analyzed models in central heating, hot water, ventilation and electricity installations for the analyzed building (own elaboration).

	Central Heating	Ventilation	Hot Water	Electricity
1.	Coal	Gravitational	Coal	Electrical grid
2.	Biomass	Gravitational	Biomass	Electrical grid
3.	Natural gas	Gravitational	Solar collectors	Electrical grid
4.	Electrical grid	Gravitational	Electrical grid	Electrical grid
5.	Heat pump powered by the electricity grid	Gravitational	Heat pump powered by the electricity grid	Electrical grid
6.	Heat pump powered by a photovoltaic installation	Gravitational	Heat pump powered by a photovoltaic installation	Electricity from a photovoltaic installation
7.	Electricity from a photovoltaic installation	Mechanical supply and exhaust	Electricity from a photovoltaic installation	Electricity from a photovoltaic installation
8.	Heat pump powered by a photovoltaic installation	Mechanical ventilation with recuperation	Heat pump powered by a photovoltaic installation	Electricity from a photovoltaic installation

When calculating the annual final energy demand, it is worth emphasizing the fact that it should be expressed as the ratio of the usable energy demand, which is determined by the heat balance of the building, to the average seasonal efficiency of the heating system.

The calculations are aimed at determining the demand for heating purposes and are given in the equation below [76,77]:

$$Q_{K,H} = \frac{Q_{H,nd}}{\eta_{H,\,tot}} \left[\frac{kWh}{year}\right] \quad (1)$$

where:

$Q_{H,nd}$—useful energy demand to heat a residential building (useful heat),

$\eta_{H,tot}$—average seasonal efficiency of the building's heating system.

An important element that is taken into account, apart from the energy intended to heat the building, is that used to prepare domestic hot water, and the energy demand for it is defined by the formula given below [76,77]:

$$Q_{K,W} = \frac{Q_{W,nd}}{\eta_{W,\,tot}} \left[\frac{kWh}{year}\right] \quad (2)$$

where:

$Q_{W,\,nd}$—demand for preparation of domestic hot water,

$\eta_{W,\,tot}$—average annual efficiency of devices preparing domestic hot water.

The above two equations allow us to determine the final energy factor (EK) of which they are components. It is important for both the auditor and the user that the EK, EU and EP coefficients are given as the number of necessary kilowatts used to heat a square meter of the building's surface throughout the year. The following equation shows us the method of calculating the required final energy EK to supply the building [76,77]:

$$EK = \frac{Q_{K,\,H} + Q_{K,\,W}}{A_f} \left[\frac{kWh}{year \times m^2}\right] \quad (3)$$

where:

A_f—heated or cooled space in the building with a specific temperature, expressed in m^2.

The primary energy demand factor is calculated on the basis of the annual energy needs for heating and domestic hot water preparation and, where applicable, cooling or ventilation of the premises and lighting in specific cases. The aforementioned EP coefficient is calculated from the following dependence:

$$EP = \frac{Q_P}{A_f} \left[\frac{kWh}{year \times m^2}\right] \quad (4)$$

where:

Q_P—annual demand for primary energy (kWh/year).

Of which the annual primary energy demand is calculated according to the formula:

$$Q_P = Q_{P,H} + Q_{P,W} + Q_{P,C} + Q_{P,L} \left[\frac{kWh}{year}\right] \quad (5)$$

where:

$Q_{P,H}$—annual primary energy demand through the heating and ventilation system for heating and ventilation,

$Q_{P,W}$—annual primary energy demand by the domestic hot water preparation system,

$Q_{P,C}$—annual primary energy demand for the space ventilation and cooling system,

$Q_{P,L}$—annual demand for a lighting system (calculated only for public buildings).

As can be seen from the above information, all indicators are directly related to each other. Having information about the height and proportions between individual indicators, we are able to relatively easily assess the overall quality of the building with the technologies used in the facility, looking through the prism of thermal modernization of partitions, the number and degree of elimination of thermal bridges, tightness of the structure as well as efficiency and efficiency of the heating installation rooms and domestic hot water.

After conducting an energy analysis of the building and carefully reading the results, the following conclusions can be reached. The main goal and criterion was to meet the requirements contained in the technical conditions of WT2021, which defined a number of requirements for the structure of the building, thermal transmittance of external partitions and energy coverage of the building's heating needs, along with meeting the needs of users related to domestic hot water. In fact, it turned out that the above-mentioned partitions not only meet the guidelines, but in most cases are of much higher quality in terms of insulation. The key element during the project implementation was to check which of the provided solutions are characterized by the best results in terms of energy demand, and above all, how renewable energy sources affect the final results of the building, together with an assessment of whether their use had a positive effect on the balance or vice versa. The seven analyzed variants could be divided into three characteristic subgroups in a simplified way. The first subgroup was characterized by the overall coverage of demand with non-renewable energy sources such as coal heating and electricity from the power grid, as well as fully electric heating from electricity from the power grid. The second subgroup was characterized by partial use of renewable energy sources. Among them, there were variants in which the object was heated by burning biomass and connecting household appliances to the power grid, burning natural gas for heating purposes and providing domestic hot water by means of solar collectors and electricity from the grid. The last variant in this subgroup was the use of an air source heat pump powered by the electricity grid. The third subgroup was characterized by the fact that both the heating of the building with the preparation of domestic hot water and electricity were fully generated by renewable energy sources.

In the case of the first variant in this subgroup, heating and domestic water were prepared by an air source heat pump powered by a photovoltaic installation. In the second case, in addition to fully electric heating, mechanical supply and exhaust ventilation were also used. The electricity of the second variant came entirely from the installation of photovoltaic cells.

When analyzing the results generated in the reports for each of the variants, it can be quickly noticed that only three out of seven proposed solutions met the guidelines, one option was close to meeting the guidelines, while the other three variants did not even come close to the stringent requirements of the WT2021 technical conditions (Figure 3). The worst solutions turned out to be those that obtained energy from the combustion of hard coal and obtained electricity from the power grid for heating purposes. The last three options managed to meet the primary energy requirements. The first variant obtained energy from the biomass combustion process, and more precisely from birch wood with very low humidity. The second solution that met the guidelines was electric heating of the building using energy from the installation of photovoltaic panels with the simultaneous use of mechanical supply and exhaust ventilation. The use of this variant made it possible to notice certain regularities. Firstly, the use of mechanical ventilation in the building generated profits thanks to the reduction of the EU's utility energy demand by 25%. Secondly, thanks to the application of the above-mentioned installation, it was also possible to reduce losses in terms of final energy (EK), which can be seen particularly easily when comparing the EK of variant VII with IV. Despite the many advantages that this installation brought, it cannot be called optimal, because the energy from the photovoltaic installation is closely related to weather conditions and, to heat the building with energy from PV panels, it would be necessary to use a fairly large number of panels. The last solution that met the guidelines was the variant which consisted of a photovoltaic installation supplying an efficient heat pump cooperating with gravity ventilation. The use of a variant containing such components obtained the best results compared to other solutions, looking through the prism of the obtained values of non-renewable primary energy (EP) and final energy (EK).

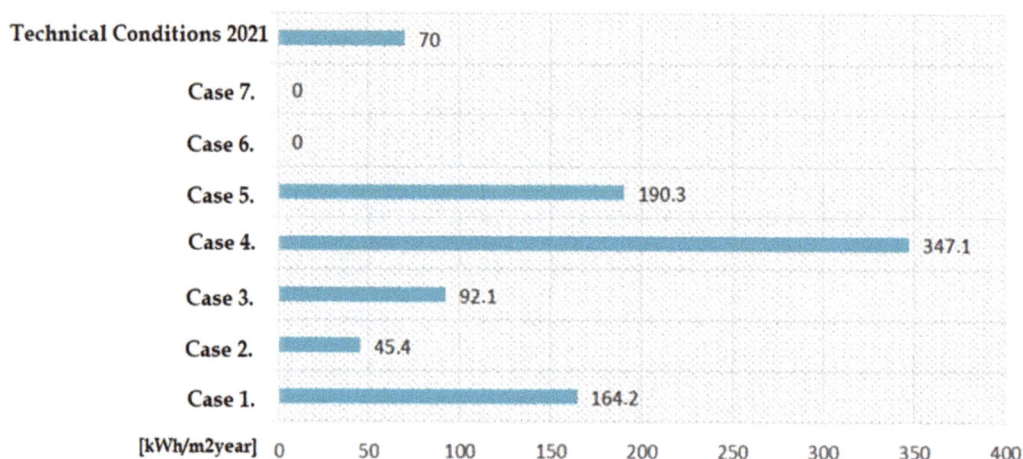

Figure 3. A collective summary of the results of the demand of the analyzed variants for EP in relation to the EP value established by the Technical Conditions for 2021 (own elaboration).

A solution that could drastically reduce EP, EK and EU coefficients would be variant VIII, which would include a compressed air source heat pump with a seasonal efficiency of 2.6, covering a total design heat load of the building of 7.5 kW in CO and DHW systems, inclusion of a mechanical ventilation installation with recuperation and powering all components and installations in the building from a 9 kW photovoltaic installation, covering the demand of the entire building at the level of 8400 kWh/year (Table 4).

Table 4. A collective summary of the results of the demand for individual energies for variant VIII (own elaboration).

Assessment of the Energy Characteristics of the Building		
Energy Performance Index	Building Being Assessed	Requirements According to Technical and Construction Regulations 2021
Annual useful energy demand indicator	EU = 58.3 kWh/m²year	
Annual final energy demand indicator	EK = 39.6 kWh/m²year	
Annual demand for non-renewable primary energy	EP = 0.0 kWh/m²year	EP = 70.0 kWh/m²year
Unit amount of CO_2 emissions	E_{CO2} = 0.0 $MgCO_2$/m²year	
Share of renewable energy sources in the annual final energy demand	U_{OZE} = 100.0 %	

In the above solution (variant VIII), we can also note that due to the 100% share of renewable energy sources in the annual final energy demand, the analyzed building is zero-emission, powered only by its own energy sources—it is a zero-energy building (ZEB).

Taking into account the entire spectrum of possibilities and information provided by this analysis, it can be stated without any doubts that the use of renewable energy sources in building heating systems and domestic hot water brings many benefits, looking through the prism of sustainable energy consumption, economics and increased comfort of life. It is worth paying attention to the fact that the use of renewable energy sources gives tangible benefits in every case of use in buildings, however, the current requirements for primary energy pose new challenges for investors, which, as it is easy to see, can only be met by renewable energy sources.

6. Discussion

One of the strategic aims of the state's energy and ecological policy is to improve the energy efficiency of the economy [5,41,58–60]. Energy efficiency is related to the area of energy use and it is particularly crucial in the process of ensuring security of energy supply, ecological security, increasing the competitiveness of enterprises and many other elements. The issue of energy efficiency is a priority, as an advance in this area is vital for the implementation of all energy policies and most environmental and climate policy objectives. The primary purpose in the area of efficiency, in addition to the targets set out in the efficiency directives [5,46–50], is currently to achieve a reduction in energy consumption compared to the projections for 2030 as a result of energy efficiency improvements. The main objective of the article was to analyze of Polish and Slovakian national policies, which are based on the European Energy Efficiency Directive 2012/27/EU, and which present new measures focused on not only increasing energy efficiency, but also increasing the well-being of households and individuals. Many organizations and countries are working to make zero-energy and resource-free buildings a common practice in new constructions by 2030 or earlier. For example, the EU requires all Member States to establish building codes that required that by 2021 (public buildings from 2019) newly erected buildings would have energy consumption close to zero, and Canada is currently developing legislation on the gradual approach to zero energy buildings (ZEBs). A key step towards net-zero energy is energy efficiency, which can drastically reduce demand by up to 80% compared to typical new designs, thus allowing moderately small renewable energy systems to deliver the remaining energy at a lower cost. The same is true for waste and water, where the emphasis is first on minimizing consumption and then on finding alternatives to achieving the net-zero target.

In order for net-zero buildings to become a common practice, specific targets should also be set in Poland and Slovakia, such as building regulations which will make it mandatory to build new buildings with a net-zero balance by 2030 [42,53,59,60,85]. To achieve this, it is essential to have technical assistance for architects, engineers and builders and innovative packages for developers, as well as research and development on material efficiency strategies (low-carbon alternatives and demand reduction). Strategies should be developed for energy-intensive buildings, such as hospitals and shopping centers. Buildings are essential in the context of the EU's energy efficiency policy, with more than 40% of final energy consumption (and 36% of greenhouse gas emissions) occurring in homes, offices, shops and other buildings. In addition, the sector has the second largest, after the energy sector itself, untapped and cost-effective energy saving potential. There are also important cobenefits of increasing the energy efficiency of buildings, such as creating new jobs, reducing energy poverty, improving health and greater energy security and industrial competitiveness. The experience of the last few years shows that Member States are increasingly making use of cohesion policy funding for energy efficiency, especially for buildings, and that they are increasingly using financial instruments. However, there is a lack of comprehensive data on the impact of this financing on energy savings in the construction sector. For a comprehensive solution to the use of RESs and the reduction of energy consumption, it is not sufficient to focus only on public buildings and buildings that, due to their use, are linked to high energy consumption. Awareness of the use of RESs in households should also play an important role in European and national policies. This not only leads to an increase in energy efficiency, but also, as recent research confirms [32–35], to a positive impact on the well-being of individuals.

Desk research and content analysis methods were used. The current situation in both countries is illustrated by case studies documenting the administrative process (Slovakia) and the calculation of energy savings (Poland) using renewable energy sources in the case of single-family houses. Case studies describe ongoing (Poland) and planned (call for proposals) (Slovakia) instruments aimed at increasing energy efficiency and increasing the well-being of residents and households. Results show that governments of both countries are undertaking several actions to increase interest of individuals and family house owners to use RESs in family houses. This outcome is supported by [33–35], who stressed the importance of quality government policies supporting RESs to increase the well-being of individuals. An analysis of European legal documents [5,40,44–51], existing financial support measures for energy efficiency in buildings [57,59,64] and different market barriers shows that the situation differs considerably between Member States in terms of the building stock, financial support measures and significant market obstacles. Although investments in the energy efficiency of buildings are rising and there are many examples of good practice for instruments that bring cost-effective energy savings, there is only limited information on the effectiveness of the different financial support measures, both at the EU and national level; there are still significant barriers to the further implementation of investments in the energy efficiency of buildings, including a lack of awareness and expertise in the financing of energy efficiency measures by all factors, high upfront costs, relatively long payback periods and (perceived) credit risk associated with energy efficiency investments and the competing priorities of the final beneficiaries.

If the EU is to improve energy efficiency by 2030 and achieve the ambition of further savings by 2050, it is necessary to increase financial support for energy efficiency in buildings. To this end, it is essential to ensure the proper implementation of the regulatory framework, to increase the availability of financial resources and to remove the main barriers. The Commission is involved in a number of initiatives and actions to reach these goals. However, given the nature of the building stock and the building sector and the responsibility of Member States to implement relevant legislation and remove national market barriers, Member States play a decisive role in ensuring that further cost-effective investments are made.

In addition, the important role of a case-by-case approach in the context of energy efficiency financing measures means that close cooperation between public authorities, financiers and the construction sector is essential. Equally important, building owners will have to convince themselves of the benefits of improving the energy performance of buildings, not only because of lower energy bills, but also in terms of increasing comfort and increasing the value of real estate. This may be one of the most serious obstacles to overcome in the pursuit of improving the energy efficiency of European buildings. However, this is supported by macroeconomic arguments and targeted incentives and information measures will be needed to change attitudes. An essential tool in this context is the building renovation action plans that Member States must establish under the new Energy Efficiency Directive. Increasing the energy efficiency of energy generation, transmission and use processes is a base of a sustainable energy policy, which is reflected in national and EU legal regulations and actions taken by various national and EU institutions. Improving energy efficiency is of great meaning for the implementation of all energy policy goals and most environmental and climate policy targets, which is why it should be a priority in modernizing the country's economy. It can be accomplished by building high-efficiency generation units, increasing the degree of application of high-efficiency cogeneration, reducing the rate of network losses in energy transmission and distribution and increasing the efficiency of energy end use. Improving the energy intensity indicators of the economy, in addition to significant economic benefits, brings measurable ecological effects (reduction of consumption of natural resources, reduction of pollutant emissions) which are not able to match the effects of any other solutions reducing the environmental nuisance of the power sector (change in the structure of energy consumption, construction of protective devices and installations, etc.).

Taking into account the whole spectrum of possibilities and information provided by this analysis, it can be said without a doubt that the use of renewable energy sources in building heating systems and hot tap water brings many benefits, from the perspective of sustainable energy consumption, the economic aspect and increased comfort of life. It is worth noting that the use of renewable energy sources gives measurable benefits in every case of use in buildings, but the current demand for primary energy poses new challenges to investors, which, as it is easy to see, can only be met by renewable energy sources. In the above solution (variant VIII), we can also notice that due to the 100% share of renewable energy sources in the annual demand for final energy, the analyzed building is zero-emission, powered only from its own energy sources—it is a zero-energy building (ZEB). Thanks to the low costs of use, greater savings are generated, and this is directly due to the improvement of energy efficiency of our building, the implementation of RESs and independence from external energy supplies by providing electricity and heat with our own energy sources.

These cases show, among other things, how investments in energy efficiency in buildings affect the well-being of households. They are the basis for prosperity and health of citizens and are the starting point for the development of innovative branches of the economy, including broadly understood energy, including distributed energy, i.e., the use of renewable energy sources. It should be remembered that by improving the energy performance of buildings, we reduce energy consumption. As a consequence, by producing our own energy, we significantly reduce the operating costs of buildings. Minimal operating costs, together with simultaneous improvement of energy and environmental efficiency and independence from external energy supplies, directly improve the security of energy production and the sense of security of residents and users of such buildings and their well-being. The sense of security and energy neutrality combined with low operating costs directly improves the well-being and comfort of life of users of zero-energy buildings.

The growing public awareness of the beneficial impact of renewable energy systems on the environment and support in the form of various programs subsidizing the implementation of new installations make the production of energy from renewable sources more and more popular and widespread.

Actions aimed at improving energy efficiency and reducing final energy consumption are undertaken by many countries across the world and the European Union. In 2012, the European Parliament and the Council of Europe published Directive 2012/27/EU, which imposes an obligation on Member States to take action to reduce final energy consumption by 1.5% per year.

Despite the still low level of belief about the profitability of using renewable energy systems, RES installations are positively perceived and recognized as a new trend in both single- and multifamily construction. The growing share of renewable energy in the national energy system affects the reduced demand for energy produced from conventional sources. This obviously translates into reduced consumption of primary energy, for example, fossil fuels. As a consequence, this translates into a reduction in the exploitation of the resources of these raw materials, and thus contributes to the protection of the natural environment.

Findings related to CO_2 are in accordance with previous studies [25,33] that pointed to the negative impact of CO_2 emissions on human health and overall individual well-being. Improving energy efficiency, reducing the operating costs of buildings and implementing the assumptions resulting from EU directives are also extremely important in terms of utility, and translate into the comfort of using buildings, their cost-free operation and the increase in the well-being of users.

The analysis of energy efficiency of buildings with the use of renewable energy sources presented in the publication refers not only to the improvement of prosperity and well-being of single- and multifamily housing. Efficiency understood as the principles of rational energy management, either regarding efficiency, where the main goal is to maximize the effect (e.g., the use of RESs, energy savings, production volume), or regarding savings—as minimization of outlays, operating costs—where it is characterized by the widest range of content covered mainly due to the fact that it concerns the relationship between effects, goals, inputs and costs. In economic theory, it is associated with the concept of the Pareto optimum, i.e., a combination of goods at which the level of utility of all market participants is maximum. Broadly understood improvement of energy efficiency refers to many aspects, e.g., construction, professional energy industry, etc. This is an aspect within the meaning of international documents in this regard, which is very important and fits into the world's energy policy until 2050, i.e., aiming at full decarbonization in relation to the use of fossil fuels and related emissions of harmful greenhouse gases and increasing the share of renewable energy sources in the electricity and heat production sector. This is extremely important not so much in the local aspect by ensuring prosperity, but also through business aspects such as the competitiveness of buildings on the real estate market—those with very high energy performance using RESs with very low costs of use and operation. Thanks to these aspects, the sector of manufacturing enterprises can produce products more cheaply, which means that the demand for products will be greater due to the lower price, without losing the quality of the product in any way, but only reducing operating costs.

In the analyzed energy policies, there are several important relationships between the individual, security in terms of feeling as well as energy, achieving a specific energy efficiency. People are exposed to the influence of various factors that can threaten their lives, health and reduce the comfort of their functioning. The issue of achieving and maintaining the optimal state of security in given conditions and its sense is therefore one of the fundamental goals of human activity. Security is not a constant and unchanging value, one of the most important aspects of activities aimed at obtaining an appropriate level of security is the systematic identification, observation, diagnosis and modification of the conditions that create the secure environment.

In this publication, we have dealt with the energy efficiency policy in Poland and Slovakia in the context of the well-being of the individual, which is directly related to the individual housing sector. The reference and connection of these practices is cross-sectoral and has a very wide meaning, affecting all sectors, not only the one related to housing construction.

As we mentioned earlier, it is defined by the global energy policy, aimed at energy transformation at a very high and demanding level, which is to lead to improved security of societies in terms of well-being as well as improved energy efficiency, lowering the operating costs of buildings and increasing the use of RESs, and changes must take place in all sectors of life (e.g., housing, services, production) to make these forecasts a reality.

7. Conclusions

- The added value of this paper is 1. comparison of the state of the art in two countries, belonging not only to European Union, but also to the Visegrád Four (meaning sharing the mutual past influenced by energy surplus), 2. application of adopted state measures in Poland and Slovakia for usage of RESs and for improving energy efficiency in a case study (combination of administrative procedures and energy saving calculation), 3. linking the expected global savings and benefits of RES use to increasing individual well-being (or, conversely, pointing out that the use of RESs in family houses is important for increasing individual well-being, which in turn leads to savings and a positive effect on increasing energy efficiency). The cases shown directly show how we can achieve energy and economic well-being. Reducing energy consumption in buildings directly translates into a reduction in operating costs, which is a measure of the economic well-being of residents. The production of electricity and heat from one's own resources and renewable energy sources allows an increase in energy and ecological welfare, along with independence from external energy supplies. These two cases described with analytical examples show how to achieve energy independence through the use of RESs, which translates directly into the economic and energy well-being of residents.
- The low energy efficiency of single-family houses results in an increase in heating costs, and thus also in the deepening of so-called energy poverty. This is a phenomenon where homeowners cannot afford energy or energy services that provide the ability to maintain the right temperature while using good-quality fuel. The energy efficiency of buildings means saving energy, and reducing energy production means not only financial savings, but also less environmental pollution.
- The use of RESs has many benefits—one of the most frequently mentioned and shown is the reduction of monthly bills for the operation of the building, and one can gain

- greater independence from energy suppliers. Thanks to the installation, as in the case of variant VIII—a heat pump and a photovoltaic installation—independence and a sense of security and self-satisfaction increase, which also increases the value of the property and its attractiveness for potential sale.
- The aim of the article is to show the possibility of achieving energy efficiency and the associated well-being of the population. Case studies show in what direction heating systems of single-family buildings should be modernized to meet the requirements of increasing energy efficiency with the use of renewable energy sources. Thus, from the perspective of national policies and strategies to increase energy efficiency and public awareness in this area, it may be crucial to focus attention on promoting the individual benefit of individuals and households in the form of increasing their individual well-being. The theoretical review of the literature on the subject and the results of empirical studies conducted by various entities in many countries show that investments in energy efficiency result not only in lowering energy costs and reducing energy take-off, but also improve air quality, living conditions and human health and well-being. In the context of strengthening the positive impact of energy efficiency on human health and well-being, it is extremely important to embed energy efficiency measures in financial support mechanisms and support them in local communities. The situation in Slovakia documents the European Union's efforts to increase the use of RESs, renovate buildings and thus reduce the energy intensity of individual housing. Due to the current subsidy call, both for the use of RESs [63] and home insulation [65], which are continuations of previous projects, it would be expected that the results on energy savings from previously supported projects would be known. Slovakia has gaps in this area as the latest data published by the Statistical Office of the Slovak Republic are for 2020. Moreover, these data refer to energy consumption in general, without the possibility to determine the energy consumption for single-family houses (and thus the possibility to identify through the trend of development the possible energy savings due to projects supporting green households or house insulation). Neither the Ministry of Economy nor the Ministry of the Environment provide information in this respect. We consider this fact to be a significant shortcoming of the administrative procedure. On the other hand, there is room for more comprehensive research on real energy savings due to state support (in this case, it would be advisable in the future either to conduct targeted interviews with beneficiaries of financial support or to carry out comprehensive data collection through a questionnaire survey).
- Taking into account the whole spectrum of possibilities and information offered by the case study and analysis of the energy efficiency of a single-family building, it can be concluded that the use of renewable energy sources in construction in heating and domestic hot water systems brings many benefits, looking through the prism of sustainable energy consumption, the economic aspect and increased living comfort.
- In this respect, however, it must be noted that in both Slovakia and Poland there is a lack of sufficient communication from the state (state policies) towards individuals and households to link increasing energy efficiency and well-being of individuals. Although we point out that the reduction of energy intensity through the use of RESs has a direct impact on the reduction of energy costs (case study Poland), it should be added that the reduction of energy intensity also represents an increase in the comfort of living, satisfaction with the quality of housing, quality of life and overall well-being. In the study, we did not investigate the perception of the owners of family houses regarding the increase in their well-being. Our intent was to show that state policies and state policy campaigns to increase energy efficiency should reflect the impact of RES use on increasing individual well-being in addition to savings on energy costs. Further research is also needed in this regard, especially through the collection of primary data from owners of those family houses in which energy efficiency improvements have already been made.

- It is worth noting that the use of renewable energy sources gives measurable benefits in every case of use in buildings, but the current demand for primary energy poses new challenges for investors, which, as can be easily seen, can only be met by renewable energy sources. A solution that could reduce EP, EK and EU coefficients could be variant VIII, which would include a compressed air source heat pump with a seasonal efficiency of 2.6, covering a total design heat load of the building of 7.5 kW in CO and DHW systems, inclusion of mechanical ventilation installation with recuperation and powering all components and installations in the building from a 9 kW photovoltaic installation, covering the demand of the entire building at the level of 8400 kWh/year (Table 4). Improving energy efficiency is clearly related to the improvement of the well-being of people using zero-energy buildings, among others, through implementation of the idea of intelligent buildings ensuring high comfort of use, safety and controlled energy consumption and high share of renewable energy in the energy balance consumed by the building, which in turn determines low operating costs of use.
- Thanks to the low costs of use, greater savings are generated, and this is directly due to the improvement of energy efficiency of our building, the implementation of RESs and independence from external energy supplies by providing electricity and heat with our own energy sources.
- In order to meet the requirements in force from 1 January 2021 for residential buildings specified in the current Regulation of the Minister competent for construction on the technical conditions to be met by buildings and their location, it is necessary to use RESs or connect them to an energy-efficient district heating network to which heat is supplied from high-efficiency cogeneration or RESs. When modernizing buildings, it is not always possible to take into account all the provisions of the above-mentioned regulation, so the use of RESs in them will be rather sporadic. To sum up, it can be clearly stated that in the next 10 years, most of the new and modernized buildings will use RESs to achieve appropriate energy efficiency indicators. The use of renewable energy sources also reduces the emission of combustion products as a result of reducing the consumption of chemical energy contained in primary fuels. The reduction of pollutant emissions (GHGs, dust, soot, BaP) released into the natural environment is proportional to the amount of non-renewable final energy replaced by RESs and the specific emissions of a given fuel consumed from a conventional source and the emissions of an alternative installation using RESs. This can be expressed by the difference between the emission of pollutants from a source and for a conventional installation (baseline state) and the emission of pollutants for an alternative installation based on RESs, which replaces this base state.
- Changes in the amount of pollutants emitted into the atmosphere when replacing conventional energy sources in the building with sources using renewable energy resources are shown in a single-family house example, consuming energy for central heating and domestic hot water preparation. Promoting and recommending the installations based on RESs in construction, in addition to improving the thermal insulation of building partitions, has not only a significant impact on increasing energy security but also achieving the required standards inside buildings with a lower operating cost. Doing so produces optimal and economically viable results in the lasting effects of non-renewable primary energy resources. Targeted and well-thought-out actions to rationalize final energy consumption for buildings should no longer be a challenge, but a necessary task in a sustainable low-carbon economy.
- Referring to the policies of both countries, it should be noted and emphasized that relevant laws and programs are being implemented to improve energy efficiency policies. The legal documents cited in the article are in response to the EU directives that are applicable in all EU countries. The referenced and cited documents such as PEP2040 and the NAPE (PL) became applicable in 2021. It is difficult to discuss their effects on the population at present, and therefore their impact on improving

social well-being. We noted a very large gap in the population's knowledge of energy efficiency in both countries. This is the added value of this paper, and it is not a fundamental purpose. A comprehensive impact of legal acts and public perception on energy efficiency in Poland and Slovakia will be possible in the near future, but not in this phase.

- The analyzed national policies to promote the use of RESs by private households have direct implications not only for individuals, households and individual well-being, but also for the business environment. The promotion of business opportunities is very significant in the context of the renewal schemes (focus on green households, house insulation, various systems of central heating, ventilation, hot water and electricity networks). The analyzed policies should help to revitalize the market for renewable energy equipment in households. At the same time, they have the potential to create a linkage between prospective production and technological capacities, subcontracting relationships, as well as the necessary service support for the installed equipment. All these steps also stimulate local employment growth, as the subsidies are allocated regionally. Last, but not least, the analyzed policies should contribute to improving the awareness and practice of RES installers and to increasing the interest in studying related fields. Thus, in this context, policies promoting the use of RESs in households not only have a positive impact on the well-being of individuals and entire households, but also on the development of business opportunities and therefore indirectly on the well-being of entrepreneurs. This area of linking energy intensity, household use of RESs and the development of entrepreneurial opportunities is not elaborated in the current literature, and represents a possible direction for future research.

- The EU has recently introduced ambitious new policies to persuade member states to take action to improve the energy efficiency of buildings. The new regulations take into account the fact that the main obstacle to building renovation is cost, including buildings in the productive sector. Currently, about 75% of buildings in the EU are energy inefficient. This means that we waste a significant portion of the electricity we use. It is possible to reduce energy waste by renovating existing buildings and using smart solutions and energy-efficient materials when constructing new buildings. Improving the energy efficiency of buildings therefore plays a key role in achieving the ambitious goal of carbon neutrality by 2050, according to the European Green Deal strategy [86].

- In summary, the analysis has shown that the study area lacks comprehensive information on the basis of which it would be possible to show in more detail how the energy saving policies of Poland and Slovakia affect the individual welfare of the population. Once the data contained in public statistics are completed, it will be possible to analyze the impact of the policies in question on the welfare of the population. However, the research gap discovered through the analysis requires further research projects to be undertaken.

Author Contributions: Conceptualization, A.B.M., M.K., K.S., M.M. and R.P.; methodology, A.B.M., M.K., K.S., M.M. and R.P.; software, A.B.M., M.K., K.S., M.M. and R.P.; validation, A.B.M., M.K., K.S., M.M. and R.P.; formal analysis, A.B.M., M.K., K.S., M.M. and R.P.; investigation, A.B.M., M.K., K.S., M.M. and R.P.; resources, A.B.M., M.K., K.S., M.M. and R.P.; data curation, A.B.M., M.K., K.S., M.M. and R.P.; writing—original draft preparation, A.B.M., M.K., K.S., M.M. and R.P.; writing—review and editing, A.B.M., M.K., K.S., M.M. and R.P.; visualization, A.B.M., M.K., K.S., M.M. and R.P.; supervision, A.B.M., M.K., K.S., M.M. and R.P.; project administration, A.B.M., M.K., K.S., M.M. and R.P.; funding acquisition, A.B.M., M.K., K.S., M.M. and R.P. All authors have read and agreed to the published version of the manuscript.

Funding: This research represents partial fulfillment and is a partial output of a scientific project of Grant Agency VEGA no. No. 1/0366/21 "Dependent Entrepreneurship in Slovakia—Reflection, Measurement and Perspectives" at the Faculty of Economics, Matej Bel University in Slovakia.

Data Availability Statement: Not applicable.

Conflicts of Interest: The authors declare no conflict of interest.

References

1. Omer, A.M. Energy, environment and sustainable development. *Renew. Sustain. Energy Rev.* **2008**, *12*, 2265–2300. [CrossRef]
2. Gillingham, K.; Newell, R.G.; Palmer, K. Energy Efficiency Economics and Policy. *Annu. Rev. Resour. Econ.* **2009**, *1*, 597–620. [CrossRef]
3. Soytas, U.; Sari, R. Energy consumption, economic growth, and carbon emissions: Challenges faced by an EU candidate member. *Ecol. Econ.* **2009**, *68*, 1667–1675. [CrossRef]
4. Chen, C.F.; de Rubens, G.Z.; Xu, X.; Li, J. Coronavirus comes home? Energy use, home energy management, and the social-psychological factors of COVID-19. *Energy Res. Soc. Sci.* **2020**, *68*, 101688. [CrossRef] [PubMed]
5. European Parliament. *Directive 2012/27/EU of the European Parliament and of the Council of 25 October 2012 on Energy Efficiency, Amending Directives 2009/125/EC and 2010/30/EU and Repealing Directives 2004/8/EC and 2006/32/EC*; European Parliament: Strasbourg, France, 2012.
6. Włudyka, T.; Smaga, M. *Instytucje Gospodarki Rynkowej*; Wydawnictwo Wolters Kluwer Polska: Warszawa, Poland, 2012; pp. 68–69.
7. Doms, M.E.; Dunne, T. Energy intensity, electricity consumption, and advanced manufacturing-technology usage. *Technol. Forecast. Soc. Chang.* **1995**, *49*, 297–310. [CrossRef]
8. Patterson, M.G. What is energy efficiency?: Concepts, indicators and methodological issues. *Energy Policy* **1996**, *24*, 377–390. [CrossRef]
9. De Lovinfosse, I. *How and Why do Policies Change? A Comparison of Renewable Electricity Policies in Belgium, Denmark, Germany, The Netherlands and the UK*; Peter Lang: Bern, Switzerland, 2008.
10. Chavanne, X. *Energy Efficiency: What It Is, Why It Is Important, and How to Assess It (Energy Science, Engineering and Technology)*; Nova Science Publishers Inc.: New York, NY, USA, 2013.
11. Di Franco, N.; Jorizzo, M. Efficiency, Energy Saving, and Rational Use of Energy: Different Terms for Different Policies. In *Innovation in Energy Systems—New Technologies for Changing Paradigms*; Ustun, T.S., Ed.; IntechOpen: London, UK, 2019; pp. 93–112.
12. Jaffe, A.B.; Newell, R.G.; Stavins, R.N. *Economics of Energy Efficiency, Encyclopedia of Energy*; Elsevier: Amsterdam, The Netherlands, 2004; Volume 2, pp. 79–80.
13. Saunders, H.D.; Roy, J.; Azevedo, I.M.L.; Chakravarty, D.; Dasgupta, S.; De La Rue Du Can, S.; Druckman, A.; Fouquet, R.; Grubb, M.; Lin, B.; et al. Energy Efficiency: What Has Research Delivered in the Last 40 Years? *Annu. Rev. Environ. Resour.* **2021**, *46*, 135–165. [CrossRef]
14. Council of the European Union. *Komunikat Komisji do Parlamentu Europejskiego, Rady, Europejskiego Komitetu Ekonomiczno-Społecznego i Komitetu Regionów Zintegrowana Polityka Przemysłowa w Erze Globalizacji Konkurencyjność i Zrównoważony Rozwój na Pierwszym Planie KOM(2010) 614*; Komunikat KE KOM(2010) 614; Council of the European Union: Brussels, Belgium, 2010.
15. Gulczyński, D. Wybrane priorytety i środki zwiększenia efektywności energetycznej. *Polityka Energetyczna—Energy Policy J.* **2009**, *12*, 173–184.
16. ODYSSEE-MURE. Definition of Data and Energy Efficiency Indicators in ODYSSEE Data Base. 2020. Available online: https://www.odyssee-mure.eu/private/definition-indicators.pdf (accessed on 15 December 2022).
17. GUS. *Energy Efficiency in Poland 2008–2018*; GUS: Warszawa, Poland, 2020.
18. Oxford English Dictionary. Available online: https://www.oxforddictionaries.com/words/the-oxford-english-dictionary (accessed on 20 December 2022).
19. Australian Centre on Quality of Life. Personal Wellbeing Index, Deakin University, Melbourne. 2012. Available online: http://www.deakin.edu.au/research/acqol/instruments/wellbeing-index/ (accessed on 15 December 2022).
20. THE WHOQOL GROUP. The Development of the World Health Organization quality of life assessment instrument: The WHOQOL. In *Quality of Life Assessment: International Perspectives*; Orley, J., Kuyen, W., Eds.; Springer: Berlin/Heidelberg, Germany, 1994; pp. 41–57.
21. Sfeatcu, R.; Cernuşcă-Miţariu, M.; Ionescu, C.; Roman, M.; Cernuşcă-Miţariu, S.; Coldea, L.; Boţa, G.; Burcea, C.C. The concept of wellbeing in relation to health and quality of life. *Eur. J. Sci. Theol.* **2014**, *10*, 123–128.
22. Steptoe, A.; Deaton, A.; Stone, A.A. Psychological wellbeing, health and ageing. PMC Article. *Lancet* **2014**, *385*, 14–20. [CrossRef]
23. Knapková, M.; Barwińska-Małajowicz, A.; Mizik, T.; Martinkovičová, M. *Time Allocation and Well-Being of University Teachers in V4 Countries during the COVID-19 Pandemic*; SIZ: Łódź, Poland, 2022; pp. 19–20.
24. Graham, C. The Economics of Happiness, World Economics, World Economics, 1 Ivory Square, Plantation Wharf, London, United Kingdom. *SW11 3UE* **2005**, *6*, 41–55.
25. Watts, N.; Amann, M.; Ayeb-Karlsson, S.; Belesova, K.; Bouley, T.; Boykoff, M.; Byass, P.; Cai, W.; Campbell-Lendrum, D.; Chambers, J.; et al. The Lancet Countdown on health and climate change: From 25 years of inaction to a global transformation for public health. *Lancet* **2018**, *391*, 581–630. [CrossRef] [PubMed]
26. Health and Wellbeing. Available online: https://www.iea.org/reports/multiple-benefits-of-energy-efficiency/health-and-wellbeing (accessed on 15 December 2022).
27. Curl, A.; Kearns, A. Housing Improvements, Fuel Payment Difficulties and Mental Health in Deprived Communities. *Int. J. Hous. Policy* **2017**, *17*, 417–443. [CrossRef]
28. Grey, C.N.B.; Tina Schmieder-Gaite, T.; Jiang, S.; Nascimento, C.; Poortinga, W. Cold homes, fuel poverty and energy efficiency improvements: A longitudinal focus group approach. *Indoor Built Environ.* **2017**, *26*, 902–991. [CrossRef]

29. Jorgenson, A.K.; Alekseyko, A.; Giedraitis, V. Energy consumption, human well-being and economic development in central and eastern European nations: A cautionary tale of sustainability. *Energy Policy* **2014**, *66*, 419–427. [CrossRef]
30. Samuel, Y.A.; Manu, O.; Wereko, T.B. Determinants of energy consumption: A review. *Int. J. Manag. Sci.* **2013**, *1*, 482–487.
31. Wang, Z.; Zhang, B.; Wang, B. Renewable energy consumption, economic growth and human development index in Pakistan: Evidence form simultaneous equation model. *J. Clean. Prod.* **2018**, *184*, 1081–1090. [CrossRef]
32. Wang, Z.; Bui, Q.; Zhang, B. The relationship between biomass energy consumption and human development: Empirical evidence from BRICS countries. *Energy* **2020**, *194*, 116906. [CrossRef]
33. Can, B.; Ahmed, Z.; Ahmad, M.; Can, M. Do renewable energy consumption and green trade openness matter for human well-being? Empirical evidence from European Union countries. *Soc. Indic. Res.* **2022**, *164*, 1043–1059. [CrossRef]
34. Ren, P.; Liu, X.; Li, F.; Zang, D. Clean Household Energy Consumption and Residents' Well-Being: Empirical Analysis and Mechanism Test. *Int. J. Environ. Res. Public Health* **2022**, *19*, 14057. [CrossRef]
35. Gordon, R.; Harada, T.; Spotswood, F. The body politics of successful ageing in the nexus of health, well-being and energy consumption practices. *Soc. Sci. Med.* **2022**, *294*, 114717. [CrossRef] [PubMed]
36. Kochanek, E. The energy transition in the Visegrad group countries. *Energies* **2022**, *14*, 2212. [CrossRef]
37. Jurgilewicz, M.; Delong, M.; Topolewski, S.; Pączek, B. Energy security of the Visegrad group countries in the natural gas sector. *J. Secur. Sustain. Issues* **2020**, *2*, 173–181. [CrossRef] [PubMed]
38. Zabojnik, S.; Hricovsky, M. Balancing the Slovak Energy Market After the Adoption of "Fit for 55 Package". *SHS Web Conf.* **2021**, *129*, 05015. [CrossRef]
39. Szmitkowski, P.; Gil-Świderska, A.; Zakrzewska, S. Electrical energy infrastructure in Poland and its sensitivity to failures as part of the energy security system. *Polityka Energetyczna-Energy Policy J.* **2019**, *22*, 59–80. [CrossRef]
40. Rozporządzenie Parlamentu Europejskiego i Rady (UE) 2018/1999 z dnia 11 Grudnia 2018 r. Available online: https://eur-lex.europa.eu/legal-content/PL/TXT/PDF/?uri=CELEX:32018R1999&from=EN (accessed on 7 November 2022).
41. USTAWA z Dnia 17 Grudnia 2020 r. o Promowaniu Wytwarzania Energii Elektrycznej w Morskich Farmach Wiatrowych. Dz.U. 2021 poz. 234. Available online: https://isap.sejm.gov.pl/isap.nsf/DocDetails.xsp?id=WDU20210000234 (accessed on 15 December 2022).
42. Ministerstwo Aktywów Państwowych. *Krajowy Plan na Rzecz Energii i Klimat (KPEiK) u na Lata 2021–2030*; Wersja 4.1 z dn; Ministerstwo Aktywów Państwowych: Warszawa, Poland, 2019.
43. Ministry of Climate and Environment. *PEP2040, Annex to Resolution No 22/2021 of the Council of Ministers of 2 February 2021 Ministry of Climate and Environment*; Ministry of Climate and Environment: Warsaw, Poland, 2 February 2021; pp. 5–6.
44. Regulation (EC) No 663/2009 of the European Parliament and of the Council of 13 July 2009 Establishing a Programme to Aid Economic Recovery by Granting Community Financial Assistance to Projects in the Field of Energy. Available online: https://eur-lex.europa.eu/legal-content/EN/ALL/?uri=CELEX%3A32009R0663 (accessed on 7 November 2022).
45. Regulation (EC) No 715/2009 of the European Parliament and of the Council of 13 July 2009 on Conditions for Access to the Natural Gas Transmission Networks and Repealing Regulation (EC) No 1775/2005 (Text with EEA Relevance). Available online: https://eur-lex.europa.eu/legal-content/EN/ALL/?uri=CELEX:32009R0715 (accessed on 7 November 2022).
46. Directive 94/22/EC of the European Parliament and of the Council of 30 May 1994 on the Conditions for Granting and Using Authorizations for the Prospection, Exploration and Production of Hydrocarbons. Available online: https://eur-lex.europa.eu/legal-content/EN/TXT/HTML/?uri=CELEX:31994L0022&from=en (accessed on 7 November 2022).
47. Directive 98/70/EC of the European Parliament and of the Council of 13 October 1998 Relating to the Quality of Petrol and Diesel Fuels and Amending Council Directive 93/12/EEC. Available online: https://eur-lex.europa.eu/legal-content/EN/ALL/?uri=CELEX%3A31998L0070 (accessed on 7 November 2022).
48. Directive 2009/31/EC of the European Parliament and of the Council of 23 April 2009 on the Geological Storage of Carbon Dioxide and Amending Council Directive 85/337/EEC, European Parliament and Council Directives 2000/60/EC, 2001/80/EC, 2004/35/EC, 2006/12/EC, 2008/1/EC and Regulation (EC) No 1013/2006 (Text with EEA Relevance). Available online: https://eur-lex.europa.eu/legal-content/EN/TXT/?uri=CELEX%3A32009L0031 (accessed on 7 November 2022).
49. Directive 2009/73/EC of the European Parliament and of the Council of 13 July 2009 Concerning Common Rules for the Internal Market in Natural Gas and Repealing Directive 2003/55/EC (Text with EEA Relevance). Available online: https://eur-lex.europa.eu/legal-content/EN/TXT/HTML/?uri=CELEX:32009L0073&from=PL (accessed on 7 November 2022).
50. Directive 2010/31/EU of the European Parliament and of the Council of 19 May 2010 on the Energy Performance of Buildings (Recast). Available online: https://eur-lex.europa.eu/legal-content/EN/TXT/HTML/?uri=CELEX:32010L0031&from=PL (accessed on 7 November 2022).
51. Regulation (EU) No 525/2013 of the European Parliament and of the Council of 21 May 2013 on a Mechanism for Monitoring and Reporting Greenhouse Gas Emissions and for Reporting Other Information at National and Union Level Relevant to Climate Change and Repealing Decision No 280/2004/EC Text with EEA Relevance. Available online: https://eur-lex.europa.eu/legal-content/EN/TXT/?uri=celex%3A32013R0525 (accessed on 7 November 2022).
52. Statistics Poland. *Fuel and Energy Economy, Energy Statistics in 2017, 2018 and 2019*; Statistics Poland: Warsaw, Poland, 2020.
53. Poland's Energy Policy 2030. Available online: https://www.gov.pl/web/ia/polityka-ekologiczna-panstwa-2030-pep2030 (accessed on 7 November 2022).
54. EU-Funded Project FinEERGo-Dom. Available online: https://fineergodom.eu/ (accessed on 7 November 2022).

55. Obwieszczenie Ministra Infrastruktury i Rozwoju z dnia 17 Lipca 2015 r. w Sprawie Ogłoszenia Jednolitego Tekstu Rozporządzenia Ministra Infrastruktury w Sprawie Warunków Technicznych, Jakim Powinny Odpowiadać Budynki i ich Usytuowanie, Dziennik Ustaw 2015 r. poz. 1422. Available online: https://dziennikustaw.gov.pl/du/2015/1422 (accessed on 7 November 2022).
56. Ustawa z Dnia 7 Lipca 1994 r.—Prawo Budowlane, Dz.U. 1994 nr 89 poz. 414. Available online: https://isap.sejm.gov.pl/isap.nsf/DocDetails.xsp?id=wdu19940890414 (accessed on 7 November 2022).
57. The Clean Air Program. Available online: https://czystepowietrze.gov.pl/czyste-powietrze/ (accessed on 7 November 2022).
58. Dyrektywa 2006/32/WE Parlamentu Europejskiego i Rady z dnia 5 Kwietnia 2006 r. w Sprawie Efektywności Końcowego Wykorzystania Energii i Usług Energetycznych oraz Uchylająca Dyrektywę Rady 93/76/ EWG—Dz. U. L 114 z 27.4.2006 r. pp. 54–64. Available online: https://eur-lex.europa.eu/legal-content/PL/TXT/PDF/?uri=CELEX:32006L0032&from=RO (accessed on 15 December 2022).
59. Integrated National Energy and Climate Plan for Years 2021–2030. Available online: https://www.economy.gov.sk/uploads/files/IjkPMQAc.pdf (accessed on 25 July 2022).
60. Greener Slovakia. Strategy of the Environmental Policy of the Slovak Republic until 2030. Available online: https://minzp.sk/files/iep/publikacia_zelensie-slovensko-aj_web.pdf (accessed on 3 August 2022).
61. Renewable Energy Statistics. Eurostat. Statistics Explained. Available online: https://ec.europa.eu/eurostat/statistics-explained/index.php?title=Renewable_energy_statistics#Share_of_renewable_energy_more_than_doubled_between_2004_and_2020 (accessed on 3 August 2022).
62. Energetika 2020. Statistical Office of Slovak Republic 2021, 166. Available online: https://slovak.statistics.sk/wps/portal/3b32dcbd-9b01-411b-a301-ee764a12cacc/!ut/p/z1/rVNNc5swEP01PoJWEp-9ERxjUuzGJhhblwxg2agOHwFimn9fufUhk4mxO9M9aCTte6vdfSvE0BqxMjmKfdKJqkxe5HnDjOeF6Vt3d9gBMPUx-A_ReOGFNgYNUPwHsAoe_wI88wmDP5-H0cpa0PuJjthld-RqZ77rOVPNDACswNPBd6bR0l5QCg69jQ8XzIHb-AMANlz_CjHEsrKruxxtqrRNcqU9KPVbOgK5vIhDkgk-gmPLu8P7xxuaUrLN0q1ip4AVDeNUSajccW4aWoJJlmTZKXadiS3a3ISOr4nFhlsVn967ouYgYA5nwCR8JI6tee54-WMC_pNLrDAwCAA-A4YEv5bmA2liLdQ-K1RQJYdQA9tgGLaONdBOI-uUKbX2iDV8xxveqG-NnOS86-r22whG0Pe9mopyr2ZVMYKvCHnVdmj9EYc2srnmxeZ-Jyg-Ct6jqKyaQn6c8B-1m36qCxNLegnWLJtQuT9lSZqZO5NZ1kmXK6LcVWh9U2xJFT9fX5kjJ7UqO_5L1vbfR7UuImmFRd-Vw_J-N-v0VN8Xz-OAx85vLqu7Xg!!/dz/d5/L2dBISEvZ0FBIS9nQSEh/ (accessed on 13 September 2022).
63. NextGenerationEU. Available online: https://next-generation-eu.europa.eu/index_en (accessed on 3 August 2022).
64. Plán Obnovy a Odolnosti Slovenskej Republiky [Recovery and Resilience Plan of Slovak Republic]. Available online: https://www.minzp.sk/poo/ (accessed on 15 July 2022).
65. Call for the Submission of Applications for Funding from the Mechanism for the Renovation of Family Houses. Available online: https://obnovdom.sk/assets/documents/requests/vyzva_1.pdf (accessed on 15 September 2022).
66. Live Economically Ministry of Transport and Construction of the Slovak Republic. Available online: https://byvajteusporne.sk/zateplovanie/ (accessed on 15 September 2022).
67. Slovak Environmental Agency. Available online: https://www.sazp.sk/ (accessed on 15 September 2022).
68. Dyrektywa Parlamentu Europejskiego i Rady 2009/125/WE z Dnia 21 Października 2009 r. Ustanawiająca Ogólne Zasady Ustalania Wymogów Dotyczących Ekoprojektu dla Produktów Związanych z Energią Zwana Dyrektywą ErP (Energy Related Products). Available online: https://www.hvacr.pl/dyrektywa-erp-dyrektywa-parlamentu-europejskiego-i-rady-2009-125-we-614 (accessed on 15 December 2022).
69. Dyrektywa Parlamentu Europejskiego i Rady 2009/28/WE z Dnia 23 Kwietnia 2009 r. w Sprawie Promowania Stosowania Energii ze Źródeł Odnawialnych. Available online: https://eur-lex.europa.eu/legal-content/PL/TXT/?uri=celex%3A32009L0028 (accessed on 15 December 2022).
70. Dyrektywa Parlamentu Europejskiego i Rady 2010/31/WE z Dnia 19 Maja 2010 r. w Sprawie Charakterystyki Energetycznej Budynków. Available online: https://eur-lex.europa.eu/LexUriServ/LexUriServ.do?uri=OJ:L:2010:153:0013:0035:pl:PDF (accessed on 15 December 2022).
71. Polska Organizacja Rozwoju Technologii Pomp Ciepła. Available online: www.portpc.pl (accessed on 10 November 2022).
72. Rozporządzenie Ministra Energii z Dnia 5 Października 2017 r. w Sprawie Szczegółowego Zakresu i Sposobu Sporządzania Audytu Efektywności Energetycznej Oraz Metod Obliczania Oszczędności Energii (Dz.U. z Dnia 13 Października 2017 r. poz. 1912). Available online: https://isap.sejm.gov.pl/isap.nsf/DocDetails.xsp?id=WDU20170001912 (accessed on 15 December 2022).
73. Rozporządzenie Ministra Infrastruktury i Rozwoju z Dnia 27 Lutego 2015 r. w Sprawie Metodologii Wyznaczania Charakterystyki Energetycznej Budynku lub Części Budynku Oraz Świadectw ich Charakterystyki Energetycznej. Available online: https://sip.lex.pl/akty-prawne/dzu-dziennik-ustaw/metodologia-wyznaczania-charakterystyki-energetycznej-budynku-lub-18176491 (accessed on 15 December 2022).
74. Michalak, P.; Szczotka, K.; Szymiczek, J. Energy effectiveness or economic profitability? A case study of thermal modernization of a school building. *Energies* **2021**, *14*, 1973. [CrossRef]
75. Szczotka, K. Analysis of the energy efficiency heating system with a heat pump. In *Problemy Inżynierii Mechanicznej i Robotyki*; Energetyka i ochrona środowiska; AGH: Kraków Poland, 2013.
76. *PN-EN ISO 12831:2006*; Instalacje Ogrzewcze w Budynkach. Metoda Obliczania Projektowego Obciążenia Cieplnego. PN-EN ISO: Warszawa, Poland, 2006.

77. *PN-EN ISO 13790:2008*; Energetyczne Właściwości Użytkowe Budynków—Obliczanie Zużycia Energii na Potrzeby Ogrzewania i Chłodzenia. PN-EN ISO: Warszawa, Poland, 2008.
78. Szczotka, K. *Badania Efektywności Energetycznej Sprężarkowych Pomp Ciepła Typu Powietrze-Woda. Wybrane Problemy Energetyki Zasobów Odnawialnych. Materiały Naukowe XXII Ogólnopolskiego Forum Zasobów, Źródeł i Technologii Energetycznych*; monografia naukowa; Ekoenergetyka: Kraków, Poland, 2015; pp. 131–147, ISBN 978-83-63318-04-8.
79. Szczotka, K.; Zimny, J.; Struś, M.; Michalak, P.; Szymiczek, J. Badania parametrów termodynamicznych sprężarkowej powietrznej pompy ciepła w celu poprawy efektywności energetycznej. *Rynek Energii* **2021**, *2*, 54–64.
80. Szymański, B. *Instalacje Fotowoltaiczne—Poradnik*; Globenergia: Kraków, Poland, 2021; ISBN 978-83-65874-00-9.
81. Zimny, J.; Michalak, P.; Szczotka, K. *Ecological Heating System of a School Building Hybrid Heating System with a Heat Pump—A Case Study*; Rynek Energii: Poznań, Poland, 2012; nr 5.
82. Zimny, J.; Michalak, P.; Szczotka, K. *Energy, Ecological and Economic Evaluation of Thermo-Modernising Project with RETScreen® Package*; Ciepłownictwo Ogrzewnictwo Wentylacja: Warszawa, Poland, 2009; 2009 R. 40; ISSN 0137-3676.
83. Zimny, J.; Michalak, P.; Szczotka, K. *Heating Installation with Heating Pump and Gas Boiler- Energy Assessment of the System by Using RETScreen® Package*; Ciepłownictwo Ogrzewnictwo Wentylacja: Warszawa, Poland, 2010; 2010 t. 41 nr 4; ISSN 0137-3676.
84. Zimny, J.; Szczotka, K. Ecological heating system of a school building—Design, implementation and operation. *Environ. Prot. Eng.* **2012**, *38*, 141–150. [CrossRef]
85. Strategia na Rzecz Odpowiedzialnego Rozwoju (SOR) do Roku 2020 z Perspektywą do 2030, uchwała Rady Ministrów z Dnia 14 Lutego 2017 r. Available online: https://www.gov.pl/web/fundusze-regiony/informacje-o-strategii-na-rzecz-odpowiedzialnego-rozwoju (accessed on 15 December 2022).
86. W Centrum Uwagi: Efektywność Energetyczna Budynków, 17.02.2020 Brussels. Available online: https://commission.europa.eu/news/focus-energy-efficiency-buildings-2020-02-17_pl (accessed on 13 December 2022).

Disclaimer/Publisher's Note: The statements, opinions and data contained in all publications are solely those of the individual author(s) and contributor(s) and not of MDPI and/or the editor(s). MDPI and/or the editor(s) disclaim responsibility for any injury to people or property resulting from any ideas, methods, instructions or products referred to in the content.

Article

Comprehensive and Integrated Impact Assessment Framework for Development Policies Evaluation: Definition and Application to Kenyan Coffee Sector

Nicolò Golinucci [1,2], Nicolò Stevanato [1,2], Negar Namazifard [2], Mohammad Amin Tahavori [2], Lamya Adil Sulliman Hussain [3], Benedetta Camilli [3], Federica Inzoli [4], Matteo Vincenzo Rocco [2,*] and Emanuela Colombo [2]

[1] Fondazione Eni Enrico Mattei (FEEM), Corso Magenta, 63, 20123 Milan, Italy; nicolo.golinucci@polimi.it (N.G.); nicolo.stevanato@polimi.it (N.S.)
[2] Department of Energy, Politecnico di Milano, Via Lambruschini, 4, 20156 Milan, Italy; negar.namazifard@polimi.it (N.N.); mohammadamin.tahavori@polimi.it (M.A.T.); emanuela.colombo@polimi.it (E.C.)
[3] Independent Researcher, 20100 Milan, Italy; lamyaadil3@gmail.com (L.A.S.H.); benedettacamilli@gmail.com (B.C.)
[4] Falck Renewables, Via Alberto Falck, 4, 20099 Milan, Italy; federica.inzoli@falckrenewables.com
* Correspondence: matteovincenzo.rocco@polimi.it

Citation: Golinucci, N.; Stevanato, N.; Namazifard, N.; Tahavori, M.A.; Sulliman Hussain, L.A.; Camilli, B.; Inzoli, F.; Rocco, M.V.; Colombo, E. Comprehensive and Integrated Impact Assessment Framework for Development Policies Evaluation: Definition and Application to Kenyan Coffee Sector. *Energies* **2022**, *15*, 3071. https://doi.org/10.3390/en15093071

Academic Editors: Chiara Martini and Claudia Toro

Received: 15 March 2022
Accepted: 20 April 2022
Published: 22 April 2022

Publisher's Note: MDPI stays neutral with regard to jurisdictional claims in published maps and institutional affiliations.

Copyright: © 2022 by the authors. Licensee MDPI, Basel, Switzerland. This article is an open access article distributed under the terms and conditions of the Creative Commons Attribution (CC BY) license (https://creativecommons.org/licenses/by/4.0/).

Abstract: The coexistence of the need to improve economic conditions and the conscious use of environmental resources plays a central role in today's sustainable development challenge. In this study, a novel integrated framework to evaluate the impact of new technological interventions is presented and an application to smallholder coffee farms and their supply chains in Kenya is proposed. This methodology is able to combine multiple information through the joint use of three approaches: supply chain analysis, input-output analysis, and energy system modeling. Application to the context of the Kenyan coffee sector enables framework validation: shading management measures, the introduction of eco-pulpers, and the exploitation of coffee waste biomass for power generation were compared within a holistic high-level perspective. The implementation of shading practices, carried out with fruit trees, shows the most relevant effects from the economic point of view, providing farmers with an additional source of income and generating $903 of work for every million of local currency (about $9k) invested in this solution. The same investment would save up to 1.46 M m³ of water per year with the eco-pulpers technology. Investing the same amount in coffee-biomass power plants would displace a small portion of production from heavy-duty oil and avoid importing a portion of fertilizer, saving up to 11 tons of CO_2 and around $4k per year. The results suggest the optimal allocation of a $100m budget, which can be affected by adding additional constraints on minimum environmental or social targets in line with sustainable development goals.

Keywords: supply chain analysis; industrial ecology; energy modeling; development policies; developing countries

1. Introduction

Over the last decades, the interest in and evidence for the numerous interconnections between energy, the environment, and society have gained increasing importance for the international community. Processes and relationships among countries are becoming global and evolving in a complex framework. In this context, it is no longer possible to consider development strategies without adopting a systemic approach where social and technological aspects are jointly taken into account. In the last decade, the recognized relevance of cross-sectoral linkages among economic sectors has driven research efforts towards the expansion of energy-economic modeling. Moreover, the 2030 Development Agenda identifies energy access as a necessary precondition for human and social promotion, as well as instrumental in fighting poverty [1].

The coexistence of the need to improve economic conditions, particularly in developing countries, and the conscious use of environmental resources play a central role in the global sustainable development challenge. To address this issue, an informed decision-making process is essential and the support of the scientific community for policymakers may be pivotal to foster innovative national development policies.

The adoption of a multidisciplinary framework would allow a comprehensive comparison and evaluation of different policy-making decisions, that can affect the environment and various energy, social, and economic sectors of the country.

Concerning agriculture, a key role is played by smallholder farmers, who produce about 80% of the food consumed in Africa and Asia [2]. Usually, these producers are characterized by a low level of productivity in the agricultural sector, which is today globally responsible for over a quarter of greenhouse gas emissions, the use of half of habitable land, and 70% of freshwater withdrawals [3].

The complexity of this problem has been recently tackled in the literature either by combining multiple models in one unique framework or by adopting a macro-economic perspective. In particular, the integrated framework of CLEWS has been presented and adopted to demonstrate the added value brought by adopting an integrated approach, able to capture multiple dimensions of sustainable development by describing the interaction between energy, water, and climate models [4–7]. Some studies take advantage of the multiple opportunities offered by integrating physical life cycle assessment and input-output models to analyze the economic, social, and environmental impacts of replacing conventional liquid fuels with alternative energy sources on the countries' economic systems [8,9]. Moreover, statistical regression analyses have been used in some studies to investigate the effect of social and environmental factors and climate adoption strategies on African farmers' revenue [10]. However, not all the complexities of sustainable development can be grasped by models: financial barriers could hamper climate-resilient investments, in particular in Sub-Saharan Africa. A proper regulatory framework is required to narrow the climate finance gap necessary for sustainable development but is difficult to be explicitly represented in models [11].

There are rare attempts that follow a comprehensive cross-sectoral analysis by deeply investigating the effects of technological innovation and new renewable energy resource introduction not only on the economy and nature but also on the energy system in detail. Therefore, to fully realize the sustainability of development opportunities in a specific sector, impact assessments should not be limited to socioeconomic and environmental indicators but incorporate explicit analysis within the energy sector of the country. The need for a framework for addressing the multi-dimensional evaluation of not only climate adoption strategies but also technological innovation on a specific supply chain emerges from the literature.

The objective of this research is the formalization and application of a ComprehensIVe and Integrated Country Study (CIVICS) framework able to assess the impact of policies considering their multidimensionality, which may be used for supporting decision-making in developing countries. The tools adopted are (i) supply chain analysis (SCA), (ii) the input–output analysis model (IOA), and (iii) energy system modeling (ESM). The SCA allows us to acquire insights into the supply chain of a specific local product considered strategic for national economic development. It permits us to focus on bottlenecks and hotspots undermining the supply chain's overall performance. This analysis allows us to identify strategies and improvement solutions that can be implemented to overcome the main issues. With the aim of determining the environmental and economic impact of the changes occurring with the adoption of the analyzed interventions, an IOA model is adopted, and to increase the level of technological detail in characterizing the energy response to changes, a dedicated ESM is needed. In particular, the capability of evaluating energy planning strategies in synergy with the analyzed economic policy, enabled by the integration among the models, represents an important added value.

The peculiarity of this research approach relies on the integration process mentioned. The adopted tools are combined in an ad-hoc developed system allowing the decision-makers to assess the multiple impacts of their national strategies and to identify them in the sustainable development framework.

This unique framework is applied in the context of developing African economies with a focus on the coffee sector in Kenya, which is one of its major economic pillars. Despite the decrease in coffee production and exports, local policymakers are concerned about supporting the coffee industry. The reason behind the decrease in the productivity of this sector are various but, in this study, the focus is on endogenous reasons that can be associated with the poor management and governance of the cooperative system and poor technological innovation. Therefore, first, a supply chain analysis is carried out to identify the potential interventions for improving productivity. Then, these interventions are applied by modeling the whole Kenyan economy, adopting a social accounting matrix developed by JRC [12] to represent all the transactions among the relevant economic agents. Finally, the role of the produced biomass from the wet processing of coffee as an energy resource is analyzed by accurately modeling the Kenyan electricity system.

The remainder of this work is structured as follows: in Section 2, the methodological approach is presented; in Section 3, the first case study is introduced and the methodology is calibrated on its peculiarities, contextually providing the lesson learned from the SCA and suggested interventions; in Section 4, results are analyzed and conclusions derived.

2. Materials and Methods

2.1. Interaction among the Tools within CIVICS Framework

Supply chain analysis (SCA) can be considered the first step of this framework. It consists of the process of investigating and studying the role and contribution of each economic agent (actors such as producers, traders, and consumers, as well as legal entities such as businesses, authorities, and development organizations) along a supply chain, that contribute directly to the generation of a final product or service. This activity involves the evaluation of every stage of the supply chain, starting from the raw materials or intermediate product acquisition and finishing downstream, after all the stages of transformation and increase in value, at the final delivery of the product to the consumer [13]. The need for such analysis can be easily understood when considering the rise of globalization and global trading. In the global supply chain, the developing countries usually play the role of supplying the raw materials to the more industrialized countries, due to a lack of know-how and expertise regarding the processing steps of a product. These countries mostly face problems affecting the performance of the supply chain which include the instability of governments and policies, corruption, labor-intensive industries, deteriorated infrastructures, the limited use of new technologies, underemployment, child labor, and the low education level of the population [14]. The fragmented market on which many supply chains of developing countries are based, alongside the low access to quality services and information for all the stakeholders of the supply chain (particularly small producers) and the informal economy somehow regulating many steps of the chain make it difficult to collect precise and accurate information to carry out a rigorous study of the supply chain of a specific product. Therefore, in this research, a customized methodological approach to SCA in developing countries, which has been developed by the authors [13] based on the steps shown in Figure 1, is adopted.

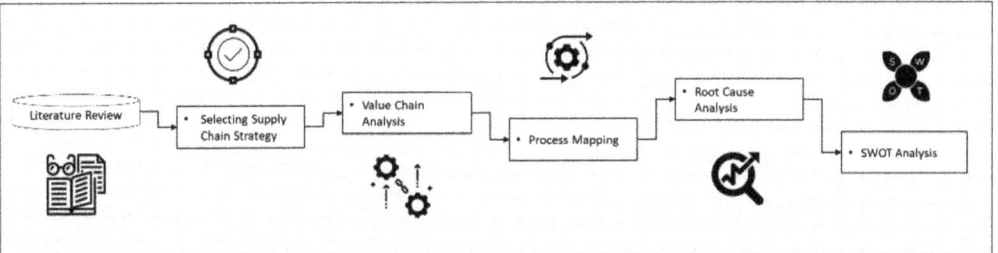

Figure 1. SCA methodological steps. Source: [13].

After investigating the bottlenecks in the supply chain of the determined local products, the identified strategies are implemented through IOA. This approach represents a suitable and comprehensive industrial ecology methodology for evaluating a structural change in a determined supply chain while considering the implications on the complex network of interlinkages among different economic sectors [15]. IOA refers to a macroeconomic analysis approach based on the study of the sectoral interrelations of an economy [16] and requires the use of input–output tables, economy-wide databases able to capture the flows of monetary value between different sectors.

The model adopted in this research is a demand-driven input-output model based on Supply and Use Tables (SUT). As it has been shown by Lenzen, the framework offered by SUT can be adopted to directly perform impact analysis [17,18]. The authors invite the reader to the A3 section of the Supplementary Material S1 for more technical details.

In this framework, it is possible to assume a change in a specific interrelation between two economic activities of a supply chain by intervening in a specific coefficient. Since the objective of this work is to evaluate the impact of a technological change related to both implementation and use, it is required to distinguish every intervention in those two steps. In both cases, there will be an impact on socio-economic factors (linked with production through the matrix of monetary exogenous coefficients f) and environmental extensions (linked with production through the matrix of physical exogenous coefficients e), respectively, F and E.

- Investment assessment: in this step, it is required to characterize all the commodities needed to have the technology produced and installed (e.g., the cost of machinery and the required training course). From a modeling point of view, this will be translated by simply adding the required commodities to the final demand vector. The investment will be handled, as shown in (1), with the current technology assessment (no subscript identifies baseline data, while subscript i identifies investment data);
- Operation assessment: in this step, it is required to describe all the cross-sectoral changes that are occurring due to the installed technology. The structural change in operation will influence, as shown in (2), how the baseline final demand is delivered (subscript o identifies data after the implementation of the intervention). These changes may be translated to the model in the following ways:
  a. Change in the use coefficients matrix (u, the use side of matrix z): a specific variation of u can reflect a change in how much input of a certain commodity is required for one unit of output (e.g., machinery, not used in the baseline, will directly increase the consumption of diesel in a certain activity).
  b. Change in the satellite account coefficients matrices (f and e): a specific variation of f or e can reflect a change in activity intensities (e.g., machinery, not used in the baseline, will directly emit an additional amount of CO_2 emissions in performing a certain activity).
  c. Change in the market share matrix (v, the make side of matrix z): a specific variation of a v coefficient represents how much of each activity is required every time a certain commodity is demanded. Therefore, a change in the v matrix

could be used to model the change in the productivity of a specific activity. In fact, productivity is how much output is produced for each unit of input, or, in the case of a demand-driven model, how much input is needed to deliver the same output (e.g., the physical productivity of coffee plants increases because of the introduced technology).

$$\Delta \underline{F_i} = \underline{\underline{f}} \, \overbrace{[(\underline{\underline{I}} - \underline{\underline{z}})^{-1} \, \underline{Y_i}]}^{X_i} - \underline{\underline{f}} \, \overbrace{[(\underline{\underline{I}} - \underline{\underline{z}})^{-1} \, \underline{Y}]}^{X}$$
$$\Delta \underline{E_i} = \underline{\underline{e}} \, [(\underline{\underline{I}} - \underline{\underline{z}})^{-1} \underline{Y_i}] - \underline{\underline{e}} \, [(\underline{\underline{I}} - \underline{\underline{z}})^{-1} \underline{Y}] \tag{1}$$

$$\Delta \underline{F_o} = \underline{\underline{f_o}} \, \overbrace{[(\underline{\underline{I}} - \underline{\underline{z_o}})^{-1} \, \underline{Y}]}^{X_o} - \underline{\underline{f}} \, \overbrace{[(\underline{\underline{I}} - \underline{\underline{z}})^{-1} \, \underline{Y}]}^{X}$$
$$\Delta \underline{E_o} = \underline{\underline{e_o}} \, [(\underline{\underline{I}} - \underline{\underline{z_o}})^{-1} \underline{Y}] - \underline{\underline{e}} \, [(\underline{\underline{I}} - \underline{\underline{z}})^{-1} \underline{Y}] \tag{2}$$

Note that a variable with one underline identifies a vector, while one with a double underline identifies a matrix. A variable in capital letters has absolute units (e.g., M$ or Gg), while one in small letters has output-specific units (e.g., M$/M$ or Gg/M$).

Where X and Y represent the total production of commodities and industrial activities and the final demand of commodities, respectively, z symbolizes the supply and use representation of the technological structure of the economy and I is the identity matrix of the same dimensions of z. The calculation is carried out through an openly available Python-based tool for performing input–output analyses, called MARIO [19].

Eventually, in order to understand the required energy system planning to align with the identified technological changes in the supply chain of the determined local products, a model of the energy system is adopted. ESM consists of the practice of building a mathematical representation of a physical energy system in order to understand its dynamics and reaction to interventions or future scenarios. It can be summarized as a discipline to support energy policy and long-term strategic energy planning decisions with insights generated by models. In particular, for this work, it is possible to narrow the discussion to engineering models for energy systems sizing, investment planning, and operation or dispatch optimization. The selected modeling framework is the open-source software Calliope [20], a "linear programming framework for spatial–temporal energy system optimization" [21]. The framework allows for a 1-year modeling horizon, works with 1 h resolution, and is based on the power nodes model, meaning that the geographical resolution of the model is left to the modeler, depending on the specific needs. A power node is created to represent a region, an area, or a building, where energy can be produced, consumed, and transferred from one another. The advantage of being able to customize the modeled power nodes is that the geographical scope and resolution representable with the framework is completely up to the necessities of the modeler and able to adapt to the availability of data, often a critical aspect when modeling systems in developing economies. In this research, the spatial resolution of our modeling is set to the national scale.

In Figure 2, a set of possible interactions between the tools, which summarize the approach, is outlined. As it is mentioned, in the presented configuration, IOA, a fit-for-purpose modeling approach of Industrial Ecology [22], acts as a bridge between a robust characterization of the supply chain under investigation and detailed modeling of the energy system. Indeed, thanks to the exchange of information between SCA and IOA, (as the outputs of the SCA are input for the IOA), it is possible to evaluate impacts at the social, economic, and environmental levels of the formulated improvement strategies. Furthermore, the integration of results between IOA and ESM permits us to formulate an energy strategy ad hoc for these interventions, addressing sustainable development objectives.

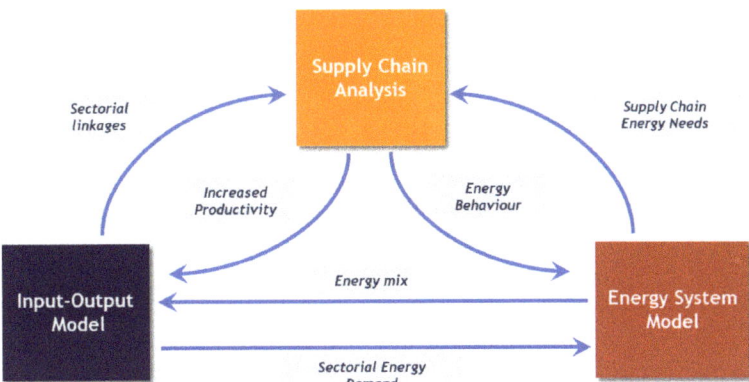

Figure 2. CIVICS Integrated Modeling Framework.

2.2. Evaluating and Comparing Interventions within the CIVICS Framework

The scope of the CIVICS framework is to evaluate economic development opportunities in a specific socio-economic context from a country perspective. At the same time, it is possible to assess how these opportunities are configured with respect to national environmental objectives. In this sense, the approach should be seen as a way to coherently compare investment opportunities within the same limiting modeling assumptions.

Each opportunity is identified by a possible technological intervention. This intervention has an impact not only on the sector in which the direct change takes place but also on its interlinked activities. In real life, these changes occur while many other interrelated activities change in magnitude or in the needed input mix. The model is a representation of an approximated reality where it is possible to isolate the effect of each specific intervention.

If an intervention is beneficial in reducing the amount of input required for delivering the same products and services that were produced and delivered in the baseline case, this means that the intervention could be used to unleash the potential for expanding the production, increasing the wages, or improving the margins. Since it is not possible to evaluate the potential effects of these potential political choices, it is preferred to build up a general economic indicator that considers the total savings triggered by each intervention with respect to the required level of investment.

The name of this indicator, defined in (3), is *Policy Return on Investment* (PRoI), and represents the expected yearly economic return on the investment from a national perspective, considering all the direct and indirect implications of changing the sectoral interdependencies on the shape of each intervention. The yearly economic return embodies not only the savings in the form of economic factors from the sector where the intervention occurs (e.g., being more productive leads to using less capital land per unit of output) but also in the form of avoided import (e.g., the new configuration implies a self-production of an organic fuel that replaces a fraction of the imported oil) and avoided internal input request (e.g., trees are introduced for their shading potential but they also produce locally consumed fruit as a by-product which substitutes a fraction of bought food).

$$\text{PRoI} = \frac{\text{Savings}_o \ [\text{M\$/y}]}{\text{Investment}_i \ [\text{M\$}]} = \text{PPBT}^{-1} \qquad (3)$$

The inverse of PRoI is the *Policy Pay-Back Time* (PPBT) and represents the number of years needed to repay the investment faced. It should be underlined that this repayment time must not be compared with entrepreneur-level repayment time, which is based on an individual investment perspective.

PRoI or PPBT are therefore used as a general economic indicator that reports the level of increase in the economic efficiency of the country involved in each intervention. Of course, there are many other possible case-specific indicators that can influence the choice between taking the investment opportunity or not.

For the sake of consistent comparison, each of these indicators should be referred to on the basis of the same functional unit. A functional unit is a quantified description of the function of a product or service that serves as the reference basis for all calculations regarding impact assessment [23]. In this case, and as a general rule, the same level of investment could be used as a functional unit to coherently compare interventions, providing useful insights for policymakers for each relevant dimension. This application of CIVICS will be referred to as *Integrated Multidimensional Analysis*.

Furthermore, this approach could be extended by adopting linear optimization techniques, which can turn into a *Policy Goal* application of CIVICS. In fact, assuming linearity between investment level and savings, therefore neglecting possible non-linear dependencies between the magnitude of the intervention and the relative costs and benefits, it is possible to build an optimization problem shaped on policy-maker objectives. For example, as can be seen by the set of inequalities in (4), it is possible to have a mix (*mix*) of interventions expressed in millions of investments that meet budget constraints and social and environmental objectives while minimizing the amount of input required to deliver the same final demand (i.e., maximizing savings).

$$\text{Max}\left(\underline{mix}^T \ \underline{PRoI}\right)$$
$$\text{s.t.} \ \underline{mix}^T \ \underline{1} \leq Budget \quad (4)$$
$$\underline{mix}^T \ \underline{i}_j \geq Minimum\_social\_or\_environmental\_objective_j \quad \forall j$$

where i_j represents the net intensity change expected with respect to social or environmental objective j. Note that all the underlined variables are vectors with dimensions as big as the number of possible interventions.

A more detailed description of the methodological approach can be found in the Supplementary Information (Supplementary Material S1).

3. Case Study

3.1. Agriculture in Kenya: Addressing the Supply Chain Analysis of the Coffee Sector

Over the last years, the Kenyan Government, in an attempt to strengthen the commitment toward sustainable development, has promoted key public investments based on four priority development pillars, namely: (i) enhancing food and nutrition security, (ii) providing affordable housing, (iii) increasing manufacturing and agro-processing, and (iv) achieving universal health coverage. The agriculture sector has a pivotal role in ushering in these sustainable economic development ambitions. Agriculture is not only central to the achievement of "a globally competitive and prosperous country with a high quality of life by 2030", but it is also expected to deliver on Kenya's global commitments, including the Sustainable Development Goals (SDGs) [24,25]. Nowadays, agricultural incomes (from crops, livestock, and fishing) account for 64% of the income sources of the poor and 53% of incomes for the non-poor (The World Bank, 2019a). Moreover, the sector establishes the industrialization framework by supplying raw materials to other industries (over 75% of industrial raw materials) and it lays the foundation of numerous off-farm activities, such as logistics and research [26]. In fact, agriculture contributed indirectly to 27% of the GDP in 2019, through the linkage with manufacturing, distribution, and other service-related sectors [27].

However, despite having one of the highest productivities in Eastern Africa, a large share of agriculture in Kenya is still prone to harvest failure (as caused by drought in 2019), being for the most part rainfed. For this reason, ongoing policy and institutional reforms are focusing on stabilizing agricultural output and reducing the risks, by supporting irrigation schemes, post-harvest losses management, and input markets.

In particular, the coffee sector in Kenya relies on a well-developed logistics hub, where all the main international traders, as well as a large pool of coffee experts, from farming to marketing, logistics, and trading are represented. However, Kenya contributes a small share to the global coffee market and accounts for 11.7% of African production. Despite the fact that coffee is still one of the strategic products for the Kenyan domestic economy, its role has been downgraded over the last decades. This decline follows a downward trend in production, which is expected to drop to a new record low for 2019–2020 (around 39,000 tons), as affected by the prolonged drought and low returns. In addition, and similarly to other coffee-producing countries, price volatility and significant fluctuations have deterred Kenyan producers and other value chain actors from making the necessary investments for increasing competitiveness, productivity, and production [28].

Average national productivity for Arabica coffee in Kenya is estimated at around 300 kg/ha of clean coffee for smallholder farms, which is low compared to average yields for Arabica worldwide (698 kg/ha) and in neighboring countries, such as Rwanda (1160 kg/ha) and Ethiopia (995 kg/ha) [29,30]. This gap may be the result of different factors such as sub-optimal or obsolete agricultural practices, the scarce availability of technical skills and knowledge, limited access to inputs and technologies (such as modern coffee varieties, chemicals, fertilizers, irrigation), and land size. At the same time, the high incidence of pests and diseases, such as coffee berry disease and leaf rust, remains a major issue, affecting cost and yields for most growers in Kenya [28,31,32].

Addressing the main outputs of the SCA, three interventions to enhance the sustainability and the value addition of the supply chain were identified for the CIVICS framework, in particular:

a. The introduction of shading management practices via trees, through intercropping in coffee plantations (shading trees);
b. The introduction of innovative water-saving pulping machinery for the wet milling process (eco-pulpers);
c. The exploitation of coffee wet-processing waste as a source of biomass for energy and fertilizer production (biomass).

These interventions have been contextualized considering Kenya's specific background, whether they have been already implemented or not in similar cases, and the existence of technologies easily available, in order to provide a set of realistic interventions. Furthermore, since the lowest level of coffee productivity is observed within smallholder production, all the interventions have been modeled as if they took place at the rural cooperative level.

3.2. Applying CIVICS Methodology to the Coffee Sector in Kenya

In order to model the Kenyan economy, it is required to represent it in such a way that economic agents' transactions could be accounted for entirely. In this research, the SUT, reported in Figure 3, is built from the information of the social accounting matrix (SAM), developed by the Joint Research Centre (JRC) [33], extended with EORA's national environmental extension for the same period (i.e., 2014) [34]. This SAM has been selected because of its recently updated data and for the characterization of household activities as a contribution to the local economy. This is very important when it is required to model the agricultural sector in a developing economy such as Kenya's, especially when analyzing the coffee sector, where smallholder production is extremely relevant.

The present structure, in the form of the observed exchanges during the year 2014, works as a baseline on which technological interventions have been modeled.

From the SCA, three possible interventions to improve the coffee sector have been identified. The technical details of the modeling of the following interventions are provided in the Supplementary Information (Supplementary Material S2), while here a general overview is provided.

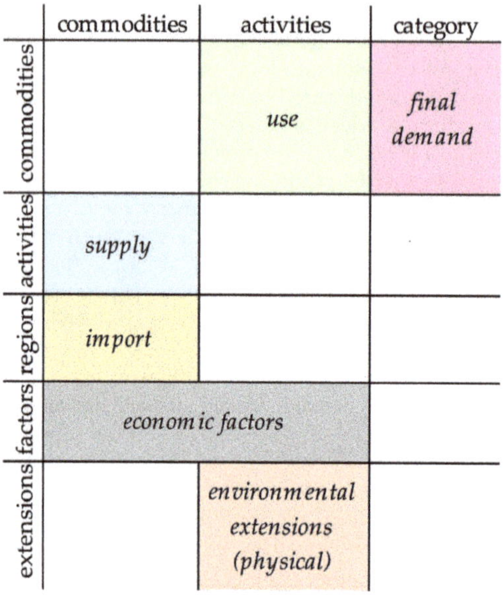

Figure 3. Structure of the SUT input–output model adopted in this research.

a. Shading management via trees

Optimal coffee-growing conditions include cool to warm tropical climates, good rainfall, and rich soils. Rising temperatures and recurrent droughts, experienced by many regions in the world as a result of climate change, represent a challenge for coffee production. Therefore, adaption practices are required in order to reduce the risks and the decline in coffee productivity. Among those, coffee shading (so-called shade-grown coffee) represents a climate-smart practice, which is gaining popularity, especially within small-holder contexts. Data sources of this intervention are reported in Table 1.

In the framework of this study, Coffee-Banana Intercropping (CBI) was considered. Research conducted in different contexts in Sub-Saharan Africa [35,36] proved that CBI systems can bring multiple benefits for smallholders, in particular:

- Increased resilience to climate change and extreme weather events;
- Increased incomes and improved food and nutrition security;
- Improved plant growth and enhanced coffee quality;
- Reduction in greenhouse gas emissions.

Potential disadvantages and barriers to the adoption of CBI were also pointed out:

- Negative impact on physical yield;
- High level of initial investment.

Table 1. Input parameters for shading tree management intervention.

Description	Value	Unit of Measure	Reference
Number of coffee plants	1800–2200 [1]	Plants/hectare	[28]
Fraction of shading trees to coffee plants	25	%	[37]
Cost of purchasing a shading banana plant	1.3	$/plant	[38]
Cost of planting a shading banana plant	0.13	$/plant	Estimation: a 10% cost over purchasing was assumed.
Banana yield	15	kg/plant	[39]
Banana price	0.065	$/kg	[39]
Reduction in physical yield (optimum level of shading)	8–15	%	[37]
Reduction in monetary yield (potential price growth)	2	%	[36]
Increase in the total soil carbon stocks	3.8	ton/ha	[36]
Reduction in required capital-machines	27	%	[40]
Growth in demand for labor	38	%	[40]
Useful life of the shading plants	20	years	Estimation: multiple banana trees emerging from the same rhizome in a couple of decades

[1] In an intercrop system, the plant population is going to be less than the actual number in Kenyan coffee monocrops, which is reported at around 2500 plants per hectare.

b. Eco-pulper for wet milling process

The pulping process is the last step of green-coffee production which takes place before drying. In the first step of the wet process, the skin and the pulp of the cherry are removed by a pulping machine, separating the pulp from the seed. Washing clears all remaining traces of pulp from the coffee seeds, which are then dried either by exposure to sunlight on concrete terraces or by passing through hot-air driers. The dry skin around the seed, called the parchment, is then mechanically removed, sometimes with polishing. This process takes place in the so-called wet mills and can affect the quality of coffee as a result of poor pulping. Losses incurred could be significant, but there are no available data to indicate their extent [28].

Pulping is normally based on a large water withdrawal and discharge, representing a risk for the sustainability of the process as well as for the communities living in the surroundings. Nowadays, available technology, the so-called eco-pulper machinery, is able to drastically reduce the impacts on water sources by minimizing water consumption and wastewater production. These machines can process up to 1 ton/h, reducing the processing time, serving several farms which can actually share the financial risk associated with the investment, and increasing the use of petroleum-based fuel. Data sources of this intervention are reported in Table 2.

Table 2. Input parameters eco-pulpers intervention.

Description	Value	Unit of Measure	References
Cost of eco-pulping machine	1430	$	[41,42]
Cost of delivery	46	$	Estimation: a 10% cost over purchasing was assumed.
Required power	1.1	kW	[41,42]
Capacity of the machine	0.5	tons of coffee/h	[41,42]
Efficiency of the machine	30	%	[41,42]
Decrease in water footprint	85	%	[41,42]
Number of smallholders to be covered by each machine	300–600	Smallholder/machine	Estimation: based on coffee fields productivity, proximity, and machinery capacity.
Productivity increase	0–2.5	%	Estimation: assumed on the basis of field interviews and expert judgments.
Carbon intensity of the eco-pulpers electricity consumption	0.27	$kgCO_2/kWh$	[43]
Useful life of the eco-pulpers	10	years	Estimation: based on similar machinery's expected life.

c. Exploiting biomass from coffee organic waste

As previously mentioned, from the SCA it has emerged that the wet processing generates waste, such as water and exhausted biomass. As also supported by observations in both Ethiopia and Kenya [44–46], waste, if not properly managed and discharged into the environment without any treatment, can affect the environment and pose a risk to communities. Data sources of this intervention are reported in Table 3 and the Logical Structure of the intervention reported in Figure 4.

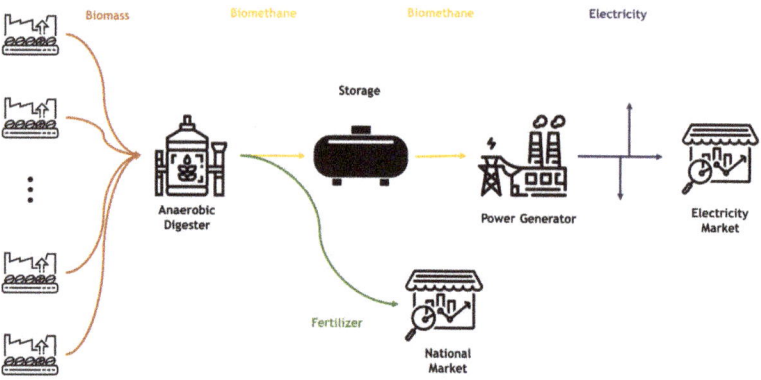

Figure 4. Structure of the "exploiting biomass" proposed implementation.

Following the principles of the circular economy, the proposed intervention aims at taking advantage of the waste biomass by feeding an anaerobic digester coupled with a biogas upgrader to produce bio-methane [47]. It is noteworthy that, in addition to the production of biogas, the anaerobic digestion of agricultural waste also produces an organic residue, namely digestate, which is rich in nutrients. If this digestate is utilized in plant production, nutrients will be reintegrated into the soil nutrient cycle, contributing to maintaining soil quality and fertility. The utilization of digestates may replace or at least reduce the use of mineral fertilizers, since they usually are rich in plant-available nutrients such as ammonium (NH_4^+), phosphate (P), and potassium (K) [48,49]. Moreover, the re-use of digestate for plant production, including coffee, is of particular interest to the Kenyan economy, being that fertilizers are massively imported into the country and on which the domestic agricultural sector relies heavily [50].

Table 3. Input parameters for biomass powerplant intervention.

Description	Value	Unit of Measure	Reference
Specific cost of biodigester	10,000	$/Nm^3/h	Estimation: assumed on the basis of field interviews and industrial players' judgments.
Specific cost of storage	0	$/Nm^3	Estimation: assumed on the basis of field interviews and industrial players' judgments.
Specific cost of generator	500	$/kW	Estimation: assumed on the basis of field interviews and industrial players' judgments.
Electricity production in one year by new plants	80	GWh	Energy modeling output (Calliope) [1]
Carbon intensity of electricity production from heavy fuel oil	0.27	kgCO_2/kWh	[43]
Efficiency of the old diesel generators to be replaced	0.4	-	Estimation: average efficiency of diesel generators.
Biomass to fertilizer rate	0.3	-	Estimation: assumed on the basis of field interviews and expert judgments.
Labor cost [2]	37.5		[51]
Size of biodigester	250	Nm^3/h	Estimation: assumed on the basis of biomass plant characteristics.
Size of generator	25,000	Nm^3	Estimation: assumed on the basis of biomass plant characteristics.
Size of storage	1	MW	Estimation: assumed on the basis of biomass plant characteristics.
Increase in use of transport commodity by cooperatives	30%	-	Estimation: assumed based on coffee fields and biomass plant proximity.
Useful life of the machines	25	years	Estimation: assumed on the basis of biomass plant characteristics.

[1] To be changed for every different number of Gensets. [2] Considering 2 technicians, one process engineer, and one electrical and power engineer per each plant.

The wet-mill process produces two different kinds of biomass waste, namely pulp (assumed to be 200% of the weight of the final green coffee production) and parchment (assumed to be 20% of the weight of the final green coffee production). The amount of waste produced refers to [52], who performed a specific analysis on the coffee industry of Kenya. The intervention proposes to collect the biomass waste at the mills level for biogas production, installing a power-producing machine in 17 mills.

The power produced by such machines is assumed to be injected into the national grid, and the fertilizer produced to enter the national market. Given the extreme seasonality of the availability of the coffee waste biomass, it is necessary to account for a storage system, in which the bio-methane is stored to allow the electricity generation to be carried out all year long. The impact of such intervention is explored with a twofold approach, taking advantage of the two modeling strategies presented in this work. Through the energy system model of the country, it is possible to assess how the national electricity system reacts to the new generating technologies, integrating them into the energy mix, and to observe how and when this energy is used, while the IOA permits us to estimate the impacts on the economy of changing the inputs to the electricity sector and avoiding the import of a part of the fertilizers required by the smallholder coffee cooperatives.

4. Results and Discussion

In order to guarantee a coherent comparative analysis, the same investment level of KES 1 million, corresponding to approximately $9k, is adopted for the analysis.

Within this modeling structure, assuming a policy goal and a set of implementation strategies, it is possible to adopt the two applications of the CIVICS framework, depicted in Section 2.2.

- *Integrated Multidimensional Analysis*: evaluate the impact of different interventions and create a set of comparable and case-specific indices. In the present case, the focus is on six indicators, which are connected to as many SDGs.
- *Policy Goal*: find an optimized mix of strategies that is compliant with policy-makers' main concerns while respecting other policy objectives. For this specific case study, a budget constraint of $100m is set. In this case, the maximization of the savings of economic production factors is first compared in the absence of further constraints and then subjected to one on green-water savings and reduction in CO_2 emissions.

Figure 5 represents an exhaustive summary of the *Integrated Multidimensional Analysis*, including six indicators by which the impact of various applied interventions is compared. In the following, a detailed description of the change in each indicator by the proposed strategies is reported.

- Required workforce: It represents the amount of additional labor, by means of required wages expenditure for $9k of investment. In this study, it can be noticed that introducing shading trees leads to the most dominant positive increase in local labor impact. In fact, the sectors associated with the harvesting and maintenance of banana trees are characterized by more labor intensiveness compared to the other interventions. This could have desirable effects in getting close to the objective depicted by SDG 8, introducing positive conditions to enable economic growth and decent jobs.
- Avoided import: Although being resilient to external shocks can play a role in improving the economic conditions of a country, it is not easy to put into practice when the considered economy is largely dependent on the import of crucial commodities (e.g., petroleum). The biomass intervention in this research permits one to decrease this dependence by the local production of a non-negligible share of imported products. Furthermore, extracting value from coffee waste, which was formerly an unexploited resource, is aligned with SDG 12, ensuring a sustainable production and consumption pattern.
- Land saving: This indicator is associated with SDG 6, which promotes the sustainable use of terrestrial ecosystems and avoids land degradation. Land-use reduction by

installing eco-pulpers is highly dependent on the assumed new productivity, influencing, importantly, the number of new inputs saved per unit of production. On the one hand, eco-pulpers allow for a more resource-efficient conversion of coffee berries into green coffee, reducing waste per unit of output. On the other hand, intercropping makes land use more efficient by exploiting banana–coffee synergies.

- Emission saving: Carbon emissions are considerably reduced by adopting biomass intervention due to the shift in electricity mix from heavy fuel oil to biomass combustion, which follows the climate action proposed by SDG 13. In fact, the activity of the highly carbon-intense heavy fuel oil is limited by substituting part of the fixed overall electricity production by means of the new—according to modeling assumptions, carbon-neutral—power production technology.
- Water-saving: This impact is extremely relevant when wet mills are substituted by eco-pulpers, allowing for the sustainable management of water introduced by SDG 6. Eco-pulpers are increasing the overall efficiency of the process (which is also true for the shading trees intervention) and heavily reducing the amount of water required per unit of processed coffee.
- PRoI: Considering shading management intervention, the main economic benefit is associated with the introduction of additional revenue-generation activity, which is the production of bananas coming out of the shading trees. This benefit compensates for the reduction in coffee productivity in cooperatives, leading to a higher *Policy Return on Investment* compared to the other strategies. On the other hand, the annual return on investment in eco-pulper intervention is mostly due to the savings coming from the direct impact of the increase in the productivity in cooperatives on the economic factors of this sector. Although the adoption of biomass resources does not increase the physical productivity of coffee, the reduction in imports of petroleum and fertilizers due to the intervention leads to an annual saving of around $20m.

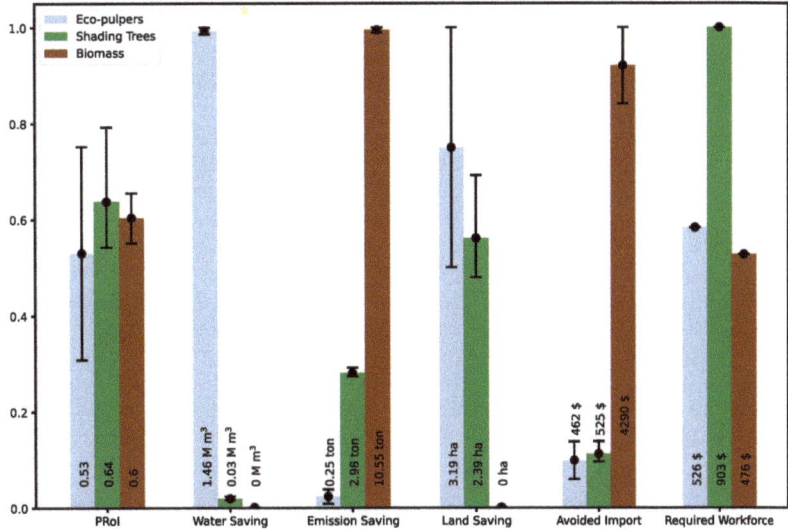

Figure 5. Comparison between performance indicators representing the net yearly gain from every intervention assuming the same level of investment (i.e., KES 1 million corresponding to nearly $9k) in average scenarios (bar height). The end of the scale corresponds to the highest value among the considered options within the sensitivity cases (error bars), while PRoI is expressed as defined in Equation (3) using KES 1 million as the denominator. MCM stands for million cubic meters.

More detailed results are represented separately in the Supplementary Information (Supplementary Material S3) of this paper where the reader can refer to the impacts on the activities and also carbon emissions in various sectors alongside the changes in the import of different commodities due to each individual intervention.

For what concerns the *Policy Goal* application of the framework, a $100 m budget has been set, allowing investments among the selected interventions, but the result may change if environmental objectives are introduced (Table 4).

Table 4. Optimization choices when running the model with only budget and physical constraints and when adding environmental objectives. The first percentage represents the budget allocation while the second compares the amount invested with the maximum possible level of investment.

	Without Env. Objectives	With Env. Objectives
Eco-pulpers	0%, 0%	1%, 34%
Shading Trees	63%, 100%	50%, 79%
Biomass	37%, 73%	49%, 98%

When no environmental objectives are set, the logic is straightforward: the only limits of the model are represented by physical boundaries, otherwise it would select only the intervention with the highest PRoI. However, since it is not possible to cover with trees more than the coffee plants, the budget is invested also in the second most profitable intervention (i.e., biomass).

The introduction of environmental constraints, in this example in the form of a minimum annual saving with respect to the baseline of circa 70 kton of CO_2 and 300 Mm^3 per year, slightly modifies the intervention choices. Now, all the biomass potential is exploited in order to reach the carbon reduction objective; similarly, the desired savings of water can be reached only by adding eco-pulpers into the interventions mix. The selected mix of interventions can be performed simultaneously and the combined results of the changes introduced by the new technologies and practices can diverge from the linear behavior assumed for finding the optimal mix.

An example of the combined effect of the intervention is represented by the employment consequences by skill level, driven by the investment of the $100 m budget if allocated as proposed by the *Policy Goal* mode including environmental objectives.

As shown in Figure 6, investments in the coffee sector trigger labor increases in other sectors all over Kenya, but the main change is associated with the increase in demand for low-skill workers in the north area of the country (see public.flourish.studio/visualisation/3338282/ accessed on 1 April 2022 for the interactive version of the map). It can be inferred that policies that aim at increasing the occupation level in the most vulnerable population share should be driven towards the coffee sector. This is particularly relevant since higher shares of unemployment are among unskilled workers and the northern part of the country is the one with the lowest wealth relative index [53]. Figure 6 shows, in particular, how from the optimal allocation of the $100 m investment, unskilled workers in the northern regions of the country are the category that benefits the most.

Furthermore, the same $100m of investment would not only positively impact the social sphere but would also benefit the economic and environmental dimensions of sustainability.

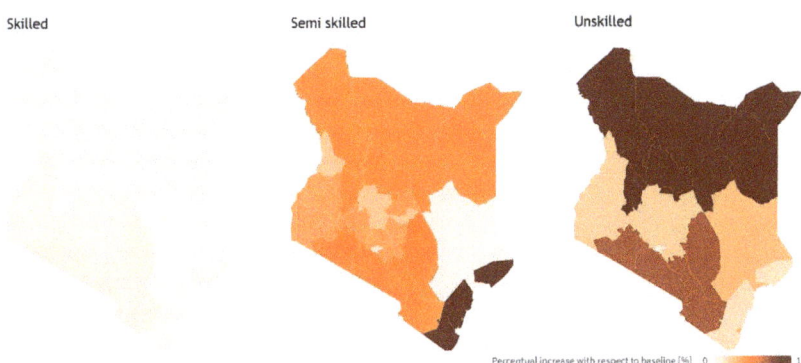

Figure 6. Labor demand induced by all the $100m mix of interventions by skill level and region.

5. Conclusions

With the aim of providing a policy support system that is scientifically solid, evidence-driven, and able to grasp the complexity related to the Water-Energy-Food Nexus and the systemic nature of the challenges involved within the sustainable development challenge, the CIVICS framework is presented in this study.

CIVICS, the ComprehensIVe and Integrated Country Study, intends to be a tool to support the decision-makers of developing countries by evaluating the impact of the proposed policies and framing them in the bigger picture of SDGs, while ensuring that the desired local outcome is achieved. In this report, some possible applications of CIVICS to different policy options for the coffee sector in Kenya are outlined in order to provide examples of the potential of the approach.

In particular, attention can be drawn to the modularity and customizability of the framework, making it flexible to the context in which it is applied, and suitable for evaluating policies that range from the national or regional level, down to very context-specific local interventions. The model is offering different tools that can be used in synergy or as stand-alone impact evaluation methods accordingly to the needs of the specific context.

In addition to that, it is worth highlighting how an interesting feature is to use CIVICS as a benchmarker between policy interventions, offering the possibility to assign a series of indicators to the proposed policies, in order to evaluate the proposals within a single framework and provide insights based on their relevance with the SDGs, or other technical and socio-economic references. Furthermore, it is possible to exploit an operational research method to identify the optimal mix of interventions under a constrained budget to meet the desired policy outcomes.

In conclusion, some key take-away messages can be derived from the presented approach. In particular, it emerged how the use of an integrated framework is pivotal to achieving the full potential of the adopted models, which gain strength and provide deeper insights when coupled with the others. The double nature of the approach guarantees the achievement of specific local goals, without overlooking international frameworks, such as Agenda 2030, as a global blueprint for inclusive development.

The goal of this work is to propose a novel methodology, reproducible in other contexts and/or geographical areas. The source code and input data are therefore released with an open license for transparency and reproducibility purposes [54]. They are made freely available in the following repository: github.com/SESAM-Polimi/CIVICS-KENYA, accessed on 1 April 2022.

Supplementary Materials: The following supporting information can be downloaded at: https://www.mdpi.com/article/10.3390/en15093071/s1, Supplementary Material S1: Detailed Materials and Methods; Supplementary Material S2: Proposed Interventions in Detail; Supplementary Material S3: Detailed results by interventions; Supplementary Material S4: Supplementary Data.

Author Contributions: Conceptualization, N.G., N.S., M.A.T., F.I., M.V.R. and E.C.; Data curation, N.G., N.S., N.N., M.A.T., L.A.S.H. and B.C.; Formal analysis, N.G., N.S., N.N., M.A.T., L.A.S.H., B.C. and M.V.R.; Investigation, L.A.S.H., B.C. and F.I.; Methodology, N.G., N.S., N.N., M.A.T., L.A.S.H., B.C., M.V.R. and E.C.; Project administration, F.I., M.V.R. and E.C.; Software, N.G., N.S. and M.A.T.; Supervision, M.V.R. and E.C.; Validation, N.G., N.S. and N.N.; Visualization, N.G., N.S. and M.A.T.; Writing—original draft, N.G., N.S., N.N., M.A.T., L.A.S.H., B.C. and F.I.; Writing—review & editing, N.G., N.S., F.I., M.V.R. and E.C. All authors have read and agreed to the published version of the manuscript.

Funding: This research received no external funding.

Institutional Review Board Statement: Not applicable.

Informed Consent Statement: Not applicable.

Data Availability Statement: The source code and input data are made available in the following repository: https://github.com/FEEM-Africa-REP/CIVICS_Kenya, accessed on 1 April 2022.

Acknowledgments: The authors thank Fondazione Eni Enrico Mattei and Politecnico di Milano for supporting the research.

Conflicts of Interest: The authors declare no conflict of interest.

References

1. UN. *Transforming Our World: The 2030 Agenda for Sustainable Development*; United Nations: New York, NY, USA, 2015.
2. Nwanze, K.F. *Smallholders Can Feed the World*; IFAD: Rome, Italy, 2017.
3. Ritchie, H.; Roser, M. Environmental Impacts of Food Production. Available online: https://ourworldindata.org/environmental-impacts-of-food (accessed on 1 April 2022).
4. Welsch, M.; Hermann, S.; Howells, M.; Rogner, H.H.; Young, C.; Ramma, I.; Bazilian, M.; Fischer, G.; Alfstad, T.; Gielen, D.; et al. Adding Value with CLEWS—Modelling the Energy System and Its Interdependencies for Mauritius. *Appl. Energy* **2014**, *113*, 1434–1445. [CrossRef]
5. Sridharan, V.; Ramos, E.P.; Zepeda, E.; Boehlert, B.; Shivakumar, A.; Taliotis, C.; Howells, M. The Impact of Climate Change on Crop Production in Uganda-An Integrated Systems Assessment with Water and Energy Implications. *Water* **2019**, *11*, 1805. [CrossRef]
6. Bazilian, M.; Rogner, H.; Howells, M.; Hermann, S.; Arent, D.; Gielen, D.; Steduto, P.; Mueller, A.; Komor, P.; Tol, R.S.J.; et al. Considering the Energy, Water and Food Nexus: Towards an Integrated Modelling Approach. *Energy Policy* **2011**, *39*, 7896–7906. [CrossRef]
7. Sridharan, V.; Broad, O.; Shivakumar, A.; Howells, M.; Boehlert, B.; Groves, D.G.; Rogner, H.H.; Taliotis, C.; Neumann, J.E.; Strzepek, K.M.; et al. Resilience of the Eastern African Electricity Sector to Climate Driven Changes in Hydropower Generation. *Nat. Commun.* **2019**, *10*, 1–9. [CrossRef]
8. Wang, C.; Malik, A.; Wang, Y.; Chang, Y.; Lenzen, M.; Zhou, D.; Pang, M.; Huang, Q. The Social, Economic, and Environmental Implications of Biomass Ethanol Production in China: A Multi-Regional Input-Output-Based Hybrid LCA Model. *J. Clean. Prod.* **2020**, *249*, 119326. [CrossRef]
9. Palma-Rojas, S.; Caldeira-Pires, A.; Nogueira, J.M. Environmental and Economic Hybrid Life Cycle Assessment of Bagasse-Derived Ethanol Produced in Brazil. *Int. J. Life Cycle Assess.* **2017**, *22*, 317–327. [CrossRef]
10. Tanimonure, V.A.; Naziri, D. Impact of Climate Adaptation Strategies on the Net Farm Revenue of Underutilised Indigenous Vegetables' (UIVs) Production in Southwest Nigeria. *Resour. Environ. Sustain.* **2021**, *5*, 100029. [CrossRef]
11. Mungai, E.M.; Ndiritu, S.W.; Da Silva, I. Unlocking Climate Finance Potential and Policy Barriers—A Case of Renewable Energy and Energy Efficiency in Sub-Saharan Africa. *Resour. Environ. Sustain.* **2022**, *7*, 100043. [CrossRef]
12. Mainar-Causapé, A.J.; Boulanger, P.; Dudu, H.; Ferrari, E. Policy Impact Assessment in Developing Countries Using Social Accounting Matrices: The Kenya SAM 2014. *Rev. Dev. Econ.* **2020**, *24*, 1128–1149. [CrossRef]
13. Adil Suliman Hussain, L.; Inzoli, F.; Golinucci, N.; Stevanato, N.; Rocco, M.V.; Colombo, E. *Supply Chain Analysis with Focus on Africa FEEM's Methodological Approach*; Milan, Italy, 2020. Available online: https://papers.ssrn.com/sol3/papers.cfm?abstract_id=3733685 (accessed on 1 April 2022).
14. Galal, N.M.; Moneim, A.F.A. Developing Sustainable Supply Chains in Developing Countries. *Procedia CIRP* **2016**, *48*, 419–424. [CrossRef]
15. Suh, S. *Handbook of Input-Output Economics in Industrial Ecology*; Springer: Berlin/Heidelberg, Germany, 2009; ISBN 9781402040832. Available online: https://link.springer.com/book/10.1007/978-1-4020-5737-3 (accessed on 1 April 2022).
16. Miller, R.E.; Blair, P.D. *Input-Output Analysis: Foundations and Extensions*; Cambridge University Press: Cambridge, UK, 2009; ISBN 1139477595.
17. Lenzen, M.; Rueda-Cantuche, J.M. A Note on the Use of Supply-Use Tables in Impact Analyses. *Sort* **2012**, *36*, 139–152.

18. Södersten, C.J.H.; Lenzen, M. A Supply-Use Approach to Capital Endogenization in Input–Output Analysis. *Econ. Syst. Res.* **2020**, *32*, 451–475. [CrossRef]
19. Tahavori, M.A.; Rinaldi, L.; Golinucci, N. SESAM-Polimi/MARIO: MARIO v0.1.0. 2022. Available online: https://zenodo.org/record/5879383#.YmIAptNBxPY (accessed on 1 April 2022).
20. Pfenninger, S.; Pickering, B. Calliope: A Multi-Scale Energy Systems Modelling Framework. *J. Open Source Softw.* **2018**, *3*, 825. [CrossRef]
21. Pfenninger, S.; Keirstead, J. Renewables, Nuclear, or Fossil Fuels? Scenarios for Great Britain's Power System Considering Costs, Emissions and Energy Security. *Appl. Energy* **2015**, *152*, 83–93. [CrossRef]
22. Shenoy, M. *Industrial Ecology in Developing Countries*; Springer: Cham, Switzerland, 2015; ISBN 9783319205717.
23. Arzoumanidis, I.; D'Eusanio, M.; Raggi, A.; Petti, L. Functional Unit Definition Criteria in Life Cycle Assessment and Social Life Cycle Assessment: A Discussion. In *Perspectives on Social LCA: Contributions from the 6th International Conference*; Traverso, M., Petti, L., Zamagni, A., Eds.; Springer International Publishing: Cham, Switzerland, 2020; pp. 1–10. ISBN 978-3-030-01508-4.
24. Boulanger, P.; Dudu, H.; Ferrari, E.; Causapé, A.J.M.; Balié, J.; Battaglia, L. *Policy Options to Support the Agriculture Sector Growth and Transformation Strategy in Kenya*; Joint Research Centre: Petten, The Netherlands, 2018; ISBN 9789279859496.
25. Government of the Republic of Kenya. The Kenya Vision 2030, The Popular Version. 2007. Available online: https://vision2030.go.ke/publication/kenya-vision-2030-popular-version/ (accessed on 1 April 2022).
26. Kenya Institute for Public Policy Research and Analysis. Kenya Economic Report 2019: Resource Mobilization for Sustainable. 2019. Available online: https://repository.kippra.or.ke/handle/123456789/2098 (accessed on 1 April 2022).
27. World Bank Group. Kenya Economic Update: Unbundling the Slack in Private Sector Investment—Transforming Agriculture Sector Productivity and Linkages to Poverty Reduction. 2019, pp. 1–86. Available online: https://openknowledge.worldbank.org/handle/10986/31515 (accessed on 1 April 2022).
28. International Coffee Organization. Country Coffee Profile Kenya. 2019. Available online: https://www.ico.org/documents/cy2018-19/icc-124-7e-profile-kenya.pdf (accessed on 1 April 2022).
29. Thuku, G.K.; Gachanja, P.; Almadi, O. Effects of Reforms on Productivity of Coffee in Kenya. *Int. J. Bus. Soc. Sci.* **2013**, *4*, 196–213.
30. Damianopoulos, R.A. *Market Adaptability, Industrial Divergence, and the Politics of Liberalization in the Kenyan and Ugandan Coffee Industries*; Queen's University: Kingston, ON, Canada, 2005.
31. Kenya Coffee Platform. Coffee Economic Viability Study. 2018. Available online: https://www.globalcoffeeplatform.org/wp-content/uploads/2021/03/Kenya-Coffee-Platform-Coffee-Economic-Viability-Study-Report-F.pdf (accessed on 1 April 2022).
32. UNCTAD Commodoities at a Glance, Special Issue on Coffee in East Africa. United Nations Conference on Trade and Development. 2018. Available online: https://unctad.org/system/files/official-document/ditccom2018d1_en.pdf (accessed on 1 April 2022).
33. Causapé, A.J.M.; Boulanger, P.; Dudu, H.; Ferrari, E.; Mcdonald, S. Social Accounting Matrix of Kenya. 2014. Available online: http://data.europa.eu/89h/bc924665-e888-4642-9b3b-297f4f6b13d3 (accessed on 1 April 2022).
34. Lenzen, M.; Moran, D.; Kanemoto, K.; Geschke, A. Building Eora: A Global Multi-Region Input-Output Database at High Country and Sector Resolution. *Econ. Syst. Res.* **2013**, *25*, 20–49. [CrossRef]
35. FAO. Agroforestry Coffee Cultivation in Combination with Mulching, Trenches and Organic Composting in Uganda. 2017. Available online: https://www.fao.org/3/ca3728en/ca3728en.pdf (accessed on 1 April 2022).
36. van Asten, P.; Ochola, D.; Wairegi, L.; Nibasumba, A.; Jassogne, L.; Mukasa, D. *Coffee-Banana Intercropping Implementation Guidance for Policymakers and Investors*; FAO: Rome, Italy, 2015; p. 10.
37. Rahn, E.; Vaast, P.; Läderach, P.; van Asten, P.; Jassogne, L.; Ghazoul, J. Exploring Adaptation Strategies of Coffee Production to Climate Change Using a Process-Based Model. *Ecol. Model.* **2018**, *371*, 76–89. [CrossRef]
38. Shading Plant Cost. Available online: https://perfectdailygrind.com/2015/06/how-much-does-it-cost-a-farmer-to-plant-a-basic-plot/ (accessed on 1 April 2022).
39. Wairegi, L.; Van Asten, P.; Giller, K.; Fairhurst, T. *Banana-Coffee System Cropping Guide*; CAB International: Wallingford, UK, 2014; ISBN 9781780644929.
40. Jezeer, R.E.; Santos, M.J.; Boot, R.G.A.; Junginger, M.; Verweij, P.A. Effects of Shade and Input Management on Economic Performance of Small-Scale Peruvian Coffee Systems. *Agric. Syst.* **2018**, *162*, 179–190. [CrossRef]
41. CAL—Coffee Machinery—Mini Eco Pulper. Available online: https://www.cal-ea.com/ (accessed on 1 April 2022).
42. Alibaba.com Eco Mini Pulper Cost. Available online: https://www.alibaba.com/product-detail/Mini-Eco-Pulper-for-coffee_135021838.html (accessed on 23 September 2020).
43. Combustion of Fuels—Carbon Dioxide Emission. Available online: https://www.engineeringtoolbox.com/co2-emission-fuels-d_1085.html (accessed on 1 April 2022).
44. Murthy, P.S.; Madhava Naidu, M. Sustainable Management of Coffee Industry By-Products and Value Addition—A Review. *Resour. Conserv. Recycl.* **2012**, *66*, 45–58. [CrossRef]
45. Ulsido, M.D.; Li, M. Solid Waste Management Practices in Wet Coffee Processing Industries of Gidabo Watershed, Ethiopia. *Waste Manag. Res.* **2016**, *34*, 638–645. [CrossRef]
46. Mwangi, R.W.; Mwenda, L.K.M.; Wachira, A.W.; Mburu, D.K. Effect of Final Processing Practices Carried out by the Coffee Cooperative Societies on the Sustainability of the Coffee Industry in Kenya. *Eur. J. Bus. Strateg. Manag.* **2017**, *2*, 60–75.

47. Surendra, K.C.; Takara, D.; Hashimoto, A.G.; Khanal, S.K. Biogas as a Sustainable Energy Source for Developing Countries: Opportunities and Challenges. *Renew. Sustain. Energy Rev.* **2014**, *31*, 846–859. [CrossRef]
48. Sogn, T.A.; Dragicevic, I.; Linjordet, R.; Krogstad, T.; Eijsink, V.G.H.; Eich-Greatorex, S. Recycling of Biogas Digestates in Plant Production: NPK Fertilizer Value and Risk of Leaching. *Int. J. Recycl. Org. Waste Agric.* **2018**, *7*, 49–58. [CrossRef]
49. Battista, F.; Frison, N.; Bolzonella, D. Energy and Nutrients' Recovery in Anaerobic Digestion of Agricultural Biomass: An Italian Perspective for Future Applications. *Energies* **2019**, *12*, 3287. [CrossRef]
50. Balié, J.; Battaglia, L.; Boulanger, P.; Dudu, H.; Ferrari, E.; Mainar Causapé, A.J. Returns to Investments in Fertilizers Production in Kenya. In *An Analysis in Support of the New "Agriculture Sector Growth and Transformation Strategy"*; Food and Agriculture Organization of the United Nations (FAO) and Joint Research C: Rome, Italy, 2019.
51. Salaries by Positions—Kenya.Paylab.Com. Available online: https://kenya.paylab.com/ (accessed on 1 April 2022).
52. Gathuo, B.; Rantala, P.; Maatta, R. Coffee Industry Wastes. *Water Sci. Technol.* **1991**, *24*, 53–60. [CrossRef]
53. Chi, G.; Fang, H.; Chatterjee, S.; Blumenstock, J.E. Microestimates of Wealth for All Low- and Middle-Income Countries. *Proc. Natl. Acad. Sci. USA* **2022**, *119*, e2113658119. [CrossRef]
54. Pfenninger, S.; DeCarolis, J.; Hirth, L.; Quoilin, S.; Staffell, I. The Importance of Open Data and Software: Is Energy Research Lagging Behind? *Energy Policy* **2017**, *101*, 211–215. [CrossRef]

Article

The Evolution of Energy Management Maturity in Organizations Subject to Mandatory Energy Audits: Findings from Italy

Annalisa Santolamazza [1,*], Vito Introna [1], Vittorio Cesarotti [1] and Fabrizio Martini [2]

1. Department of Enterprise Engineering, "Tor Vergata" University of Rome, 00133 Rome, Italy
2. DUEE-SPS-ESE Laboratory, Italian National Agency for New Technologies, Energy and Sustainable Economic Development (ENEA), Lungotevere Thaon di Revel, 76, 00196 Rome, Italy
* Correspondence: annalisa.santolamazza@uniroma2.it

Abstract: Promoting energy efficiency is a key element of the strategic commitment of the European Community. Prominent among the binding measures established by the 2012 Energy Efficiency Directive to further this vision is the requirement for large companies to conduct energy audits every four years. After receiving the second cycle of energy audit reports in December 2019, a new description of the energy situation of Italian companies was made available. This presented the previously inaccessible possibility of comparing the two situations reported in 2015 and 2019 to assess the development of energy efficiency practices in organizations subject to the legislative obligation of energy audits in the country. To this end, in collaboration with the Italian National Agency for New Technologies, Energy and Sustainable Economic Development (ENEA), a project was initiated with the aim of developing the tools and methodologies necessary to assess in more detail the evolution that has occurred in the four years since 2015. In this paper, the findings of the analysis conducted on a significant sample of companies in Italy are presented. Through the design of a maturity model to assess the degree of progress achieved in a company's energy management, the results of the two situations were analyzed. The analysis was deepened by assessing the progress achieved in different aspects of Energy Management: "Strategic Approach", "Awareness, Competence, and Knowledge", "Methodological Approach", "Organizational Structure", "Energy Performance Management and Information System", and "Best Practices". Furthermore, the observed variations were statistically tested using a pairwise *t*-test to make statistical inferences about the maturity of the total population of Italian enterprises under legislative obligation. The results have shown an increase in overall energy management maturity in each maturity dimension.

Keywords: energy efficiency; energy audit; energy management; maturity model; energy efficiency directive

Citation: Santolamazza, A.; Introna, V.; Cesarotti, V.; Martini, F. The Evolution of Energy Management Maturity in Organizations Subject to Mandatory Energy Audits: Findings from Italy. *Energies* **2023**, *16*, 3742. https://doi.org/10.3390/en16093742

Academic Editor: Abu-Siada Ahmed

Received: 4 April 2023
Revised: 17 April 2023
Accepted: 19 April 2023
Published: 27 April 2023

Copyright: © 2023 by the authors. Licensee MDPI, Basel, Switzerland. This article is an open access article distributed under the terms and conditions of the Creative Commons Attribution (CC BY) license (https:// creativecommons.org/licenses/by/ 4.0/).

1. Introduction

Energy efficiency promotion, the utilization of renewable sources, and pollutant emission reduction are crucial components of the European Community strategy. In October 2012, the Energy Efficiency Directive (EED) was established with the aim of achieving a 20% reduction in energy use before 2020 [1], energy efficiency target uploaded by the 2018/2002 directive to 32.5% by 2030 (relative to 1990 levels) [2], defining a balanced collection of binding measures and recommendations. The EED created a framework of measures to promote energy efficiency and ensure that European goals are met, as well as facilitate future advancements in energy efficiency after 2020. Article 8 of the framework requires affected companies to conduct energy audits. Large and/or energy-intensive organizations must provide these audits every four years and include details such as facility location, corporate characteristics, manufacturing processes, and finished products. In 2014, the Italian government implemented the EED through Legislative Decree No. 102/2014, expanding the requirement to include energy-intensive enterprises. The Italian

National Agency for New Technologies, Energy and Sustainable Economic Development (ENEA) is responsible for managing and implementing the EED framework in Italy and uses the "Audit102" web portal to collect energy audits.

Energy audits are deemed the first step in increasing energy efficiency within an organization [3,4]. An energy audit, according to the European technical standard EN 16247, is a systematic and comprehensive analysis of the energy performance of an organization, equipment, systems, or processes. The purpose of the energy audit is to identify energy savings opportunities and propose solutions to improve energy efficiency, reduce energy consumption, and reduce greenhouse gas emissions. The energy audit process typically includes a preliminary phase to gather information about the facility and stakeholders' objectives, field work to collect data, analysis to identify areas of energy waste and inefficiency, identification of energy saving opportunities and recommendations, and reporting [5]. The results of an energy audit can help owners and managers make informed decisions about investing in energy-efficient upgrades and can ultimately lead to significant energy and cost savings [4,6].

In this context, the present paper reports relevant research findings aimed at gaining insight into the current situation and the changes undergone by companies subject to the legislative obligation of conducting mandatory energy audits. Indeed, a research project was conducted in collaboration with ENEA, with the aim of developing the necessary tools and methodologies to assess in more detail the evolution that has occurred in Italian companies subject to mandatory energy audits.

In particular, in the present paper, the following research questions have been asked:

- **RQ1**: Has there been a change in the energy management practices of organizations subject to mandatory energy audits in the timeframe between the two mandatory energy audit cycles?
- **RQ2**: In reference to organizations subjected to mandatory energy audits, which energy management areas have undergone changes in the timeframe between the two mandatory energy audit cycles?

The novelty of this study lies in both the topic and the specifics of the research. Indeed, to the authors' knowledge, it is the first time that organizations subject to mandatory energy audits have been analyzed in terms of the evolution of their energy management good practices, focusing on aspects such as the development of energy monitoring system [7,8], the implementation of EnPIs (Energy Performance Indicators) [9], and awareness and training programs for personnel [9]. Moreover, the significance of the research also lies in the scope of the analysis since it regards data from over 340 organizations which led mandatory energy audits, with varying initial conditions and dimensions.

The structure of the paper is as follows: Section 2 introduces the background, describing the use of maturity models to evaluate energy management aspects; Section 3 describes the research methodology used; Section 4 presents the findings; Section 5 presents the discussions; finally, Section 6 concludes the paper, highlighting the main results and interesting insights for future developments.

2. Background

The tool chosen as the most suitable to be able to evaluate the change of energy management practices in companies is the Maturity Model. In fact, maturity models are widely used to assess the organizational status of the company in a specific area and support the identification of potential areas for improvement. The concept of corporate maturity was conceived in 1979 by Philip Crosby in the work titled "Quality is free" [10] with the purpose of providing a tool for corporate management to measure, and therefore control, the degree of quality management in the organization. The proposed instrument was the "Quality Management Maturity Grid (QMMG)". Subsequent to its first formulation, the concept of maturity has evolved over time, thanks to the interest of both academics and practitioners. Nowadays, the sectors in which maturity models are applied have expanded, from project management to security management and sustainability [11]. For example, a

2012 literature review identified 237 articles regarding the research on maturity models, covering more than 20 different domains [12].

Becker, Knackstedt, and Pöppelbuß provided a clear definition of a maturity model in 2009 [13]: "A maturity model consists of a sequence of maturity levels for a class of objects. It represents an anticipated, desired, or typical evolution path of these objects shaped as discrete stages". Therefore, a maturity model is used to represent an evolutionary path for certain entities that may be represented by organizations or processes [13,14].

Maturity models are also tools suitable for the knowledge transfer process, since they can define a specific improvement path based on an assessment of current conditions and their comparison with relevant best practices [15]. They can often be configured in self-assessment mode, allowing professionals and organizations to identify key areas for improvement and actions to be taken.

Moreover, maturity models can be defined by different levels (or stages) of maturity and by several structuring dimensions. Dimensions give a systematic representation of the field of interest and should be defined so that they are distinct and representative of all aspects of the activity/process for which maturity is being evaluated [16]. Therefore, maturity models can be one-dimensional, multi-dimensional, or even hierarchical through the use of subdimensions [17].

Finally, the main characteristics common to all maturity models are as follows [11]:

- Model structure—this can be "continuous" or "in stages". For models in stages, each maturity level is considered as the basis for the next level. In continuous models, the approach to improvement is based on the development of processes' capacities and is ongoing and flexible [11,16,17].
- Methodology of analysis—this refers to the manner used to evaluate the organization's maturity.
- Reference to international standards—this can be beneficial for an organization that already has applied an international standard to choose to use a maturity model that is based on the same standard but, in contrast, other organizations could benefit more from using a maturity model not tailored to a specific standard.
- Mode of assessment—this refers to the operational procedures used to conduct the evaluation. Most models are characterized by the presence of questionnaires with closed questions or grids. The number of questions is a compromise between a thorough evaluation and the aim to appeal to even less structured and less experienced organizations. Moreover, the possibility of self-assessment is an effective way to allow even less aware organizations to obtain an overall assessment of their maturity.
- Results of the assessment—this refers to the differences in terms of results provided. They may vary according to the degree of detail of the assessment (e.g., a simple number or a more structured report). Often, the results of the assessments are supported by graphical tools to better convey the concept.

During the past decade, various models have been developed to evaluate the maturity of organizations in energy management. These models differ in terms of their structure, analysis methodology, reference to international standards, mode of assessment, results of assessment, and domain. The most widely used model structure is staged, which is easier for less mature organizations to understand. However, Carbon Trust developed both a staged and a continuous structured model in 2011 [18]. Moreover, different assessment methods have been used, such as workshops [19], interviews [20–22], and questionnaires. The last typology is the most used since it enables organizations to self-assess their performance independently, with varying degrees of depth (i.e., the number of questions ranges from 15–20 [23–25] to 40–60 [11,18,26]). Furthermore, most models analyze single sites, but Finnerty et al. have focused their attention on evaluating the maturity of multisite organizations, defining a self-guided assessment comprising sections for both the specific site and the overall organization [27,28]. Çoban and Onar had a similar focus but used a fuzzy methodology to implement the assessment [29]. Moreover, Wehner et al. defined a maturity model from a staged structure for the energy efficiency initiatives adopted by logistics service providers [30], while Benedetti et al. have focused on the management of specific energy assets such as compressed air systems [15]. Finally, Jin et al. recently proposed a maturity model to analyze the Chinese situation as well as that of other emerging economies [31].

While different attempts to define models to assess the maturity of organizations in energy management can be identified in the scientific literature, in this article, as described by the following section, a new specific maturity model was defined in collaboration with ENEA to evaluate how the dissemination of best practices in energy management has evolved in companies submitted to mandatory energy audits [32].

3. Research Methodology

3.1. Summary of the Research Methodology

The research methodology used to evaluate the development of energy management aspects in Italian companies, required to conduct mandatory energy audits, involved a series of steps.

Firstly, a maturity model was designed to assess the changes in critical characteristics of energy management in these companies. Subsequently, maturity assessment questionnaires were delivered and collected from a significant sample of companies. The collected data was then analyzed using main statistical tools, such as descriptive statistics and inferential statistics, to identify patterns and trends in the data. Overall, this methodology enabled the researchers to gain a comprehensive understanding of the evolution of energy management practices in Italian companies and to draw meaningful conclusions based on the analysis of the collected data (Figure 1).

Figure 1. Summary of the research methodology used.

3.2. Design of the Energy Management Maturity Model

The definition of the maturity model followed these methodological steps: definition of the structure of the model, definition of analysis methodology, and definition of assessment procedures. The most common structure found for the models examined was the staged one, which was evaluated as the most suitable to allow one to carry out an assessment of the evolution of maturity in the energy management of companies.

The proposed model was based on a model already developed by the authors, which has, however, been heavily modified to make it suitable for the purpose of the project, taking into account legislative and regulatory changes, changing the number of maturity dimensions, questions, and associated answers [11].

It was decided to use 5 levels, the most common in existing models, as a good compromise between the need for differentiation and the ease in the recognition of the actual behaviors:

- Level 1—Elementary

Energy consumption is not considered relevant. In the organization, the energy performance of the organization has never been evaluated.

- Level 2—Occasional

There is a tentative interest in the organization towards the issue of energy consumption. Generally, there is a lack of adequate commitment and support from above, and energy efficiency is pursued in an occasional manner. The preliminary collection of consumption data and energy costs might start.

- Level 3—Project-based

A first strategy is identified and targets are set. Typical of this stage is the execution of an energy audit or the identification of specific opportunities for improvement. The collection and evaluation of energy data is systematized.

- Level 4—Management

The company is led toward the development of an Energy Management System with an adequate information system and monitoring and the development of a plan of activities to achieve efficiency targets.

- Level 5—Optimized

Inside the organization, an Energy Management System is present and continuously optimized, with the support of top management and the full involvement of the entire organization. In the case of a model in stages it is necessary to establish the operational mode to assess within the companies the achievement of different maturity levels (e.g., whether to reference to dimensions, targets, or processes such as the processes of ISO 50001).

In the proposed model, key aspects of energy management within an organization have been defined and used to create 6 dimensions. Each level may contain aspects related to the different dimensions of maturity. Below, the six maturity dimensions identified are listed (Figure 2):

- Strategic approach (i.e., energy policy, measurable objectives, responsibilities, and action plan) (SA);
- Awareness, competence, and knowledge (i.e., knowledge of the energy market, self-generation systems, capability to manage relationships with energy suppliers and services, equipment and materials providers, knowledge of the energy consumption structure of the site, analytical and statistical tools and methods of financial analysis) (ACK);
- Methodological approach (i.e., the consistency, continuity, and systematization of planned actions) (MA);
- Organizational structure (i.e., relations within the organization and the approach used to define and coordinate tasks) (OS);
- Energy performance management and Information Systems (i.e., measurement system, data collection, analysis and reporting, energy performance indicator definition) (EPMIS);
- Best practices (i.e., standardization and optimization of activities and processes that have an impact on the energy performance of the organization, such as maintenance and usage of machines and systems, purchase, design, and plant modifications, risks and opportunities assessment) (BS).

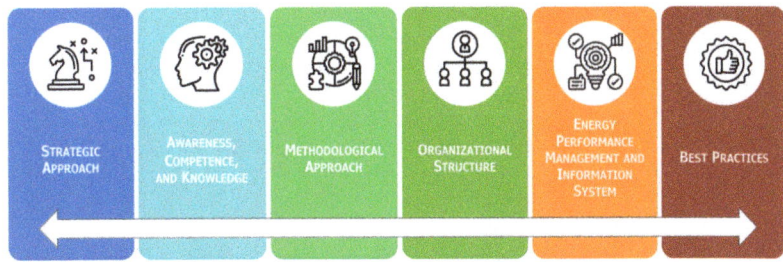

Figure 2. Representation of six maturity dimensions of the maturity model.

The assessment method chosen for the proposed maturity model was the self-assessment guided through a questionnaire. The reasons for this are related to the intention to reduce the risk of misunderstandings due to personal interpretations that could skew the results of the assessment while also enabling remotely data collection via web platform to promote data collection. For each level, a number of questions associated with each dimension have been identified, resulting in a total of 48 questions:

- 12 questions for Level 2;
- 14 questions for Level 3;
- 15 questions for Level 4;
- 7 questions for Level 5.

Since the first level is an elementary stage, it is not associated with any questions. From levels 2 to 5, questions are associated with a series of responses to characterize the specific level (the number of responses is equal to 4 for the first three levels, from second to fourth, while it is equal to 2 for the last level). Each question is also associated with maturity dimensions, as displayed in Table 1.

Table 1. Association of each question of the maturity model and maturity dimensions.

Maturity Dimension	Associated Questions
Methodological Approach	Level 2: Q04 Level 3: Q18; Q19 Level 4: Q27; Q36; Q37; Q38 Level 5: Q47
Strategic Approach	Level 2: Q01; Q02; Q05 Level 3: Q13; Q23 Level 4: Q28; Q39 Level 5: Q43
Best Practices	Level 2: Q11; Q12 Level 3: Q25; Q26 Level 4: Q32; Q33; Q34; Q41 Level 5: Q48
Awareness, Competence, Knowledge	Level 2: Q03; Q10 Level 3: Q16; Q17 Level 4: Q35; Q40 Level 5: Q45
Energy Performance Management and Information Systems	Level 2: Q07; Q08; Q09 Level 3: Q20; Q21; Q22; Q23 Level 4: Q30; Q31 Level 5: Q42
Organizational Structure	Level 2: Q06 Level 3: Q14; Q15; Q24 Level 4: Q29 Level 5: Q44; Q46

The organization that answers the questionnaire must choose the answer that better reflects their situation. The answers are defined so that if an answer is true, the previous ones are true. As a result, the score for each response can be calculated cumulatively. In order to enable companies to assess how their approach to energy management has evolved in the years between the two mandatory energy audit cycles, two answers are given for each question:

- The first one, representative of the situation prior to the conduction of the energy audit (2015);
- The second one, representative of the situation after the conduction of the second mandatory energy audit (2020–2021).

The presentation of the results is achieved through three indicators:

- The global maturity index, a number between 1 and 5, which summarizes the overall level of maturity of the organization;

- The degree of coverage of the different levels;
- The development of maturity in different dimensions.

Thus, in accordance with the definition of the model, for every indicator two evaluations are made: the first representative of the situation before the conduction of the first energy audit and the second representative of the situation after the conduct of the second energy audit cycle.

To improve the effectiveness of the model, its first draft was tested on a small sample of companies by first letting each company answer the questions autonomously and then establishing an interview with it. Thus, it was possible to verify the adequacy of the results obtained and the ability of the tool to capture the changes undergone by the companies over the years and identify the causes. Moreover, the questionnaire was shared during several meetings with trade associations and ENEA. Their feedback was collected to assess the comprehensibility of the questions and the reliability of the results.

To provide benefits to the companies undergoing the maturity assessment, we decided to create a report describing the results of the analysis, highlighting the variation in global maturity index and each maturity level and dimension, also suggesting the most crucial areas to prioritize for their energy management improvement.

3.3. Data Collection

In the first months of 2021, with the collaboration of ENEA, the questionnaire for the maturity model was published in online form in a private section of the same portal used by Italian companies to submit their mandatory energy audit (https://audit102.enea.it/, accessed on 20 March 2023). The delivery and collection of the maturity assessment questionnaires was a success, making it possible to establish relevant results.

The number of companies in the database thus developed was 411 at the end of 2021. Of this initial sample, 68 companies were discarded for two main reasons:

- Companies that did not answer all the questions in the questionnaire;
- Companies that carried out the questionnaire by answering for the "first audit cycle" referring to a closer timeframe.

Thus, the sample analysed comprised 343 companies.

In Figure 3 it is possible to observe the distribution of the companies in the sample in relation to the main economic sectors (that is, the NACE code [33]).

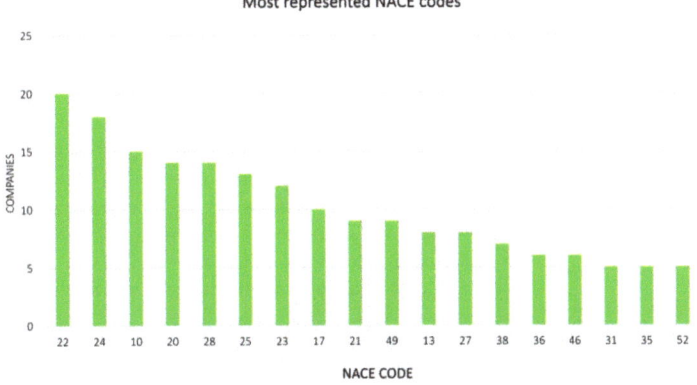

Figure 3. Display of the most represented NACE codes in the collected sample.

3.4. Data Analysis

To analyze the results of the maturity questionnaire given to the sample of Italian companies, the main descriptive statistics were selected with the aim of identifying the following information [34]: distribution, mean, and standard deviation.

Subsequently, inferential statistics tools were used to expand the study.

For each company, having obtained two non-independent values for all the different maturity parameters defined in the construction of the maturity questionnaire (one relating to the situation prior to the first energy audit in 2015 and one relating to the current situation), a statistical test for paired data was carried out. The aim was to assess whether in the period between the first mandatory energy audit and the second mandatory energy audit cycle the performance in terms of energy management had improved in Italian companies. Therefore, it was investigated whether it was possible to assess in a statistically significant manner that, for the parameters quantified by the maturity questionnaire, the average of the population represented by Italian companies that have undergone mandatory energy audits in compliance with the legislative obligation had increased.

First of all, the difference for each of the pairs of paired observations was calculated. The hypothesis to be tested statistically is as follows:

$$H_0 : \mu_2 - \mu_1 = \delta \leq 0 (variable\ not\ increased) \tag{1}$$

$$H_1 : \mu_2 - \mu_1 = \delta > 0 (variable\ increased) \tag{2}$$

with μ_2 e μ_1, respectively, the average values of the examined maturity index relative to the situation after the second audit cycle (2020–2021) and relative to the situation before the first energy audit cycle (2015).

If the test resulted in the rejection of the null hypothesis ($H_0 : \delta \leq 0$) it would be possible to conclude that from 2015 to the following situation, the specific maturity indicator examined for companies subjected to mandatory energy audit has improved with a significance level of 5%. Therefore, this test was carried out by analysing differences between the situation before the 2015 energy audit cycle and the situation after the second mandatory energy audit cycle, in relation to different variables:

- Global maturity index;
- Degree of coverage of maturity levels;
- Level of coverage of maturity dimensions.

Finally, to further explore the analysis and identify which specific aspects have changed more significantly and which, on the contrary, have remained more stable over the years, the variation of each individual question of the questionnaire was observed to assess how much individual requirements have been met.

It should be noted that to be able to exercise statistical inference to draw statistically valid conclusions about the entire population of companies that have complied with the legislative obligation, the following assumptions are valid (in relation to the analysis of paired data) [23,24]:

- The sample of companies is assumed to be statistically representative of at least the entire population of companies subject to the legislative obligation;
- Subjects (in this case companies) must be independent. The measurements of one subject must not influence the measurements of the others;
- Paired measurements must be obtained from the same subject;
- The measured differences must have a normal distribution, or the central limit theorem must still be valid (sample size > 30–40 elements).

4. Results

4.1. Analysis of the Global Maturity Index

Figure 4 shows the comparison of the global maturity index distributions at the time of the first audit (2015) and now (2021), after the second deadline of the legislative obligation. As can be seen, the distribution of the maturity index has changed, moving to the right, signifying an increase in the overall level of maturity of companies in energy management. In Figure 5 the distribution of the change in the overall maturity index is also shown.

Specifically, in the first cycle, the average value of the maturity index of the companies in the sample was 2.27 (with a standard deviation of 0.85), while thereafter the average value was 3.19 (with a standard deviation of 0.88).

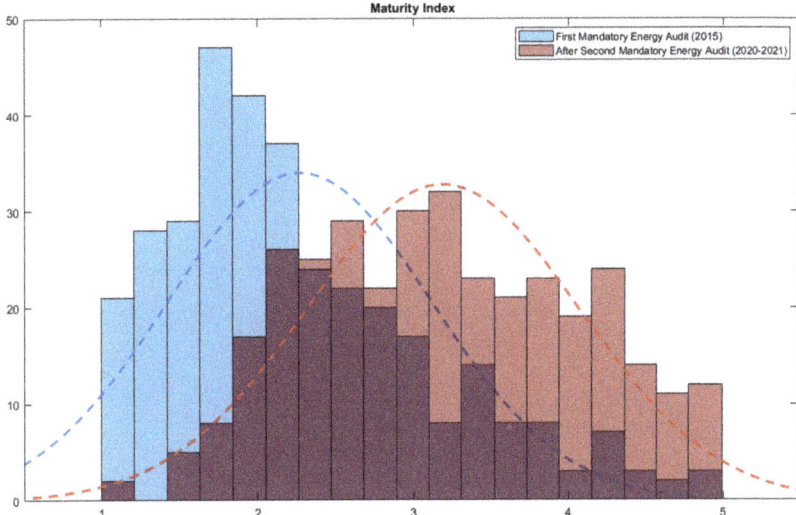

Figure 4. Comparison of the distribution of the global maturity index in 2015 and after the second round of energy audits (2020–2021).

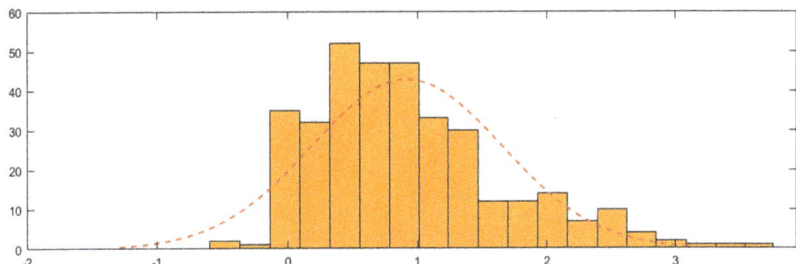

Figure 5. Distribution of the difference in the overall maturity index between the two instances (first audit cycle vs. second audit cycle).

In order to verify the actual statistical significance of the observed variation, a paired t-test was then carried out in relation to the maturity index found in the two observed situations.

A t-test was performed for paired data to assess whether it was possible to conclude that the population of companies in 2021 had a higher average overall maturity index than previously in 2015. The p-value resulting from the analysis is less than 0.001 (2.68×10^{-72}), so it is possible to conclude that the maturity index of the companies subjected to the legislative obligation has increased in these years with a significance level of 0.05.

4.2. Analysis of Maturity Levels

The analysis can be deepened by looking at how different levels of maturity have evolved (Figure 6). Each level has increased in coverage by an average of 20–25%.

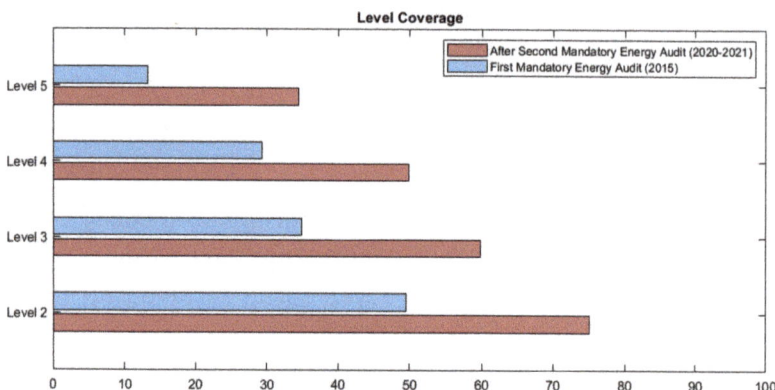

Figure 6. Comparison of level of coverage of the individual levels in 2015 and after the second round of energy audits (2021).

Figure 7 compares the box plots of degree of coverage of the individual levels in 2015 and after the second round of energy audits (2020–2021).

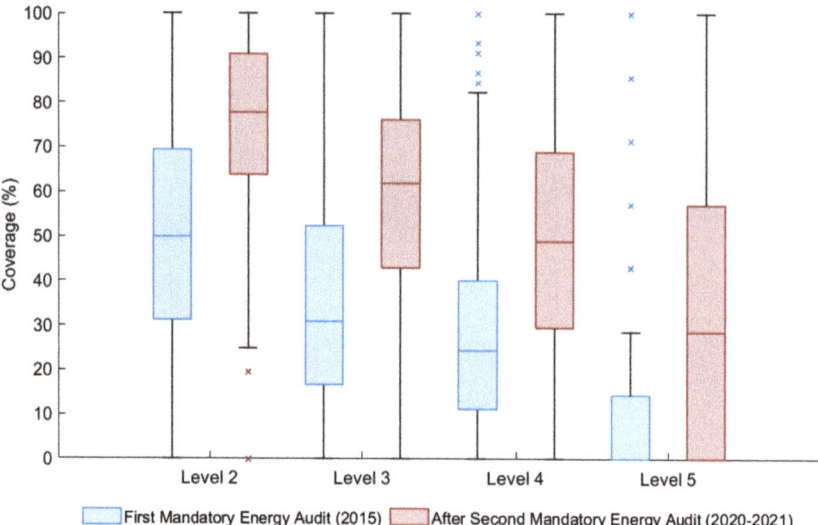

Figure 7. Box-plot comparison of the degree of coverage of the individual levels in 2015 and after the second round of energy audits (2020–2021).

In Figure 8 a summary of the comparisons of maturity level coverage in 2015 and after the second cycle of mandatory energy audits (2020–2021) is shown.

To verify the actual statistical significance of the apparent variation observed, a paired t-test was performed on variations in the degree of coverage of the levels. The results are reported in Table 2.

As shown in Table 2, all p-values resulting from paired t-test were less than 0.001, so it is possible to conclude that all maturity levels coverage for companies subjected to the legislative obligation have increased in these years with statistical significance.

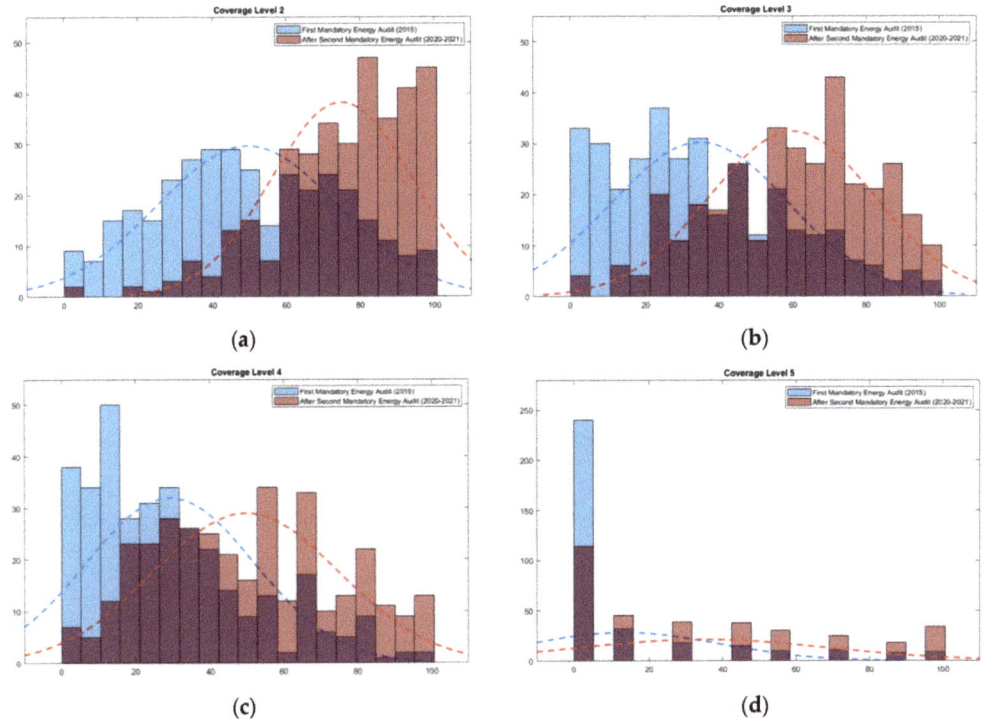

Figure 8. Summary of comparisons of maturity levels' coverage in 2015 and after the second round of energy audits (2021): (**a**) distribution comparison for Level 2 coverage; (**b**) distribution comparison for Level 3 coverage; (**c**) distribution comparison for Level 4 coverage; (**d**) distribution comparison for Level 5 coverage.

Table 2. Statistical analysis of the coverage of maturity levels for the sample: mean value and standard deviation of each maturity level with respect to the first and second mandatory energy audits and p-value for the t-test for the variations between two periods for each maturity level.

Maturity Level	Mandatory Energy Audit Cycle	Mean	Standard Deviation	p-Value for t-Test
Level 2	First	49.59	24.45	6.52×10^{-80}
	Second	75.00	18.96	
Level 3	First	34.90	24.00	5.88×10^{-75}
	Second	59.89	22.42	
Level 4	First	29.36	22.66	4.96×10^{-58}
	Second	49.84	24.96	
Level 5	First	13.29	25.69	6.62×10^{-35}
	Second	34.49	34.23	

4.3. Analysis of Maturity Dimensions

The analysis continues by examining the variations found in the different dimensions of maturity. On average, all dimensions increased by about 20% in their coverage (Figure 9).

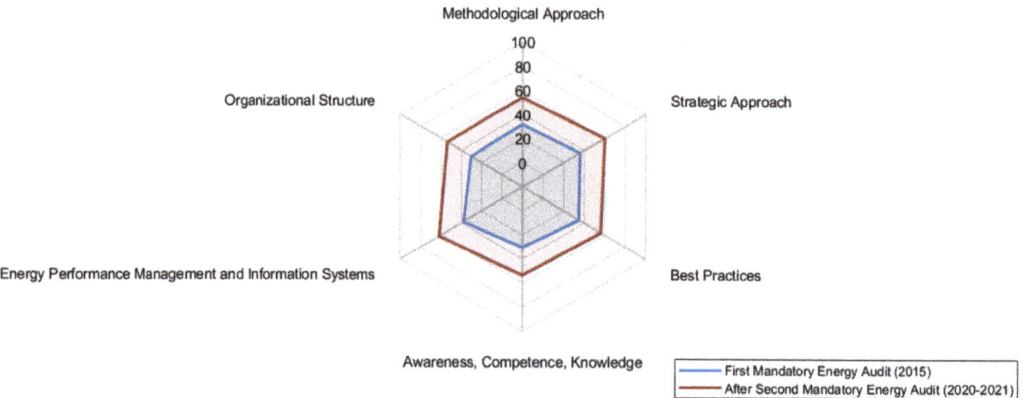

Figure 9. Comparison of level of coverage of the individual dimensions in 2015 and after the second round of energy audits (2020–2021).

Figure 10 compares box plots of the degree of coverage of the different dimensions in 2015 and after the second round of energy audits (2021).

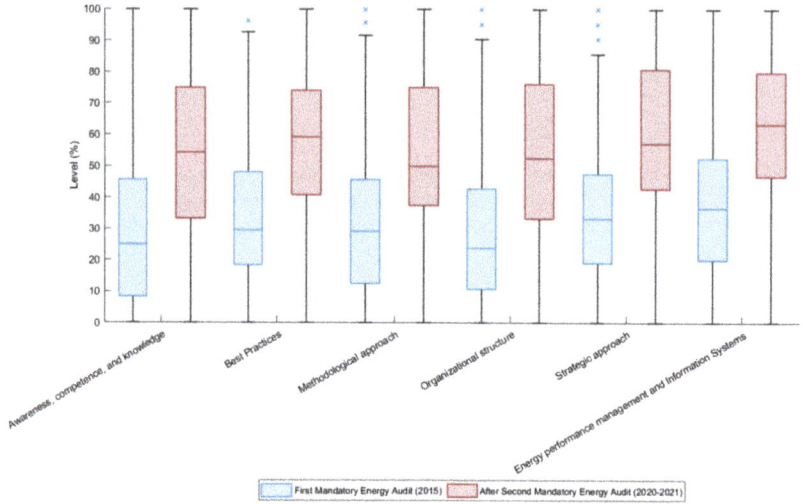

Figure 10. Box-plot comparison of the level of coverage of the individual dimensions in 2015 and after the second round of energy audits (2020–2021).

Observing the variation of the dimensions in the sample (Figure 10), it can be seen that the dimensions relating to "Energy Performance Management and Information Systems" and "Strategic Approach" are the dimensions that have seen the greatest change in the sample.

Figure 11 shows a summary of comparisons between the coverage levels of Maturity Dimensions in 2015 and after the second round of energy audits (2020–2021).

To verify the actual statistical significance of the apparent variation observed, a paired t-test was carried out concerning the variations in the degree of coverage of the maturity dimensions. The results are reported in Table 3.

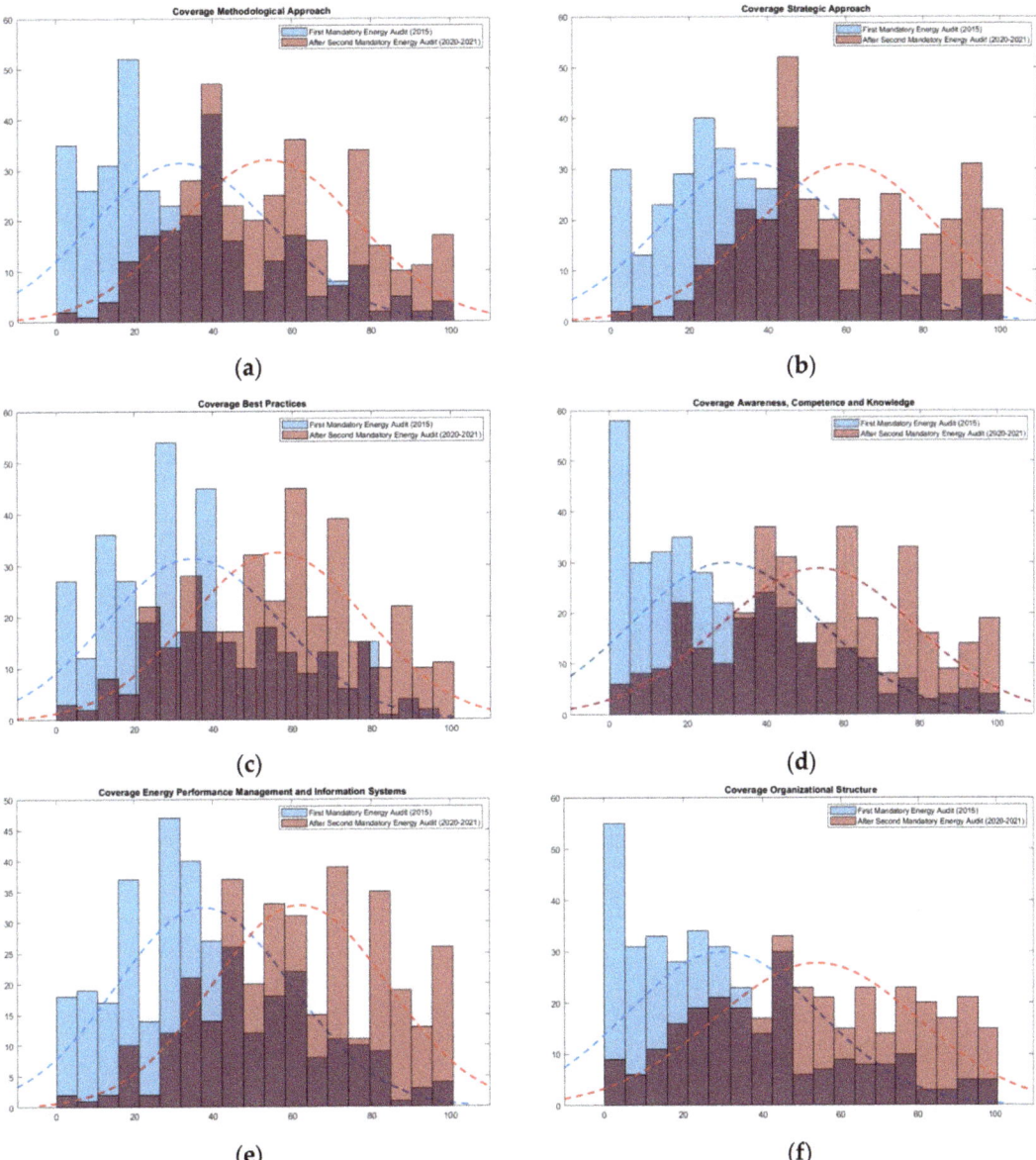

Figure 11. Summary of comparisons of maturity dimensions' coverage in 2015 and after the second round of energy audits (2020–2021): (**a**) distribution comparison for Methodological Approach; (**b**) distribution comparison for Strategic Approach; (**c**) distribution comparison for Best Practices; (**d**) distribution comparison for Awareness, Competence, Knowledge; (**e**) distribution comparison for Energy Performance Management and Information Systems; (**f**) distribution comparison for Organizational Structure.

As shown by Table 3, all p-values resulting from paired test-t were less than 0.001, so it is possible to conclude that all maturity dimensions coverage for companies subjected to the legislative obligation increased in these years with statistical significance.

Table 3. Statistical analysis of the coverage of maturity dimensions for the sample: mean value and standard deviation of each maturity dimension with respect to both the first and second mandatory energy audits and the *p*-value for the *t*-test for the variations between two periods for each maturity dimension.

Maturity Dimension	Mandatory Energy Audit Cycle	Mean	Standard Deviation	*p*-Value for *t*-Test
MA [1]	First	31.92	22.65	1.71×10^{-63}
	Second	54.15	22.67	
SA [1]	First	36.17	23.54	1.24×10^{-69}
	Second	60.23	23.55	
BP [1]	First	34.89	22.22	9.20×10^{-69}
	Second	56.28	22.25	
ACK [1]	First	30.02	25.04	4.88×10^{-66}
	Second	53.39	25.05	
EPMIS [1]	First	37.69	22.25	1.52×10^{-70}
	Second	62.04	22.27	
OS [1]	First	30.22	26.01	2.84×10^{-62}
	Second	53.98	26.04	

[1] MA: Methodological Approach; SA: Strategic Approach; BP: Best Practices; ACK: Awareness, Competence, Knowledge; EPMIS: Energy Performance Management and Information Systems; OS: Organizational Structure.

In general, all maturity dimensions also showed significant improvements, demonstrating an overall improvement in the practices with which companies that have complied with the energy audits obligation manage energy.

4.4. Analysis of the Individual Requirements (Analysis for Each Question)

After finding an actual change in the level of coverage of all levels of maturity management and dimensions, we proceeded to investigate further.

In order to identify which specific aspects have changed more significantly and which, on the contrary, have remained more stable over the years, we have proceeded to analyze in detail the variation of answers for each question, observing how the individual requirements of the maturity dimensions are being satisfied.

It is possible to compare the initial and current situation of energy management in relation to the "Awareness, Competence, Knowledge" dimension in Italian companies, observing the most widespread answers for each question (Figure 12).

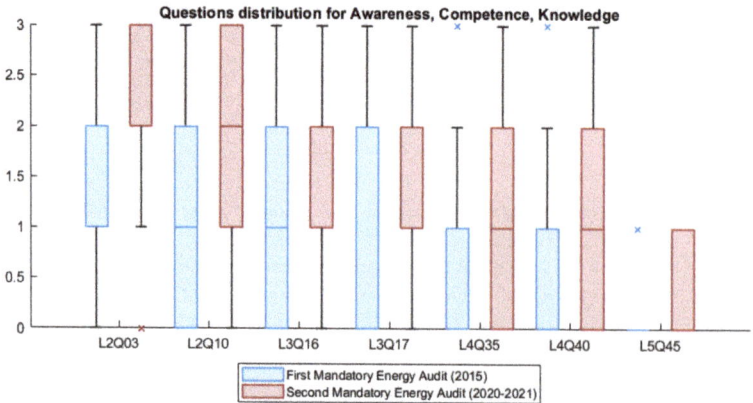

Figure 12. Box-plot of sample answers for each question—Dimension "Awareness, Competence, Knowledge".

- Lev 2—Question 03: promotion of energy efficiency within the organization.

In the past, the organization had conducted promotional activities in a sporadic or ad hoc manner to raise staff awareness. However, currently, systematic ad hoc initiatives are being implemented both internally and externally to the organization, ensuring that all staff are aware of the importance of energy efficiency.

- Lev 2—Question 10: technical knowledge of the energy aspects of personnel responsible for energy management.

Previously, the level of technical knowledge varied among different companies. Some companies had no specific knowledge or training, while others had almost sufficient knowledge. Currently, the situation is still variable. Some companies have limited and heterogeneous knowledge but are expecting to activate a training program soon. For other companies, the level of knowledge is adequate and they maintain it through periodic training activities.

- Lev 3—Question 16: technical training (energy procurement, energy production/transformation, energy use, innovative technologies) offered to personnel responsible for energy management.

Previously, there was a lack of adequate specific training: some companies were about to start a training program, while others had covered only a few topics or trained only some of the concerned staff. Currently, the situation is quite similar, with some companies starting a training program or only covering main issues.

- Lev 3—Question 17: type of management training (economic-financial evaluation of energy projects, energy audits, methods and tools for consumption analysis, information systems for energy management, energy management systems) offered to energy management personnel.

Previously, the situation was very varied: the staff had not yet received adequate specific training, were about to start a training program, or the training had covered only some topics or only part of the staff concerned. Currently, a training program is about to start or the training has covered only some issues or only part of the staff concerned.

- Lev 4—Question 35: operational training on energy management (good practices related to energy use, maintenance, etc.).

Previously, there was no initiative in the direction of operational training or a training plan had been defined but had not yet been started. Currently, the situation is variable, with some companies having a formal training plan that is not yet complete, covering only some roles or aspects, while others have not started any initiative in this direction.

- Lev 4—Question 40: organization awareness.

Previously, the organization had not adequately addressed the awareness of its commitment to energy efficiency by staff and their role and responsibilities in achieving the objectives. However, currently, some companies are carrying out a series of activities to fully achieve this aspect, while for others, it is still not adequately addressed.

- Lev 5—Question 45: continuous training on energy efficiency.

In the past, only a few companies evaluated the training needs of their employees, planned, implemented, and periodically verified the effectiveness of training activities for energy management. Furthermore, only a few companies had planned to update their training programs periodically with respect to technological innovations. Currently, this happens more frequently than in the past, but it is still limited. The analysis highlights that there is still a wide margin for improvement in personnel training. The development of an energy management system can help organizations pay more attention and systematicity to training on energy issues.

It is possible to compare the initial and current situation of energy management in relation to the "Best Practices" dimension in Italian companies, observing the most widespread answers to the questionnaire for each question (Figure 13).

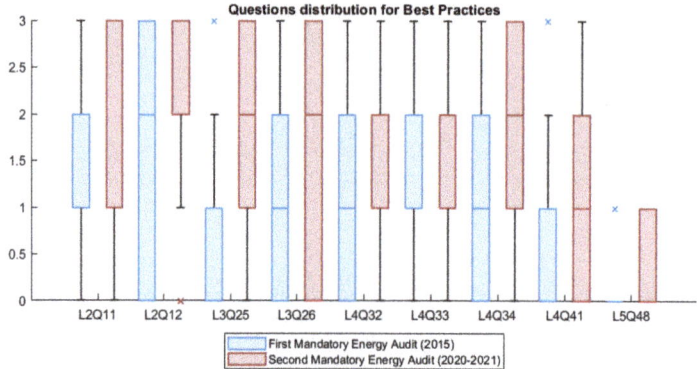

Figure 13. Box-plot of sample answers for each question—Dimension "Best Practices".

- Lev 2—Question 11: use of incentive tools in the energy field available to the organization to promote the financing of energy efficiency interventions.

In the past, companies were generally interested in promoting energy efficiency interventions but had not evaluated specific tools such as "Energy Efficiency Certificates". Now, some companies are fully aware of these tools and systematically consider their suitability when assessing the feasibility of efficiency projects. However, there is still room for improvement in this area to help companies turn energy audit opportunities into real savings.

- Lev 2—Question 12: self-production of energy.

Previously, some companies had not addressed self-production of energy, while others had conducted preliminary or structured analyses. Today, most companies have conducted at least a preliminary analysis and will undertake interventions if the assessment of costs and benefits is positive.

- Lev 3—Question 25: search for energy efficiency opportunities (e.g., through energy audits).

In the past, energy efficiency opportunities identified through energy audits had not led to the implementation of good practices for the use or maintenance of utilities and energy systems. Now, some companies have implemented new procedures for at least one of these functions, indicating greater confidence in audit tools and a willingness to identify feasible improvement interventions related not only to system modifications but also to methods of use and maintenance.

- Lev 3—Question 26: risk analysis related to energy supply.

Before, some companies had never done this type of analysis, while others had done it but were struggling to act further. Now, the situation is varied, with some companies still not conducting this analysis, while others have taken preventive measures and developed emergency plans in case of energy supply interruption. This highlights a weakness in the companies' approach to analyzing energy supply risks and offers room for improvement.

- Lev 4—Question 32: identification and planning of good practices for the use of the organization's plants and machinery.

Previously, some companies had no initiatives in this area, while others had good practices for some activities but were not well-documented or implemented regularly. Few companies were starting to systematically identify good practices.

Currently, there are good practices for some relevant activities that are not always documented or implemented regularly. Some companies are in the process of systematically and thoroughly identifying good practices for all relevant activities that significantly impact energy use. There is room for improvement which can be achieved through the development of an energy management system.

- Lev 4—Question 33: identification and planning of good practices for the implementation of maintenance activities of the organization's plants and machinery.

Both previously and currently, good practices exist for some relevant activities, but are not always documented or implemented regularly. Some companies are in the process of systematically and thoroughly identifying good practices for all relevant activities that significantly impact energy use. The same considerations as in the previous point apply.

- Lev 4—Question 34: identification and planning of good practices for the implementation of the design and purchase of plants, machinery, and services.

Previously, some companies had no initiatives in this area, while others only had good practices for some activities, sometimes not well documented or implemented regularly. Few companies were starting to systematically identify good practices.

Currently, some companies have systematically and thoroughly identified good practices for all relevant activities that significantly impact energy use and they are regularly checked and updated. However, there is still room for improvement.

- Lev 4—Question 41: risk assessment and opportunities for energy performance.

Previously, some companies had never conducted an energy risk analysis, while others were still in the process of developing it. Currently, more companies have conducted a preliminary risk analysis and established related preventive and corrective actions.

- Lev 5—Question 48: research implementation and updating of good practices for the organization's significant energy-related activities (source/service acquisition, design, installation, modifications, use, and maintenance of machinery and equipment).

Previously, only a few companies systematically researched, documented, and implemented good practices that were consistently followed by all employees. Fewer companies regularly reviewed and updated these good practices for continuous improvement based on suggestions from staff at all levels of the organization.

Currently, this happens more frequently than before, but still to a limited extent.

It is possible to compare the initial and current situation of energy management in relation to the "Energy Performance Management and Information Systems" dimension in Italian companies, observing the most widespread answers for each question (Figure 14).

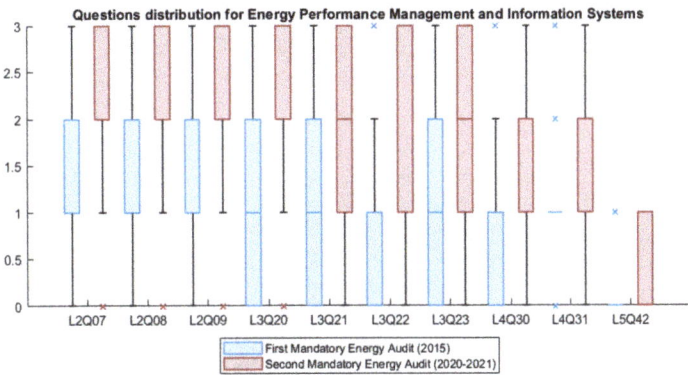

Figure 14. Box-plot of sample answers for each question—Dimension "Energy Performance Management and Information Systems".

- Lev 2—Question 07: analysis of cost and consumption data.

    In the past, companies only conducted occasional analysis of cost and consumption trends over time in case of anomalies or, for more energy-conscious companies, setting targets to follow. Today, companies periodically compare current and historical data to identify anomalies and set targets based on benchmark data. Some companies also conduct specific analyses on cost and consumption ratios and specific consumption in case of anomalies. This shift represents a significant improvement from an occasional to a systemic approach to energy performance control.

- Lev 2—Question 08: data collection for energy source costs and consumption.

    Previously, companies collected data on energy source costs and consumption annually or in each billing period. Nowadays, companies collect and report data several times a year and some companies even identify and report additional information necessary for understanding the consumption data (i.e., production units and working hours). This represents a significant improvement from previous practices.

- Lev 2—Question 09: energy tariff analysis methods.

    Previously, companies compared rates of different suppliers and conducted rate checks annually to identify the best rate for at least some energy sources, sometimes with the help of external professionals. Today, this practice is common for all main energy sources and the person responsible for the purchase selects the appropriate tariff structure with input from other managers (i.e., production manager). This reflects an increased attention to the choice of tariff.

- Lev 3—Question 20: development of an energy measurement system.

    Previously, some companies had no system to collect data on energy consumption, while others had at least defined methods for collecting data and set up a permanent measurement system for the main functional areas. Nowadays, at least the data collection has been defined and a permanent data collection and recording system has been set up for the main functional areas. Some companies have even established a detailed permanent data collection and recording system that covers the main significant processes and uses. This improvement is correlated with the attention paid to the measurement of consumption data in the recommendations of the second mandatory audits issued by ENEA.

- Lev 3—Question 21: measurement of energy drivers.

    Previously, some companies had not addressed the issue of identifying energy drivers, while others had carried out systematic analyses to identify them, but only for measurement points. Nowadays, the most relevant energy drivers have been measured, but only a few companies have proceeded to introduce them into the permanent measurement system with consumption energy. This indicates companies' interest in understanding the causes of variation in energy consumption over time.

- Lev 3—Question 22: analysis of energy consumption data.

    Previously, data analysis was performed only at the global system level, seldom deepening the analysis. Today, at least the contribution of main functional areas and the temporal trend of consumption for each measurement point are periodically analyzed. More established companies conduct a periodic analysis that systematically takes into account consumption recorded with respect to monitored energy drivers. This represents a significant improvement that serves as the basis for understanding energy consumption dynamics and identifying anomalies and opportunities to reduce consumption.

- Lev 3—Question 23: Energy Performance Indicators (EnPIs).

    In the past, some companies did not use any energy performance indicators while others had specific EnPIs for main functional areas. Nowadays, almost all companies use global-level EnPIs, which consider energy drivers that affect performance. Some companies also use specific EnPIs for the main functional areas and energy processes/uses. Significant improvement can easily be related to the requirement of evaluating energy performance indicators defined by mandatory energy audits.

- Lev 4—Question 30: energy consumption forecasting.

Previously, there was usually no consumption forecasting methodology or global consumption was predicted based solely on historical data. Currently, global consumption is predicted using historical data or forecast models that consider energy drivers (e.g., multivariable regression analysis). Although the improvement is limited in this case, the ability to predict consumption using complex models is crucial for organizations to maintain control over energy performance.

- Lev 4—Question 31: periodic consumption control.

In the past, some companies did not carry out consumption checks, while others experimented with control strategies for significant functional areas/systems in terms of energy consumption. Nowadays, periodic checks based on historical consumption are carried out and some companies are experimenting with control based on consumption forecasting through models that consider energy drivers. This improvement aligns with the previous point.

- Lev 5—Question 42: the information system for energy management.

Previously, only a few companies had an adequate information system for energy management that covered all areas/systems/services relevant to energy purposes, was integrated with the company's information system, and was subject to periodic reviews and adjustments. Nowadays, this is more common, but still to a limited extent.

It is possible to compare the initial and current situation of energy management in relation to the "Methodological Approach" in Italian companies, observing the most common answers to the questionnaire for each question (Figure 15).

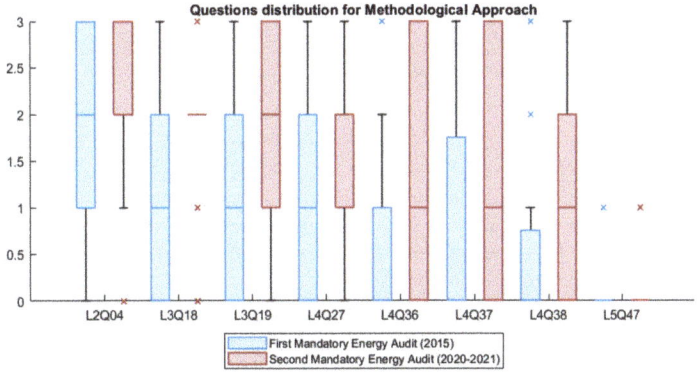

Figure 15. Box-plot of sample answers for each question—Dimension "Methodological Approach".

- Lev 2—Question 04: regarding attitude towards energy efficiency opportunities.

Previously, the situation was very diverse: opportunities were not frequently sought, but there was a general interest if they were randomly identified. However, now opportunities are taken more promptly when there is a positive quantitative assessment of the related costs and benefits, or whenever they arise.

- Lev 3—Question 18: energy audits frequency.

Previously, the situation was varied, with some companies never conducting audits and others conducting them with a frequency of 4 years. Currently, energy audits are conducted periodically with a frequency greater than that required by law.

- Lev 3—Question 19: energy saving opportunities periodically identified (e.g., through energy audits).

Previously, the situation was diverse: many companies had never identified savings opportunities, while others had done so, but only with a summary cost/benefit assessment and, occasionally, a technical-economic feasibility analysis. Now, opportunities are reported in a list that provides a description and a summary cost/benefit assessment for each of them and sometimes an implementation plan is prepared for positively evaluated opportunities for which there is financial availability. This new approach is a direct consequence of the practices introduced with energy audit.

- Lev 4—Question 27: development of an energy management plan.

Previously, activities were not always defined. Currently, activities are defined but are not always formalized and shared outside of the people responsible for implementation.

- Lev 4—Question 36: non-conformities management.

Previously, there was no initiative in this direction or at most there were methods of managing non-conformities applied in an infrequent or irregular manner. Currently, non-conformities are managed in a regular and adequate manner.

- Lev 4—Question 37: internal audits (inspections).

Previously, the situation was diverse: some companies had never conducted any, but for others, the management of internal audits almost always took place on a regular and adequate basis. Currently, internal audits are performed on a regular and adequate basis.

- Lev 4—Question 38: energy management system (e.g., according to ISO 50001 standard).

Previously, most of the time there was no real Energy Management System, or work was underway to develop it. Currently, there is either no real Energy Management System or there is a management system that is fully and continuously implemented over time. This indicates that in several companies, the growing sensitivity to energy management has led to the decision to develop a real management system, even if there is more room for improvement.

- Lev 5—Question 47: visibility of the organization.

It can be said that both previously and currently, few companies have been perceived and taken as a point of reference in the field of energy management. However, the work of these companies in this area is often cited as best practice and there are requests for presentations of their energy management system. This result seems to indicate that even organizations that have introduced the energy management system perceive they have further room for growth.

It is possible to compare the initial and current situation of energy management in relation to the "Organizational Structure" dimension in Italian companies, observing the most widespread answers to the questionnaire for each question (Figure 16).

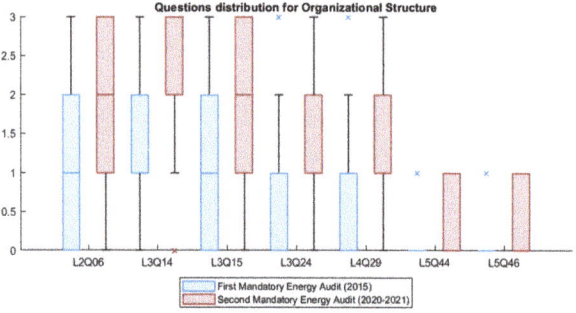

Figure 16. Box-plot of sample answers for each question—Dimension "Organizational Structure".

- Lev 2—Question 06: the Energy Manager.

Before, the situation varied in terms of the existence of an Energy Manager within companies. Some companies did not have one, while others had one informally or separately from the rest of the company. Currently, some companies have a formal energy manager, but not all, which sometimes limits their ability to involve staff from other areas as needed. This is still a significant improvement, but there is room for further progress.

- Lev 3—Question 14: sharing of the energy issues within the organization.

Previously, most managers recognized the importance of energy management, but some saw it outside of their responsibilities, while others were only reactive when involved in specific projects. Nowadays, most managers are convinced of its importance and are proactive in reducing consumption or encouraging others to do so. This is a fundamental change as it requires the full participation of the organization for significant and continuous improvements.

- Lev 3—Question 15: operational participation of the organization.

Previously, the Energy Manager operated autonomously within the organization. Now, the Energy Manager works with external experts and has varying degrees of systematic involvement with other managers within the organization. This is a crucial improvement in line with the previous point.

- Lev 3—Question 24: internal communication.

In the past, there was little to no contact between the Energy Manager and the departments/areas that used energy, or only ad hoc meetings were held with representatives of different areas. Currently, meetings are held with representatives from various areas with varying degrees of consistency, depending on the company. Improved communication is a crucial step towards greater involvement of the entire organization.

- Lev 4—Question 29: responsibility and tasks for energy management within the organization.

Previously, there was little awareness of the impact of different roles on energy consumption, and only a few key figures were identified, but without specific tasks and responsibilities. Now, some companies have identified the key figures and their impact on energy consumption and defined specific tasks and responsibilities to achieve energy efficiency. This is another crucial improvement in line with the previous points.

- Lev 5—Question 44: attitude of the organization in energy management.

Previously, only a few companies viewed energy management as a strategic element and implemented measures to continuously and efficiently. Now, more companies view energy management as important, but it still happens to a limited extent.

- Lev 5—Question 46: external communication on energy management.

Previously, only a few companies considered it important to disclose information on their energy performance and established an external communication plan. Today, more companies do so, but it is still limited.

It is possible to compare the initial and current situation of energy management in relation to the "Strategic Approach" dimension in Italian companies, observing the most common answers to the questionnaire for each question (Figure 17).

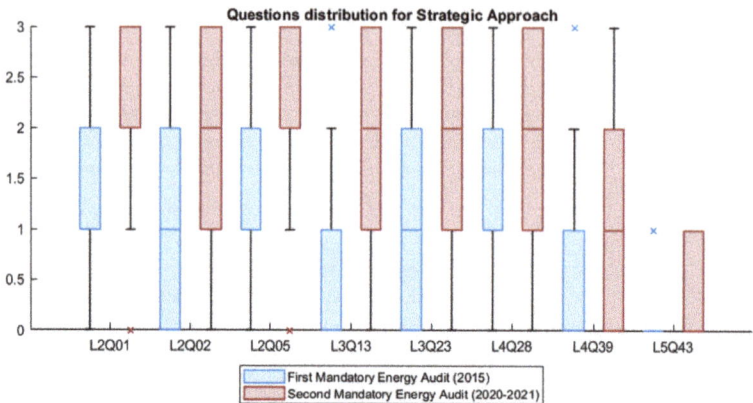

Figure 17. Box-plot of sample answers for each question—Dimension "Strategic Approach".

- Lev 2—Question 01: the issue of an organization's energy consumption.

Previously, some companies did not prioritize energy management and only took occasional measures to address it. Today, the importance placed on energy management is growing in companies, but there is still room for improvement in the systematic reduction in consumption and costs.

- Lev 2—Question 02: the organization's energy policy.

Previously, some companies did not have an energy policy, while others, while having it, did not share it widely. Now, most companies have an energy policy, but its dissemination and adoption are still variable, indicating a need for improvement.

- Lev 2—Question 05: organization's investment policies.

Previously, energy-saving investments were only considered if they were significantly cheaper than investments in the organization's core business. Now, there is greater awareness of the importance of energy-saving investments, but not all companies evaluate them on the same level as core business investments.

- Lev 3—Question 13: energy goals set by the organization.

Previously, energy goals were either nonexistent or established only at a global level. Now, some companies have defined specific energy goals for different levels and areas of the organization, indicating significant improvement with room for growth.

- Lev 3—Question 23: energy performance indicators (EnPIs).

Previously, some companies did not use any EnPI, while others defined some specific EnPIs for the main functional areas. Today, most companies use EnPIs at the global level, with some using more specific indices for functional areas and energy processes.

- Lev 4—Question 28: management control over the organization's energy performance.

Previously, management only periodically checked energy costs against the budget, without much discussion. Now, some companies periodically discuss energy performance reports to verify goal achievement and define improvement action plans, demonstrating greater emphasis on understanding the relationship between budget trend and company actions.

- Lev 4—Question 39: review of the Energy Management System (EMS).

Previously, management did not have direct involvement in the periodic review of EMS, but this was planned for the near future. Now, some companies have involved management in EMS review, but the timing and review methods are still not standardized, indicating significant room for improvement.

- Lev 5—Question 43: alignment of the energy management system with the strategic objectives of the organization.

Previously, only a few companies periodically defined and described the organization's energy-related strategic objectives at various levels, with the drive for the entire organization to work towards achieving them and producing measurable results. More companies are doing so, but to a limited extent, indicating a need for improvement.

4.5. Correlation Analysis between Global Maturity Index and Its Initial Value

Following the analyses conducted, an additional statistical analysis is reported to determine if there is a correlation between the final global maturity index and its initial value (i.e., first mandatory audit cycle), in order to understand if the initial level of maturity can influence its development. A preliminary analysis was conducted through the correlation diagram in Figure 18.

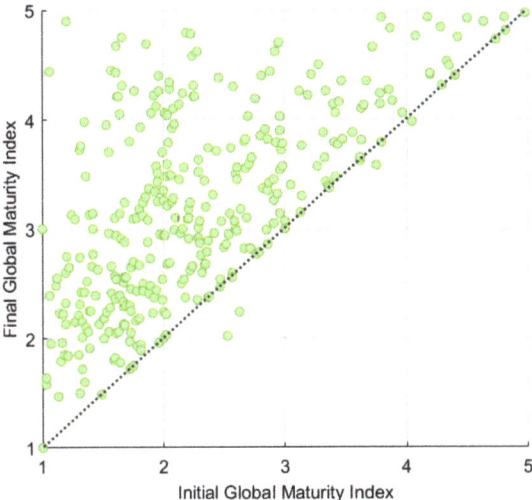

Figure 18. Scatter plot for the correlation between the initial and final global maturity index.

The dotted line identifies companies that have not undergone any variations in their global maturity index. The entire area above the line is widely occupied, although it can be observed that the number of companies that achieved indexes up to 3.5–4 starting from 1–2 is higher compared to those with bigger increases for the same starting range, as expected.

Moreover, the existence of correlation between the improvement of the overall maturity indicator and its initial level has been investigated through a statistical analysis.

The verification of the existence of correlation could be quantitatively conducted through the measurement of the main performance parameters (linear correlation coefficient R or Pearson's index, R^2 and observed p-value). The resulting p-value for the analysis was less than 0.001; therefore, it can be concluded that there was a negative correlation between the improvement of the overall maturity indicator and its initial level, indicating that energy management maturity has increased more over the years for initially less advanced companies, who have been able to take advantage of the opportunity given by energy audits to improve this aspect.

Considering that the practices corresponding to the systematic use of energy audit generally coincide with the practices corresponding to maturity level 3 of the model, it is interesting to note how even companies starting from non-elementary levels have had improvements, thus consolidating the good practices related to the use of such a tool.

5. Discussion

From the analysis it emerged that on average, companies that have complied with the obligation of mandatory energy audit have increased the maturity in their energy management.

Indeed, by examining the changes that have taken place and taking into account the significance of the different levels of maturity, it can be inferred that on average, Italian companies conducting mandatory energy audits have made progress in consolidating their approach to energy consumption. Specifically, focusing on the maturity possessed by the organizations, there has been a shift away from an "Occasional" or "Project-based" approach towards a more systemic approach, where companies are developing their own strategies for reducing energy consumption and costs by setting specific objectives and using energy audits' tools. However, when it comes to organizations starting from a higher maturity level, such as "Management", there is a wide range of progress. Although in 2015 there were fewer companies actively working towards developing a comprehensive energy management system, there is now a growing interest among companies, though this interest is at different stages of development. This consideration is also supported by the correlation analysis between the variation of maturity index and its initial value, which indicates that initially less advanced companies have generally achieved the greatest improvements.

Moreover, looking at the changes that have occurred in terms of maturity dimensions, the two most developed dimensions have been "Strategic Approach" and "Energy Performance Management and Information System", which are also the ones with the highest initial values.

The improvement in the "SA" dimension shows the growth of support from top management in the development of actions pertaining energy efficiency, which could be the result of the fact that the audit obligation has brought the energy issue to the attention of management. On the other hand, the improvement in the "EPMIS" dimension highlights the general improvement in the measurement system for collecting, analysing, and reporting all the organisation's energy performance data. This result could be easily explained by the need to collect reliable data to carry out the energy audit and by the stimulus to the development of the measurement system provided by the guidelines formulated by ENEA in view of the second cycle of mandatory energy audits and the percentage coverage thresholds of the measurement and/or monitoring plans indicated therein.

On the other hand, the analysis of single requirements shows that generally there is still lack in terms of preparation and training for energy management personnel and organizational awareness on the topic ("ACK" dimension), identification, and implementation of operational procedures for design, maintenance, and purchase and other such activities with impact on energy efficiency ("BP" dimension) and internal communication and clear responsibility for energy management personnel in the organization ("OS" dimension). In particular, it appears that the less developed requirements are the one associated with maturity levels 4 and 5. Indeed, this happens also for the "MA" dimensions, where the answers showed how organizations lack the structure and managing activities typically associated to the presence of an energy management system. It can be noted that all these practices and characteristics which appear to still be underdeveloped are not aspects that are not hugely affected by an energy audit execution.

Actually, the identified areas of improvement are usually achieved through the development of an energy management system; therefore, in order to achieve further progress, it is important to promote the stabilization of the observed improvement.

To foster energy audits benefits and take the first steps in this direction, organizations could enact several measures. For example, during the efficiency opportunity identification phase of the energy audit, personnel participation could be fostered to ensure the identification of the largest number of efficiency measures and their importance while also increasing the organization' awareness on the theme. The detailed list of identified measures should also be kept and reviewed. Moreover, organization could be made part of the data analysis phase, thus improving their knowledge of analysis tools in order to foster their independent future use for both control and budgeting activities [35].

6. Conclusions

In this paper, the findings of an analysis on the evolution of energy management in organizations subject to mandatory energy audits have been described.

To start, a maturity model was created to evaluate how energy management practices changed in companies. The questionnaires were distributed to a significant number of companies to assess the variation of their maturity level. The analysis conducted on this significant sample of companies (343 enterprises) confirmed the model's ability to discriminate different realities in terms of maturity and identify areas in which improvement has occurred. The data collected were analyzed using statistical techniques to gain a thorough understanding of how energy management practices have evolved in Italian companies and draw meaningful conclusions from the data analysis. For all observed variations (global maturity index, each level coverage, and each dimension coverage), their statistical significance was demonstrated, highlighting the increase in all energy management maturity areas. The average value of the global maturity index of the companies in the sample varied from 2.27 to 3.19, whereas each maturity level increased its coverage, on average, by at least 20% (up to 24% for Levels 2 and 3) and each maturity dimension increased its coverage up to 20%, on average.

Furthermore, the possible correlation between the improvement of the global maturity index and the initial maturity level was also analyzed to understand whether the initial maturity level could influence its development. A statistically significant relationship was observed, with the largest improvements obtained by companies starting from a more elementary maturity level at the onset of the obligation. This finding is reasonable considering that these companies, being in more elementary situations in terms of maturity, could easily take advantage of the opportunity provided by mandatory energy audits to evolve towards energy efficiency.

Moreover, it should be noted that the energy maturity requirements most improved are the ones that can be more affected by an energy audit execution such as EnPIs establishment and control, energy measurement campaign and information system, and efficiency opportunity identification.

In conclusion, the level of energy management maturity of companies which have undergone mandatory energy audits has increased, thus suggesting the positive contribution provided by this measure. The findings from the maturity model can support the legislator's assessment, identifying areas where it could be more useful to develop supportive policies.

In relation to the results described, two limitations should be noted. The adhesion to the compilation of the questionnaire by organization subject to mandatory energy audit was voluntary. Another possible limitation of the study lies in the choice of the self-assessment tool, which leaves companies free to autonomously evaluate the answers to be included in the questionnaire.

Further development of this study could see the replication of the analysis on other countries, analyzing the differences among European organizations and investigating the possible correlation to specific national legislative actions. Moreover, it would be important to repeat the assessment of the evolution of maturity indicators after the third mandatory energy audit cycle. Finally, further research will investigate the evaluation of the variation of EnPIs and their correlation with the variation of the maturity indicator.

Author Contributions: Conceptualization and methodology, A.S. and V.I.; formal analysis, A.S.; writing—original draft preparation, A.S.; writing—review and editing, A.S., V.I., F.M. and V.C.; visualization, A.S.; supervision and project administration, V.I.; funding acquisition, F.M. All authors have read and agreed to the published version of the manuscript.

Funding: This work is part of the Electrical System Research (PTR 2019–2021), implemented under Programme Agreements between the Italian Ministry for Economic Development and ENEA, CNR, and RSE S.p.A.

Data Availability Statement: Data is contained within the article.

Conflicts of Interest: The authors declare no conflict of interest.

References

1. Directive 2012/27/EU of the European Parliament and of the Council of 25 October 2012 on Energy Efficiency, Amending Directives 2009/125/EC and 2010/30/EU and Repealing Directives 2004/8/EC and 2006/32. *OJ* **2012**, *L315/1*, 1–56. Available online: https://eur-lex.europa.eu/eli/dir/2012/27/oj (accessed on 22 March 2023).
2. Directive (EU) 2018/2002 of the European Parliament and of the Council of 11 December 2018 Amending Directive 2012/27/EU on Energy Efficiency (Text with EEA Relevance). 2018, Volume 328. Available online: https://eur-lex.europa.eu/eli/dir/2018/2002/oj (accessed on 22 March 2023).
3. Hercé, C.; Biele, E.; Martini, C.; Salvio, M.; Toro, C. Impact of Energy Monitoring and Management Systems on the Implementation and Planning of Energy Performance Improved Actions: An Empirical Analysis Based on Energy Audits in Italy. *Energies* **2021**, *14*, 4723. [CrossRef]
4. Schleich, J. Do Energy Audits Help Reduce Barriers to Energy Efficiency? An Empirical Analysis for Germany. *IJETP* **2004**, *2*, 226. [CrossRef]
5. *EN 16247-1:2022*; Energy Audits—Part 1: General Requirements. European Committee for Standardization (CEN): Brussels, Belgium, 2022.
6. European Investment Bank; Revoltella, D.; Kalantzis, F. *How Energy Audits Promote SMEs' Energy Efficiency Investment*; Publications Office of the European Union: Luxembourg, 2019; ISBN 978-92-861-4228-4.
7. Bonfá, F.; Benedetti, M.; Ubertini, S.; Introna, V.; Santolamazza, A. New Efficiency Opportunities Arising from Intelligent Real Time Control Tools Applications: The Case of Compressed Air Systems' Energy Efficiency in Production and Use. In Proceedings of the Energy Procedia, Hong Kong, China, 22–25 August 2018; Elsevier Ltd.: Amsterdam, The Netherlands, 2019; Volume 158, pp. 4198–4203.
8. Salvatori, S.; Benedetti, M.; Bonfà, F.; Introna, V.; Ubertini, S. Inter-Sectorial Benchmarking of Compressed Air Generation Energy Performance: Methodology Based on Real Data Gathering in Large and Energy-Intensive Industrial Firms. *Appl. Energy* **2018**, *217*, 266–280. [CrossRef]
9. *ISO 50001:2018*; Energy Management Systems—Requirements with Guidance for Use 2018. International Organization for Standardization (ISO): Geneva, Switzerland, 2018.
10. Crosby, P.B. *Quality Is Free: The Art of Making Quality Certain*; McGraw-Hill: New York, NY, USA, 1979.
11. Introna, V.; Cesarotti, V.; Benedetti, M.; Biagiotti, S.; Rotunno, R. Energy Management Maturity Model: An Organizational Tool to Foster the Continuous Reduction of Energy Consumption in Companies. *J. Clean. Prod.* **2014**, *83*, 108–117. [CrossRef]
12. Wendler, R. The Maturity of Maturity Model Research: A Systematic Mapping Study. *Inf. Softw. Technol.* **2012**, *54*, 1317–1339. [CrossRef]
13. Becker, J.; Knackstedt, R.; Pöppelbuß, J. Developing Maturity Models for IT Management. *Bus. Inf. Syst. Eng.* **2009**, *1*, 213–222. [CrossRef]
14. Mettler, T.; Rohner, P.; Winter, R. Towards a Classification of Maturity Models in Information Systems. In *Management of the Interconnected World*; D'Atri, A., De Marco, M., Braccini, A.M., Cabiddu, F., Eds.; Physica-Verlag HD: Heidelberg, Germany, 2010; pp. 333–340. ISBN 978-3-7908-2403-2.
15. Benedetti, M.; Bonfà, F.; Bertini, I.; Introna, V.; Salvatori, S.; Ubertini, S.; Paradiso, R. Maturity-Based Approach for the Improvement of Energy Efficiency in Industrial Compressed Air Production and Use Systems. *Energy* **2019**, *186*, 115879. [CrossRef]
16. Fraser, P.; Moultrie, J.; Gregory, M. The Use of Maturity Models/Grids as a Tool in Assessing Product Development Capability. In Proceedings of the IEEE International Engineering Management Conference, Cambridge, UK, 18–20 August 2002; IEEE: Cambridge, UK, 2002; Volume 1, pp. 244–249.
17. Lahrmann, G.; Marx, F. Systematization of Maturity Model Extensions. In *Global Perspectives on Design Science Research*; Winter, R., Zhao, J.L., Aier, S., Eds.; Springer: Berlin/Heidelberg, Germany, 2010; pp. 522–525.
18. Carbon Trust Energy Management—A Comprehensive Guide to Controlling Energy Use. In *Energy Management—A Comprehensive Guide to Controlling Energy Use*; Association for Decentralised Energy: London, UK, 2011.
19. Ngai, E.W.T.; Chau, D.C.K.; Poon, J.K.L.; To, C.K.M. Energy and Utility Management Maturity Model for Sustainable Manufacturing Process. *Int. J. Prod. Econ.* **2013**, *146*, 453–464. [CrossRef]
20. Qiang, R.; Jiang, Y. Enterprise Energy Saving and Emission Reduction Level Assessment Research Based on the Capability Maturity Model. In Proceedings of the 2009 International Conference on Management and Service Science, Beijing, China, 20–22 September 2009; IEEE: Beijing, China, 2009; pp. 1–5.
21. Curry, E.; Conway, G.; Donnellan, B.; Sheridan, C.; Ellis, K. A Maturity Model for Energy Efficiency in Mature Data Centres. In Proceedings of the SMARTGREENS 2012—1st International Conference on Smart Grids and Green IT Systems, Porto, Portugal, 19–20 April 2012; pp. 263–267.
22. Curry, E.; Conway, G.; Donnellan, B.; Sheridan, C.; Ellis, K. Measuring Energy Efficiency Practices in Mature Data Center: A Maturity Model Approach. In Proceedings of the Computer and Information Sciences III—27th International Symposium on Computer and Information Sciences, Paris, France, 3–4 October 2012; ISCIS 2012. pp. 51–61.

23. EDF Climate Corps EDF Smart Energy Diagnostic Survey. 2015. Available online: http://edfclimatecorps.org/edf-smart-energy-best-practices-survey (accessed on 20 March 2023).
24. Jovanović, B.; Filipović, J. ISO 50001 Standard-Based Energy Management Maturity Model—Proposal and Validation in Industry. *J. Clean. Prod.* **2016**, *112*, 2744–2755. [CrossRef]
25. Prashar, A. Energy Efficiency Maturity (EEM) Assessment Framework for Energy-Intensive SMEs: Proposal and Evaluation. *J. Clean. Prod.* **2017**, *166*, 1187–1201. [CrossRef]
26. O'Sullivan, J. *Energy Management Maturity Model (EM3)—A Strategy to Maximize the Potential for Energy Savings through EnMS*; Sustainable Energy Authority of Ireland: Dublin, Ireland, 2012.
27. Finnerty, N.; Sterling, R.; Coakley, D.; Keane, M.M. An Energy Management Maturity Model for Multi-Site Industrial Organisations with a Global Presence. *J. Clean. Prod.* **2017**, *167*, 1232–1250. [CrossRef]
28. Finnerty, N.; Sterling, R.; Coakley, D.; Contreras, S.; Coffey, R.; Keane, M.M. Development of a Global Energy Management System for Non-Energy Intensive Multi-Site Industrial Organisations: A Methodology. *Energy* **2017**, *136*, 16–31. [CrossRef]
29. Çoban, V.; Onar, S.Ç. Energy Management Maturity Model Based on Fuzzy Logic. In *Intelligent and Fuzzy Techniques in Big Data Analytics and Decision Making*; INFUS 2019. Advances in Intelligent Systems and Computing; Kahraman, C., Cebi, S., Cevik Onar, S., Oztaysi, B., Tolga, A., Sari, I., Eds.; Springer: Cham, Switzerland, 2020; Volume 1029.
30. Wehner, J.; Taghavi Nejad Deilami, N.; Altuntas Vural, C.; Halldórsson, Á. Logistics Service Providers' Energy Efficiency Initiatives for Environmental Sustainability. *IJLM* **2022**, *33*, 1–26. [CrossRef]
31. Jin, Y.; Long, Y.; Jin, S.; Yang, Q.; Chen, B.; Li, Y.; Xu, L. An Energy Management Maturity Model for China: Linking ISO 50001:2018 and Domestic Practices. *J. Clean. Prod.* **2021**, *290*, 125168. [CrossRef]
32. Santolamazza, A.; Introna, V.; Cesarotti, V.; Martini, F.; Toro, C. Proposal of an Energy Management Maturity Model to Assess the Progress Achieved through Mandatory Energy Audit Application in the Eu Energy Efficiency Directive Context. In Proceedings of the Summer School Francesco Turco, Bergamo, Italy, 9–11 September 2020.
33. *EUROSTAT NACE Rev. 2*; Office for Official Publications of the European Communities: Luxembourg, 2008; ISBN 978-92-79-04741-1.
34. Montgomery, D.C.; Runger, G.C. *Applied Statistics and Probability for Engineers*; John Wiley & Sons, Inc.: Hoboken, NJ, USA, 2018.
35. Cesarotti, V.; Di Silvio, B.; Introna, V. Energy Budgeting and Control: A New Approach for an Industrial Plant. *Int. J. Energy Sect. Manag.* **2009**, *3*, 131–156. [CrossRef]

Disclaimer/Publisher's Note: The statements, opinions and data contained in all publications are solely those of the individual author(s) and contributor(s) and not of MDPI and/or the editor(s). MDPI and/or the editor(s) disclaim responsibility for any injury to people or property resulting from any ideas, methods, instructions or products referred to in the content.

Article

Evaluation of Energy Performance Indicators and Energy Saving Opportunities for the Italian Rubber Manufacturing Industry

Matteo Piccioni [1], Fabrizio Martini [2], Chiara Martini [2] and Claudia Toro [2,*]

[1] Department of Astronautical, Electrical and Energy Engineering (DIAEE), "Sapienza" University of Rome, Via Eudossiana 18, 00184 Rome, Italy; matteopiccioni1@gmail.com

[2] DUEE-SPS-ESE Laboratory, Italian National Agency for New Technologies, Energy and Sustainable Economic Development (ENEA), Lungotevere Thaon di Revel, 76, 00196 Rome, Italy; fabrizio.martini@enea.it (F.M.); chiara.martini@enea.it (C.M.)

* Correspondence: claudia.toro@enea.it

Abstract: The objective of this work is the energy characterisation and evaluation of the energy efficiency potential of the rubber manufacturing industry in Italy, exploiting the detailed data included in energy audits by large and energy-intensive companies. This sector is divided into two sub-activities: the manufacture of rubber products and the production of tyres. Existing studies are focused mainly on tyre production, and there is a lack of quantitative evaluation of energy indicators that can provide guidance for improving process efficiency. In this work, updated global and specific energy performance indicators (EnPIs) related to the production process and to the auxiliary and general services are defined and evaluated. At the same time, targeted actions and interventions to improve the energy efficiency of the sector are analysed, showing the role of different intervention areas and their cost-effectiveness. The analysis is based on 100 Italian mandatory energy audits of the sector collected according to Art.8 EU Directive 27/2012. The applied methodology made it possible to calculate specific energy performance indicators by considering the overall and sub-process energy consumption of different production sites. Based on a detailed database containing real data from recent energy audits, this study provides an up-to-date and reliable benchmark for the rubber industry sector. In addition, the analysis of energy audits allows the identification of the most effective energy efficiency interventions for the rubber industry in terms of cost-effectiveness and payback time.

Keywords: energy efficiency; EnPIs; rubber industry; energy audit

Citation: Piccioni, M.; Martini, F.; Martini, C.; Toro, C. Evaluation of Energy Performance Indicators and Energy Saving Opportunities for the Italian Rubber Manufacturing Industry. *Energies* **2024**, *17*, 1584. https://doi.org/10.3390/en17071584

Academic Editor: François Vallée

Received: 24 January 2024
Revised: 26 February 2024
Accepted: 13 March 2024
Published: 26 March 2024

Copyright: © 2024 by the authors. Licensee MDPI, Basel, Switzerland. This article is an open access article distributed under the terms and conditions of the Creative Commons Attribution (CC BY) license (https://creativecommons.org/licenses/by/4.0/).

1. Introduction

Energy efficiency plays a crucial role in a clean energy transition of the industrial sector. At worldwide level, industry accounts for 37% of final energy use and 24% of carbon emissions [1]. Making production processes more efficient and rationalizing the use of energy resources are the main objectives of the European Commission's approach to the issue of energy efficiency in production processes.

The Energy Efficiency Directive 2012/27/EU (EED) [2] (and the 2023 directive amendment [3]) is a key element of Europe's energy legislation. The 2023 revision makes the energy efficiency target binding for EU countries, which should collectively reduce energy consumption by 11.7% in 2030, compared to the 2020 reference scenario projections.

According to Article 8 of the EED, since 2015, large companies have been obliged to carry out an energy audit on their production sites every four years. An energy audit is defined as "a systematic procedure having the purpose of obtaining adequate knowledge of the current energy consumption profile of a building or group of buildings, of an industrial or commercial operation or installation, of a private or public service, by identifying and quantifying cost-effective energy saving opportunities, and reporting the findings".

Energy Audits (EAs) are a key tool to assess the existing energy consumption and identify the whole range of opportunities to save energy in a productive site. The recast of the EED changed the obliged parties, who will no longer be defined according to company size but according to a consumption threshold of 10 TJ per year (2.8 GWh).

In Italy, the Directive was transposed with the Legislative Decree 102/2014, later amended by Legislative Decree 73/2020. The recast of the Directive, published in October 2023, must be transposed within the next two years. According to Art. 8 of Legis. D. 102/14, two categories of companies are obliged to carry out energy audits on their production sites: large enterprises and energy-intensive enterprises. The second category is represented by enterprises voluntarily applying for tax relief on the purchased electricity and registering in the list of the Environmental Energy Services Fund (CSEA, a government agency on electricity). These companies present large energy consumptions (in absolute terms and relative to their internal costs), and they must be part of some specific industrial sectors (mainly Annexes 3 and 5 of EU Guidelines 2014/C 200/01 [4]).

Pursuant to Article 8 of Legis. D.102/2014, by the deadline of December 2019, 11,172 energy audits of production sites relating to 6434 companies [5] have been sent to ENEA, the Italian National Energy Agency in charge of the management of the scheme. The manufacturing sector represents 53% of total energy audits, and plastics and rubber manufacturing (according to the Nomenclature of Economic Activities, the sector code is NACE C22) is the second sector by importance, with around 8% of the total.

Eurostat data for 2021 [6] show that this sector is responsible for about 3.2% of energy consumption in relation to the total manufacturing industry in the European Union. Italy is in line with this figure, with a slightly higher share of 2.9%. Italy is the third country in Europe, after Germany and France (27.7% and 16.4% respectively), as the sector is important in terms of energy consumption, with a value of 15.7%. The manufacture of rubber and plastic is an important sector in Italy, which is the second EU Member State considering the national share of the total sectoral value added (14.5%, second only to Germany, which corresponds to 32.4%) [7].

Rubber is a strategic material, and rubber products are essential in several sectors and applications, such as automotive, construction, aerospace, food, pharmaceutical, oil and gas, etc. Energy is needed for all the phases of the rubber manufacturing cycle, and it represents one of the main costs in this sector [8] and is thus closely related to the competitiveness of enterprises. The rubber sector (NACE Group C22.1) has two distinct parts: the manufacture of rubber tyres and tubes, the retreading and rebuilding of rubber tyres (C.22.1.1), and the manufacture of other rubber products (C22.1.9). The rubber industry is an important sector in Italy: in 2020, the number of enterprises in the sector was 2301, and the added value was 23.9% of the NACE C22 total [9]. As for rubber production, in 2022, Italy was the fourth largest producer in the EU, after Germany, France and Spain [10]. This report includes information provided by Assogomma, the National Association of Rubber Producers. The companies in four Northern regions, namely Lombardy, Veneto, Emilia Romagna and Piedmont, account for 82% of companies. Lombardy alone has 45% of employees and 55% of companies.

This study aims to calculate the energy performance indicators for the Italian rubber manufacturing industry, exploiting the information provided by the mandatory energy audits collected by ENEA in the years 2019–2022. About 100 energy audits related to rubber processing production sites have been analysed. These represent the totality of audits related to sites of large or energy-intensive companies uploaded to the ENEA portal in the four-year period 2019–2022. In Italy, mandatory energy audits can only be prepared by subjects and/or bodies certified by accredited bodies, such as E.S. Co. (Energy Service Company), E.G.E. (Energy Management Expert) and Energy Auditors certified according to the specific and related technical regulations.

The first step toward effective energy management of production processes requires measuring and benchmarking energy-efficiency performance [11]. Several studies propose methods for the development of key performance indicators (KPIs) [12] and classification methods for different types of indicators [13,14]. In [15], a review of existing studies is presented, providing an overview of indicator typologies, methodological issues, and applications for energy performance evaluation and improvement. The authors show that existing studies are mainly dedicated to the development of indicators for specific industrial sectors with different energy intensities, such as steel [16], pulp and paper [17], aluminium [18], food [19], textiles [20] and engineering [21]. At the sector level, they depend on various parameters such as activity level, structure and maturity of energy efficiency [22]. Despite several efforts to standardise the use of indicators to compare energy efficiency across countries and sectors [23], only applied to a limited number of energy-intensive industries have been investigated using Energy Performance Indicators (EnPIs) [24]. The European IPPC Bureau provides a key contribution to creating, reviewing, and updating Best Available Techniques (BAT) reference documents (BREFs), which analyse more than 52,000 installations across Europe covered by the Industrial Emissions Directive (2010/75/EU) [25]. These documents are the European reference for the consumption of specific agro-industrial activities or cross-cutting issues such as energy efficiency, industrial cooling systems or emissions from storage with relevance for industrial manufacturing in general. BREFs provide a set of EnPIs, but without country-specific information.

Regarding specific studies related to the calculation of energy indicators for rubber manufacturing, few studies can be found in the literature. Particularly with regard to tyre production, specific energy consumption indicators are presented in [26,27], respectively, a study from the Thai Ministry of Energy and a study from the U.S. Department of Energy on rubber and plastic plants. An analysis of energy use in the tyre manufacturing industries is presented in [28], showing that electric motors account for a major share of the total energy consumption, followed by pumps, heaters, cooling systems and lighting. In Stankevičiūtė's study [29], on the other hand, an industrial rubber manufacturing company is chosen as a case study, and energy flows and energy management are studied in order to suggest ways to improve energy performance in tyre manufacturing processes. An analysis of energy use and savings in the Malaysian rubber-producing industries is presented in [9]. Decarbonisation options for the Dutch tyre industry have been evaluated in [30], highlighting that the most promising solutions are related to natural gas substitution by using biomass boilers, electric boilers, hybrid boilers or hydrogen boilers.

The literature review has shown that existing studies are focused mainly on tyre production and or specific sites, and there is a lack of extensive and in-depth studies on the evaluation of energy indicators of the different production sub-processes for both tyre production and general rubber products, that can provide guidance for improving process efficiency. The novelty of this work was to calculate up-to-date energy indicators for the sector as a whole (rubber products and tyre production) and to deepen the energy analysis of production processes. This made it possible to better frame and assess efficiency options for the sector.

In the present work, the benchmark indicators' definition was made using a comprehensive methodology proposed in [31] and already successfully applied to several productive sectors, such as ceramic [32], oil refinery [33], and pharmaceutical [34]. This methodology made it possible to calculate specific EnPIs by considering the overall and sub-process energy consumption of different production sites, including relevant variables. Based on a major database containing real data from recent energy audits, this study was able to provide an up-to-date and reliable benchmark for the rubber industry sector. In addition, the analysis of energy audits has enabled the identification of the most effective energy efficiency interventions for the rubber industry, thus allowing an assessment of the sector's overall savings potential.

The research questions (RQs) addressed in this study are:

- RQ1—Is it possible to quantify the total, electric, or thermal-specific energy consumption of the rubber manufacturing sector?

- RQ2—What is the quantitative and qualitative information on energy efficiency interventions in this sector?

EnPIs can be calculated based on economic data (e.g., value-added from production) or physical data (e.g., tonnes or cubic metres of products). This paper provides a contribution in the context just described: first, it analyses a specific sector, rubber, having a strategic role not often investigated; second, it provides EnPIs information at country level, which can be compared with available benchmarks, if any.

EnPIs can play a fundamental role in identifying effective Energy Performance Improvement Actions (EPIAs). Despite the multiple benefits of the adoption of EPIAs, there are several barriers related to their implementation and thus, an energy efficiency gap exists [35–37]. The paper puts together EnPIs information with the data associated with EPIAs described in energy audits, both relative to implemented measures and proposed measures. The EPIAs listed in the energy audit in the rubber sector are described, providing the cost-effectiveness of different areas of intervention.

The paper is structured as follows: Section 2 includes details about data collection, processing and categorization, Section 3 describes the main results in terms of definition of plant energy models, calculation of EnPIs and EPIAs analysis, and Section 4 illustrates the main conclusions.

2. Materials and Methods

2.1. Data Collection and Pre-Processing

In the following work, energy audits from two Italian industrial sectors, classified according to their respective NACE Code, were analysed. The study aims to achieve three main objectives through the analysis of the energy audits received by ENEA in the second audit cycle starting from 2019.

The first step of the procedure is to define a reference "Plant Energy Model" [38], according to ENEA methodology, associated with the rubber manufacturing sector production processes. The definition of the plant energy model is essential to associate the consumption of energy carriers with the different phases of the process, as well as the auxiliary operations and general services. The second and main objective of the study is the computation of first and second-level benchmark energy performance indicators (EnPi). The purpose of defining energy performance indices is to identify reference values that allow companies to appropriately plan their energy policy. EnPis is the most used tool for benchmarking energy performance. Energy consumption benchmarking, both internal (through historical/trend analysis) and external (comparison with other companies in the sector), is a powerful tool for evaluating performance, identifying critical issues and any improvement measures and therefore improving energy efficiency.

The main purpose of energy audits is to identify energy efficiency solutions by evaluating them through a cost-benefit analysis. Finally, both the energy efficiency solutions carried out by the companies and those proposed in the energy audits are analysed, making it possible to identify the areas where it is most advantageous to invest in terms of energy saving based on the economic factors in the audits.

The data is categorised according to the European NACE [39] (Nomenclature of Economic Activities) classification. Table 1 shows the descriptions of the sectors analysed according to the NACE classification and the number of audits collected for each category. This is the total number of audits uploaded by Italian obligated companies, large and energy-intensive enterprises, to the ENEA portal in the four-year period 2019–2022 for the sectors of interest and represents the sample of data analysed for this study.

Table 1. The number of audits for each subcategory of NACE 22.1.

NACE Code	Description	Number of Audits
C22.1.9	Manufacture of other rubber products	92
C22.1.1	Manufacture of rubber tyres and tubes and rebuilding of rubber tyres	6

The first sector is the manufacturing of other rubber products, which consists of a large number of companies with a high product heterogeneity. For this reason, they have been grouped into sub-categories such as gaskets, compounds, rubber hoses or belts.

The second sector relates to the manufacture of tyres and inner tubes. As can be seen from the data in Table 1, for this sector, only six energy audits were available. However, these companies, mainly large companies, have a very high annual production, supplying tyres both for normal vehicles and for the most prestigious motoring circuses.

To fulfil the obligations established by Legislative Decree 102/2014, ENEA demands companies to compile two main documents. The first one consists of the energy audits report while the second document is a spreadsheet that collects consumption data for each direct energy carriers present in the different plants (mainly electricity and natural gas). The latter is customised for each NACE Code category and is regularly updated with the contribution of the specific Trade Associations.

For each energy carrier, consumption is then broken down according to three main categories:
- Main Activities: only processes closely related to the company's "core business".
- Auxiliary Systems: all those processes transforming the incoming energy carrier into different energy carriers used in the main functional areas, such as compressed air plant or thermal power plant.
- General Systems: Secondary activities are not directly connected with the core business. Examples of general services are air conditioning, lighting, heating, etc.
- For each production site, the following information was available:
- Site identification data, name, city of residence, VAT number, NACE code.
- General details of facilities and machinery used, floor plans, etc.
- Electrical and thermal load curves and consumption for the individual areas.
- Excel file summarising consumption according to the plant energy model.

To process the indicators, an in-depth data cleaning phase of the databases was necessary, with the dual objective of increasing the quality of the data and understanding the main production processes within the various factories. The most critical issues were found in companies with the lowest production levels, where the quality of energy audits was significantly lower than in larger companies. The analysis of the energy audits revealed two main recurring problems:
- No breakdown of energy consumption, reporting only total consumption for each energy carrier.
- Distribution of consumption according to the structure of the site and facilities (Department X, or Line Y) instead of a correct breakdown according to the production process.

These problems led to a reduction of analysable data, often only allowing the calculation of global energy performance indicators, i.e., referring to the total consumption of the site. For NACE sector C22.1.9, it was also necessary to group the different companies according to their main products. Two main clusters were defined and elaborated: the first, which is more numerous, refers to companies that produce gaskets and O-rings, while the second includes companies that produce raw compounds. For NACE sector C22.1.1, no clustering was necessary due to the homogeneity of the data set.

2.2. Data Analysis

Production Processes and Energy Carriers in the Rubber Manufacturing Industry

In the first phase of the work, an electrical and thermal model for the rubber manufacturing sectors was developed. This model is fundamental for the subsequent definition of first and second-level energy indicators and constitutes a useful tool to assist companies in reporting and compare their consumption. A detailed description of the methodology applied in this work can be found in [31].

Due to the high heterogeneity of the final products in the NACE C.22.1.9 sector, it was necessary to cluster the dataset in two groups:
- Cluster A: gasket/seal and O-ring manufacturers.

- Cluster B: raw compounds manufacturers,

while the NACE Sector 22.1.1 did not require any clustering.

The results of NACE C22.1.9 Cluster A and NACE C22.1.1 will be the subject of this article.

Figure 1 shows the flow chart identified for NACE Sector C22.1.9 and the breakdown of direct and indirect energy carriers involved in the process.

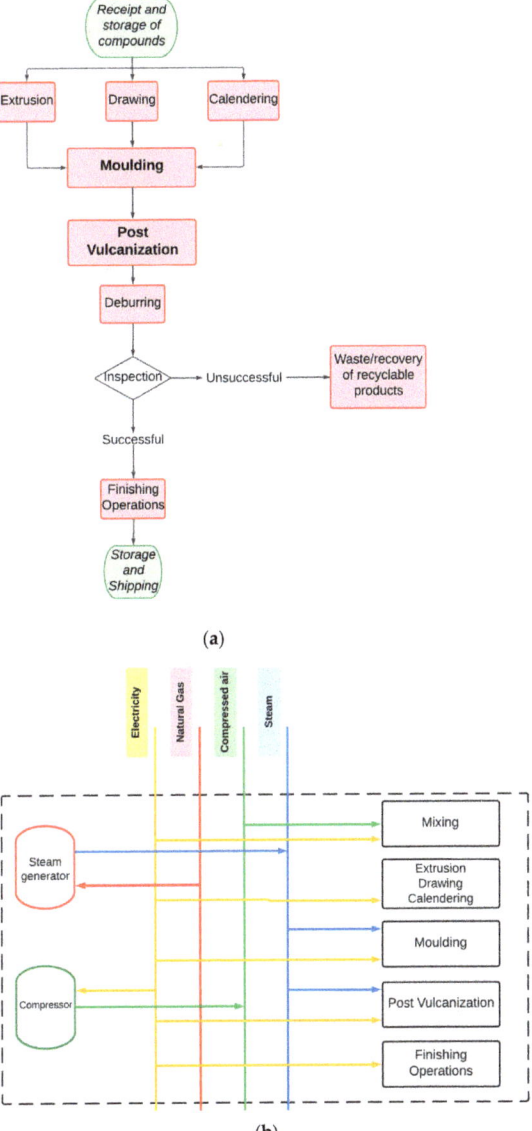

Figure 1. NACE C22.1.9: Gasket manufacturers cluster. (**a**) Flow chart for gasket manufacturers; (**b**) Breakdown of direct and indirect energy carriers. Source: Authors' elaboration.

The raw material generally used is the raw compound, which is initially processed through various phases and then undergoes moulding, the most energy-intensive process in the process. In this phase, the vulcanisation of the rubber takes place, i.e., a chemical-physical process at high temperature (about 160 °C to 200 °C) [40] necessary to change the behaviour of the compound from plastic to elastic. Moulding generally takes place by compression, placing the rubber in a special machine and then closing the heated press to give it a specific shape and the desired mechanical properties. Another method is injection moulding, where the rubbery material is initially melted and plasticised and then injected under high pressure into a melt form, after which the material is allowed to solidify into the desired shape. The next phase is called post-vulcanization, the second most intensive, in which the material is heated in an oven to remove the stresses caused by the previous processing. To complete the process, the gasket is first cleaned of burrs formed in the moulding process phase and then undergoes testing and finishing prior to storage and shipment.

Figure 1b shows the main energy carriers used by the process, namely electricity, natural gas, compressed air, and steam. Compressed air is mainly used for the pneumatic drive of the machines, while steam is used for the moulding and curing processes, which are the most energy-intensive processes.

Figure 2 shows both the flow chart identified for NACE sector C22.1.1 and the breakdown of direct and indirect energy carriers in the plant.

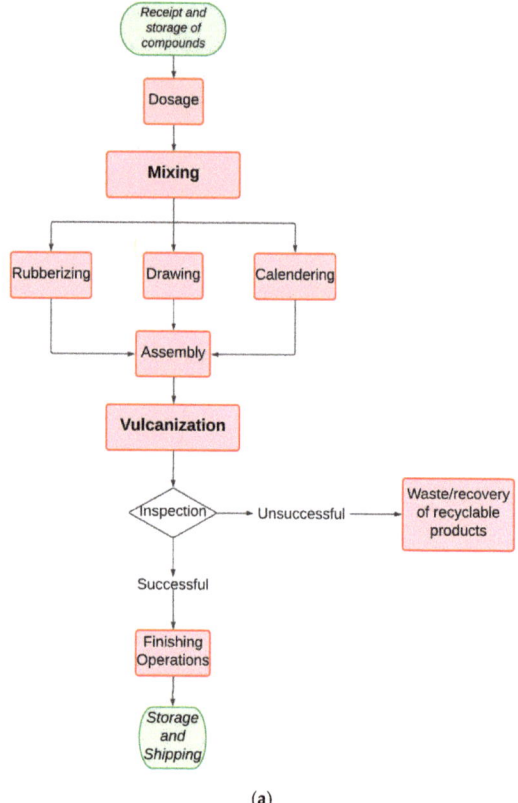

(a)

Figure 2. *Cont.*

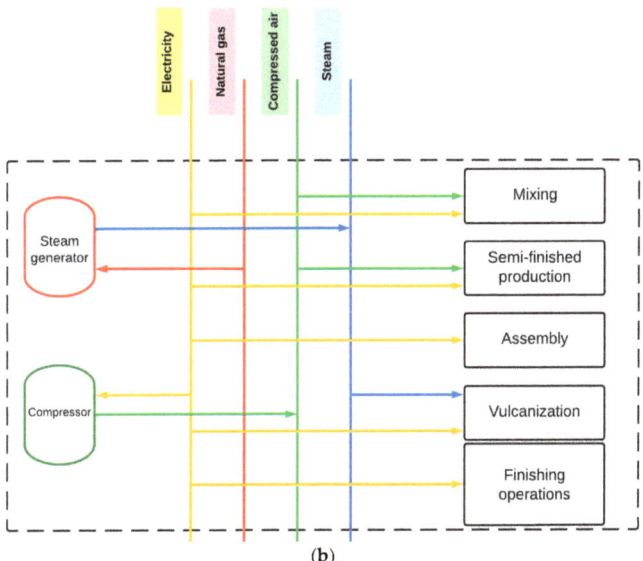

Figure 2. NACE C22.1.1 (**a**) Flow chart for tyre manufacturers; (**b**) Breakdown of direct and indirect energy carriers. Source: Authors' elaboration.

The process starts with the production of the raw compounds. Raw materials used are multiple and can have up to 45 components. Both natural and synthetic rubber can be used in addition to further substances such as carbon black, plasticisers, sulfur, and silica, as well as additives that are inserted in specific parts of the process. Generally, the raw materials can be stored in a "hot chamber" in an air-conditioned warehouse at 45–55 °C. To produce the compound, the raw materials are initially minced and dosed appropriately by means of weighing belts and then fed into the mixer, generally known as Banbury. The process is exothermic and reaches temperatures up to 100 °C. The product of the mixing process is then cooled by means of special batch-off tanks and reduced to rubber sheets. Subsequently, the compound sheets are used to produce the semi-finished products, which are generally rubber wire beads or rubber sheeting, which are then assembled by special machines, called packing machines, to create the 'raw' tyre. The subsequent phase is vulcanisation, which is the most energy-intensive phase of the plant from the thermal energy point of view. It is based on the use of a mould to compress and heat the raw tyre produced to give it the desired properties, using both electricity and steam. Specific additives called vulcanisers are also used, which ensure the formation of so-called crosslinks within the three-dimensional rubber lattice, i.e., links between the polymer chains that give the material the necessary stability, elasticity, and mechanical properties [41]. As with the gaskets, checks are subsequently carried out to ensure the tyre strength and reliability, and then the finishing operations are performed.

3. Results and Discussion

3.1. Plant Energy Model

As previously stated, the first step in calculating energy consumption indicators is to identify a reference "Plant Energy Model" [38] for the production processes. The plant energy model must be created for each energy carrier identified, and the plant must be subdivided into homogeneous functional areas. Based on the breakdown of the processes presented in the previous section and the information contained in the energy audits, different models were developed for each direct energy vector used in the processes, i.e., electricity and natural gas, as shown in the following pictures. The information presented in this section represents the context for the calculation of first and second-level EnPIs. At the same time, the breakdown of energy consumption by energy carrier is useful for a better understanding of the energy efficiency interventions identified in energy audits and analysed in this study. The electricity and natural gas models developed for sectors NACE C22.1.9 and NACE C22.1.1 are shown in Figures 3 and 4 respectively.

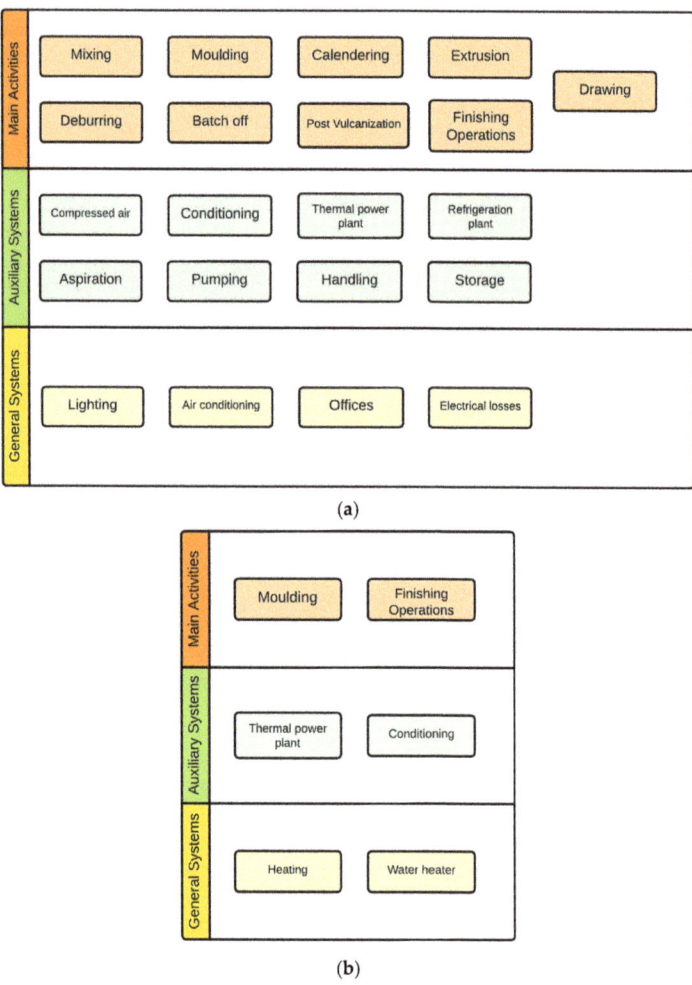

Figure 3. NACE C22.1.9: Gasket manufacturers cluster. (**a**) Plant Energy Model: Electricity; (**b**) Plant Energy Model: Natural Gas. Source: Authors' elaboration.

From the analysis of energy consumption reported in the energy audits, it was possible to evaluate the weight of the different energy carriers. Natural gas accounts for more than a third of the total energy consumption for NACE C22.1.9, while for NACE C22.1.1, it accounts for 50%. The distribution of electric energy consumption among the three main functional areas and among the main process activities for NACE C22.1.9 is reported in Figure 5. The quality of the data collected in the energy audits was not sufficient to allow a similar analysis for thermal energy.

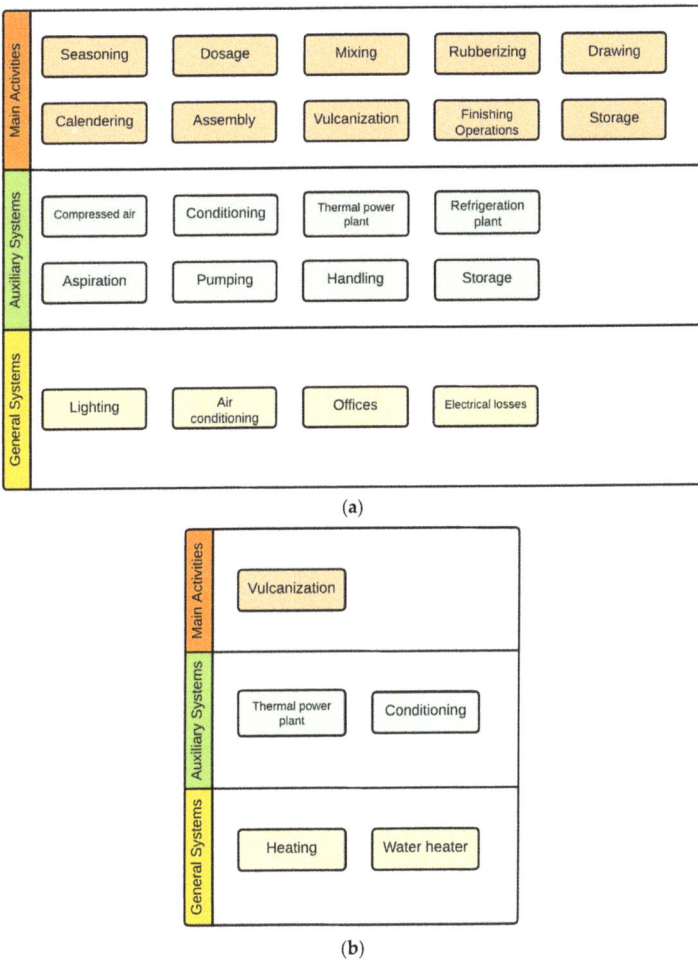

Figure 4. NACE C22.1.1 (**a**) Plant Energy Model: Electricity; (**b**) Plant Energy Model: Natural Gas. Source: Authors' elaboration.

Figure 5a shows how the main activities represent the most energy-intensive category of the plant, accounting for about 2/3 of the total electricity consumption, followed by auxiliary services consuming $\frac{1}{4}$ of the total, and finally, general services. It is also useful to analyse how electricity consumption is distributed among the different main activities, shown in Figure 5b. Moulding uses approximately 60% of the total electricity, with the remaining operations not reaching 10%. Given the high incidence of consumption in the moulding process, the computation of benchmark energy performance indicators and the identification of the best opportunities for energy efficiency will be focused on this phase of the process. Note the presence of an unidentified consumption category due to the volume of data in the summary reports of low quality, which led to a non-categorization of the latter.

The distribution of electricity consumption among the three main functional areas for NACE C22.1.1 is shown in Figure 6a. Figure 6b shows how the core activities have the largest percentage of electricity consumption; in particular, vulcanisation and mixing account for 2/3 of the electricity consumption for the core activities, which is the reason energy performance indicators and energy efficiency interventions will focus on these phases. A similar pie chart for thermal energy would see a very high percentage of consumption devoted to vulcanisation; unfortunately, in most cases, consumption due to steam production for vulcanisation was not specifically allocated, making a similar analysis not possible.

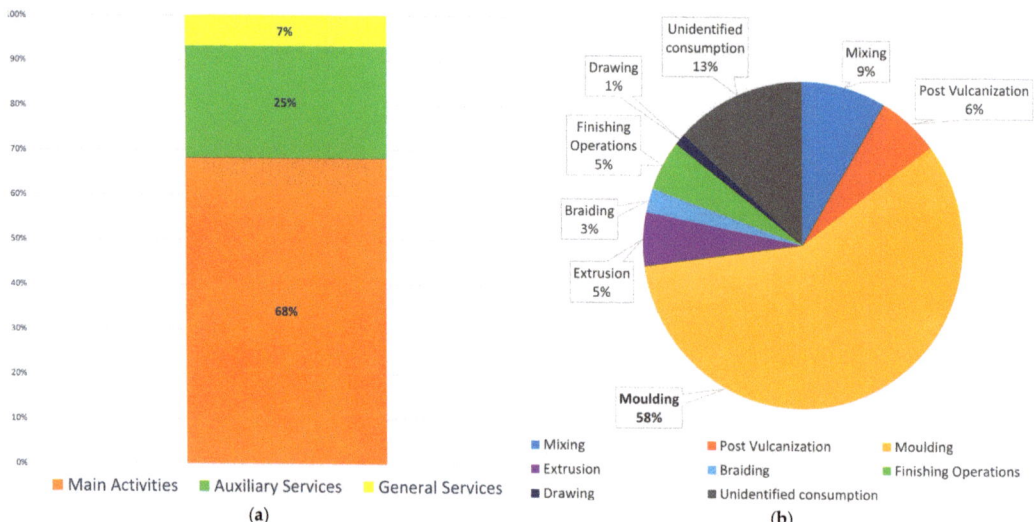

Figure 5. NACE C22.1.9: Gasket manufacturers cluster. (**a**) Electricity consumption distribution for functional macro-areas; (**b**) Electricity consumption distribution for main activities. Source: Authors' elaboration on Microsoft Excel.

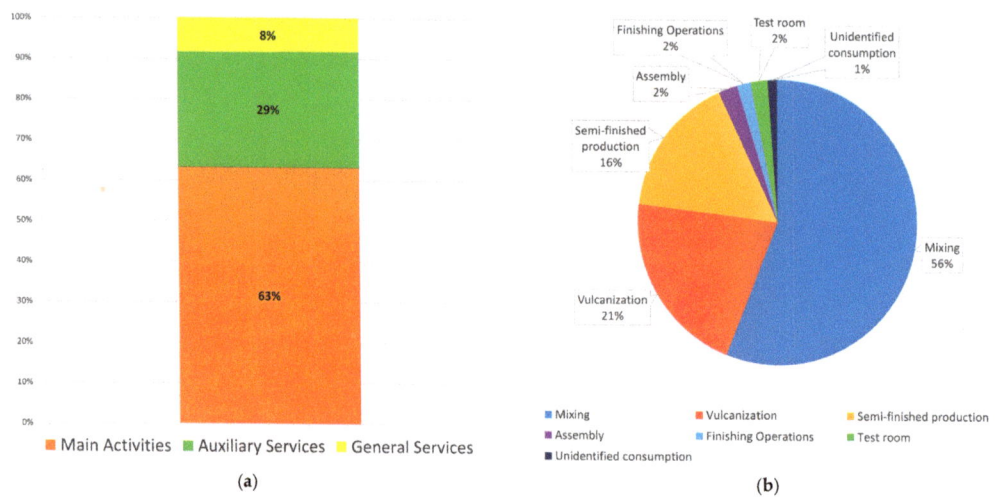

Figure 6. NACE C22.1.1: (**a**) Electricity consumption distribution for functional macro-areas; (**b**) Electricity consumption distribution for main activities. Source: Authors' elaboration on Microsoft Excel.

The energy carriers used are the same as in the NACE C22.1.9 sector, with electrical energy being used first for the mixing process, as shown in Figure 5b, and most of the thermal energy used for the vulcanisation process.

3.2. Benchmark EnPIs

3.2.1. First Level Benchmark EnPIs

The energy audit database as described in the previous paragraph was used to calculate updated energy performance indicators for the NACE C22.1.9 and C22.1.1 sectors. The data presented in this section and in the following one answer to the first research question, on the values of EnPIs.

EnPIs have been introduced by the ISO 50001:2018 standard [42], where they are described as a combination of processes efficiency, energy consumption, and management of energy sources and their end use [40].

Energy performance indicators were calculated as specific consumption (total, electric or thermal), having the energy driver as the denominator:

$$\text{EnPI} = \frac{\text{Energy consumption}}{\text{Energy driver}}$$

The denominator is specific for each energy performance indicator, using the relevant variables available in the database such as tons of final products. When available, specific uses, such as normal cubic metres of compressed air, were used. To assess the reliability of the computed benchmark EnPIs, a coefficient of variation is introduced, which represents the ratio between the standard deviation and the mean of the calculated EnPIs.

A regression analysis was performed for each EnPI, and the evaluation was conducted using two statistical parameters, namely R^2 and p-value. The coefficient of determination R^2 gives an indication of the strong predictive ability of the correlation model. In simpler terms, R^2 represents the difference between the values of the dependent variable that can be accounted for by the variation in the independent variable. The higher the value of R^2, the better the ability of the independent variables to predict the values of the dependent variable. Conversely, low values of R^2 indicate a low predictivity of the regression model, mainly since the dependent variable analysed also relies largely on other parameters that were not considered by the model. The linear correlation can be considered strong if $R^2 > 0.5$ and moderate if $R^2 > 0.25$. The p-value is used to check the statistical significance of the estimated coefficients or the estimated model. Models with p-values ≤ 0.05 are considered statistically significant.

Figures 7 and 8 show the results of the regression analyses, while Table 2 shows the calculated statistical parameters.

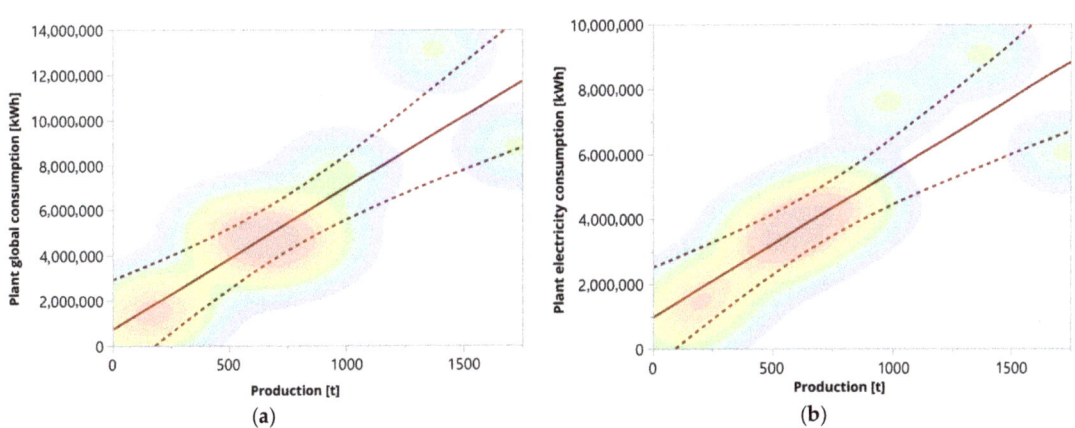

Figure 7. NACE C22.1.9: Gasket manufacturers cluster. (**a**) Global regression model; (**b**) Electric regression model. Source: Authors' elaboration on JMP statistical software V.17.

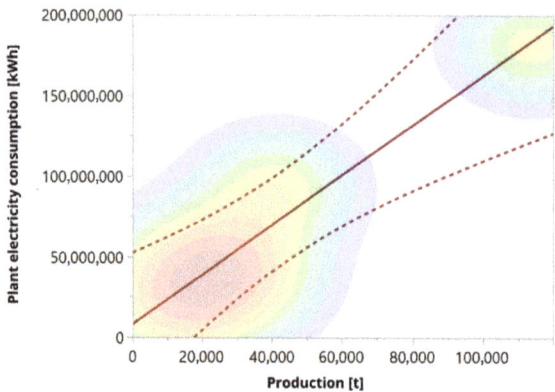

Figure 8. NACE C22.1.1. Electric regression model. Source: Authors' elaboration on JMP statistical software V.17.

Table 2. Statistical parameters of first-level regression analyses.

Sample Sites	Regression Model	p-Value	R^2
NACE C22.1.9	Global	<0.0004	0.773
	Electric	<0.0003	0.747
NACE C22.1.1	Electric	<0.0081	0.930

For NACE C22.1.9, regression analyses of the global and electrical models show a good correlation, both from an R^2 and from an intercept point of view, which is positive, indicating the reliability of the model even for low outputs.

The thermal model is not present as it is not entirely satisfactory; in fact, it has a much lower R^2 and a negative intercept. In any case, the amount of thermal energy used depends on the type of machinery present, which is mainly for moulding, leading to a high level of randomness in consumption. For this reason, it is advisable to relate the consumption of the plant using mainly the global energy model.

In addition to the regression analysis, the average values of the EnPIs, referring to the consumption of the entire plant, for the electrical, thermal (consumption of natural gas and other fuels), and global (electrical + thermal) models, are elaborated and reported in Table 3. The calculated indicators confirm the greater weight of electrical consumption compared to thermal consumption, so the sector's efficiency must come through a reduction in this consumption. However, a reduction in natural gas consumption is still important in decarbonising the sector. As pointed out in the literature review, no indicators related to the energy consumption of the production of rubber products were found, so a comparison with the calculated values is not possible.

Table 3. Results of the first level EnPIs calculations.

Sample Sites	EnPI (kWh/t)	$EnPI_{avg} \pm st.dev$	Reliability
NACE C22.1.9	EnPI_el	6.270 ± 1.335	Medium/High
	EnPI_th	1.365 ± 1.160	Low
	EnPI_gl	7.656 ± 1.895	Medium/High
NACE C22.1.1	EnPI_el	1.764 ± 631	Medium/High
	EnPI_th	3.963 ± 2.727	Low/Medium
	EnPI_gl	5.727 ± 3.280	Medium

For the NACE C22.1.9 sector, the electrical and global models are reliable, with a standard deviation between 20% and 25% of the average. The thermal model, however, is not reliable, and it is not an indication of how an increase in quality is needed in the reporting of thermal consumption by the reporters of the audits. For NACE sector C22.1.1, similar results have been found for the electrical and global models, while the thermal model is more reliable compared to the previous sector. This is mainly due to an increase in the quality of the data in the energy audits.

For NACE sector C22.1.1, only the regression analysis performed for the electrical model is reported. Those for the thermal and global models are not shown due to the high randomness of the results obtained. This is mainly due to the low number of companies subjected to mandatory energy audits in the sector, which results in low reliability, particularly referring to the p-value. On the other hand, the high quality of the data allowed for more accurate analyses, especially for second-level EnPIs.

The average values shown in Table 3 can be compared with the few values reported in previous literature studies. In particular in [26] a global EnPI of about 5000 kWh/t is presented, while in [27] global EnPI of US tyre factories of about 3000 kWh/t is reported.

3.2.2. Second Level Benchmark EnPIs

The added value of the present work was to delve deeper into the energy analysis of production processes by going on to evaluate the second-level indicators that are a fundamental tool for understanding the ways in which the production site can be made more efficient, and that must guide the choice of interventions.

In this section, second level EnPIs obtained for the main activities will be illustrated, for the moulding and post vulcanization phases for NACE sector C22.1.9 (gasket cluster) and for mixing and curing phases for NACE sector C22.1.1 will be analysed.

Figures 9 and 10 show the regression analyses performed for NACE C22.1.1, while Table 4 shows the associated statistical parameters.

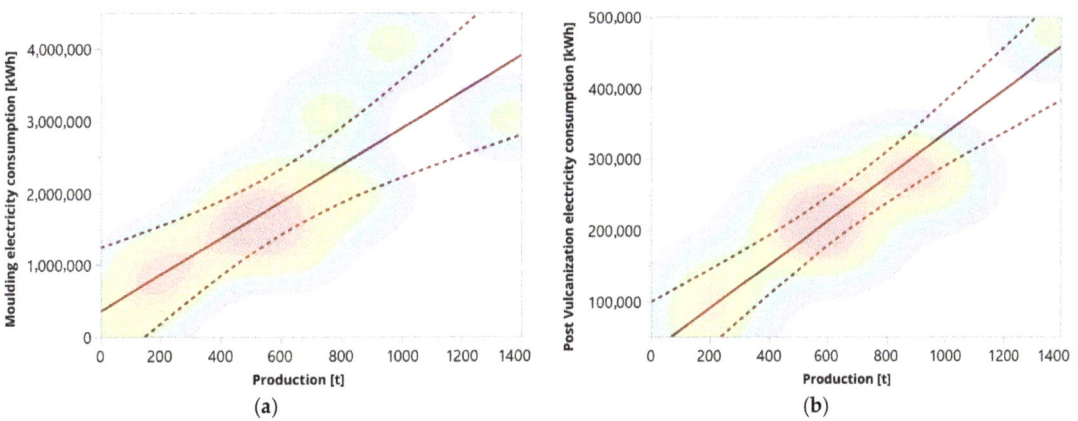

Figure 9. NACE C22.1.9: Gasket manufacturers cluster. (**a**) Electric regression model for moulding; (**b**) Electric regression model for post-vulcanization. Source: Authors' elaboration on JMP statistical software V. 17.

For NACE C22.1.9, the regression shows a good correlation, determined by the high R^2. For the moulding process, even though the consumption data are evenly distributed over the range of existence of the production, it has a significantly lower randomness than post-curing. This is probably due to the different types of presses used to perform the process, as injection presses generally consume more than compression ones. Despite the very high R^2 of post-curing, the relative regression line has a negative intercept, indicating that the model is not totally reliable for low production levels.

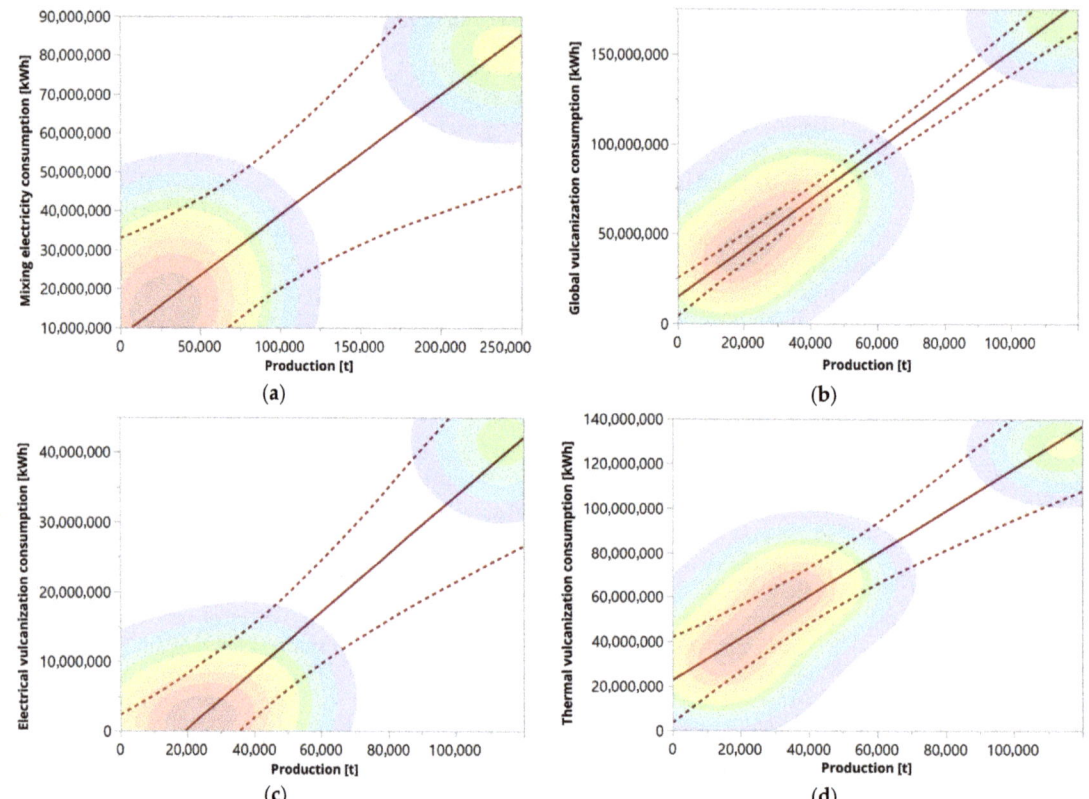

Figure 10. NACE C22.1.1. (**a**) Electric regression model for mixing; (**b**) Global regression model for vulcanisation; (**c**) Electric regression model for vulcanisation; (**d**) Thermal regression model for vulcanisation. Source: Authors' elaboration on JMP statistical software V. 17.

Table 4. Statistical parameters of second-level regression analyses.

Phase	Regression Model	p-Value	R^2
NACE C22.1.9	Moulding	<0.0014	0.696
	Post-vulcanization	<0.0002	0.917
NACE C22.1.1	Mixing	<0.0226	0.955
	Vulcanization$_{gl}$	<0.0020	0.995
	Vulcanization$_{th}$	<0.0029	0.964

For the mixing phase of the NACE C22.1.1 sector, the data density is not homogeneous over the entire range of production. This results in a set of sites with similar production levels, which generate a leverage effect on the regression, decreasing the quality of the analysed model, which can only be considered reliable mainly for low and high production levels. Like the calculation of EnPIs, the regression for the vulcanisation phase allows a very accurate consumption model to be obtained, particularly when referring to the global analysis. The randomness of electrical and thermal consumption due to the different types and performances of the ovens is, in fact, cancelled out with the global model, where the model's confidence lines lie almost completely on the same regression line.

The average second level EnPIs, referred to the consumption of individual phases for the electrical, thermal, and global (electrical + thermal) models are presented in Table 5.

Better results have been obtained for the moulding and post-vulcanization phases (both electrical) for NACE C22.1.9, while for NACE C22.1.1, the specific consumption related to mixing (electrical) and vulcanization (overall and thermal) phases were calculated.

Table 5. Results of the second level EnPIs calculations.

Phase		EnPI (kWh/t)	$EnPI_{avg} \pm st.dev$	Reliability
NACE C22.1.9	Moulding		3.270 ± 984	Medium
	Post-vulcanization		370 ± 89	Medium/High
NACE C22.1.1	Mixing		515 ± 206	Medium
	$Vulcanization_{gl}$		1.913 ± 322	High
	$Vulcanization_{th}$		1.775 ± 435	Medium/High

For NACE sector C22.1.9, the calculated EnPIs have a fair level of reliability. It also shows that moulding is by far the most energy-intensive phase of the plant, with an EnPI an order of magnitude higher than post-vulcanization. For moulding, the values obtained can be compared with the previous ENEA study [40], where a value of 4300 kWh/t is reported. The higher value reported in the previous study is due to the fact that the reference's IPE is specific to injection moulding (more energy-intensive than the compression moulding included in the present study).

For the NACE C22.1.1 sector, the overall EnPI identified for the vulcanisation phase has a lower coefficient of variation, supporting the reliability of the model. The thermal model also shows good confidence, while the electric indicator is not reported due to its high coefficient of variation. Conversely, the electrical model for the mixing phase is reliable, as reported in Table 5.

For the mixing phase, it is possible to compare the value obtained with the value of 770 kWh/t reported in [29] related to the analysis of a Swedish plant in 2000. Moreover, the results reported in [40] are available, which, however, are for rubber manufacturing in general and not specifically for tyre production, and in fact, report an EnPI of about 200 kWh/t. The one calculated is higher, probably due to additional raw materials and additives used in the mixing phase of tyre manufacturing.

3.3. Energy Efficiency Interventions

This session includes information on the second research question concerning qualitative and quantitative information on energy efficiency measures.

Energy audits include detailed information on the implemented and proposed EPIAs. The measures identified are described in the audits both qualitatively and quantitatively, reporting the achievable energy and economic savings, the investment required to carry out the measure, and any economic indicators such as NPV and Payback Period. Before starting the analysis, the data preparation phase involved the classification of interventions according to 17 main areas, shown in Table 6.

Table 6. Classification of energy efficiency interventions.

Classification of Interventions	
Air conditioning	Electrical Systems
Aspiration	General
Building Envelope	Other areas
Cogeneration	Power factor correction
Compressed Air	Renewable energy production
Cooling systems	Thermal power plant/Heat Recovery
Electric Motors/Inverters	Transport

3.3.1. Analysis of Energy Efficiency Interventions for the NACE C22.1.9 Sector

In the NACE C22.1.9 sector, 406 energy efficiency measures were identified in audits, amounting to about 10 ktoe of annual energy saving, corresponding to a total investment of EUR 32 million. On the other hand, 66 interventions were carried out by companies in previous years, with 724 tons of annual energy saving and EUR 3 million invested.

The comparison between the distribution of the Interventions identified and implemented is included in Figure 11. Figure 11b demonstrates that about one-third of implemented EPIAs are associated with lighting, which was made efficient mainly through LED relamping, followed by Compressed air and Production lines interventions. In identified interventions, the most populated area is compressed air, with interventions on the search/elimination of leaks and replacement of compressors. More details on the interventions carried out on production lines in NACE C22.1.9 are provided in Table 7.

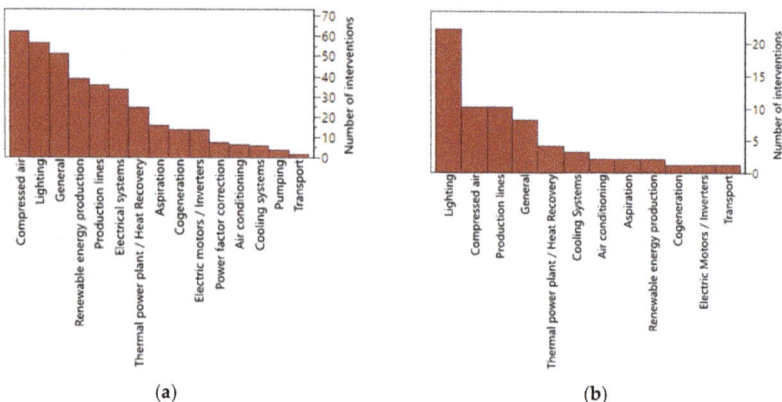

Figure 11. NACE C22.1.9. Distribution according to the main areas of the interventions identified and carried out. (**a**) Identified interventions. (**b**) Interventions carried out. Source: Authors' elaboration with JMP statistical software V. 17.

Table 7. NACE C22.1.9. Main interventions carried out on production lines.

NACE C22.1.9		
Main Activities		
Production Phase	**Solution**	**Description**
Moulding	Press insulation	Installation of insulation planes on outside walls of presses
	Presses replacement	For very old systems, it is directly recommended to replace the presses with the latest of the latest generation, which are more insulated and efficient.
	Optimisation of press start-up time	By analysing the electrical load curves of the plant, is it possible to determine margins for postponing the switch on of pressed to avoiding wasting electricity.
	Optimising press heating	Replacement of electric heating elements for heating the moulds with more modern and efficient technologies such as inductive heating.
Mixing	Mixer replacement	For very old systems, we directly recommend replacing the mixer with new models. An example is the change from a machine working with direct current to one working with alternating current, driven by an asynchronous motor.
	Mixer efficiency	Correct metering of the amount of chilled water required to operate the machine can lead to considerable energy savings.
Drawing	Drawer replacement	For very old systems, we directly recommend replacing the drawer with more efficient models of the latest generation.
Seasoning in hot chamber	Seasoning efficiency	Transition from obsolete raw material heating systems such as AHU heating elements to innovative systems equipped with heat pumps.

An important parameter in assessing the effectiveness of an energy efficiency intervention is the Cost Effectiveness, defined as:

$$\text{Cost Effectiveness} = \frac{\text{Investment } [\text{\euro}]}{\text{Yearly energy savings [toe]}}$$

This parameter makes it possible to assess, depending on the type of intervention, the necessary investment to save one toe per year. The lower the Cost Effectiveness, the more advantageous the energy efficiency intervention will be. It is particularly important to relate cost effectiveness to parameters such as investment and payback period. In Figure 12, the relationship between these parameters is highlighted in a bubble plot, which presents Cost Effectiveness and Simple Payback Period, both averaged, in ordinates and abscissae, respectively. The average investment realised can instead be assessed by scaling the bubbles, classified according to the different main areas of intervention defined by ENEA.

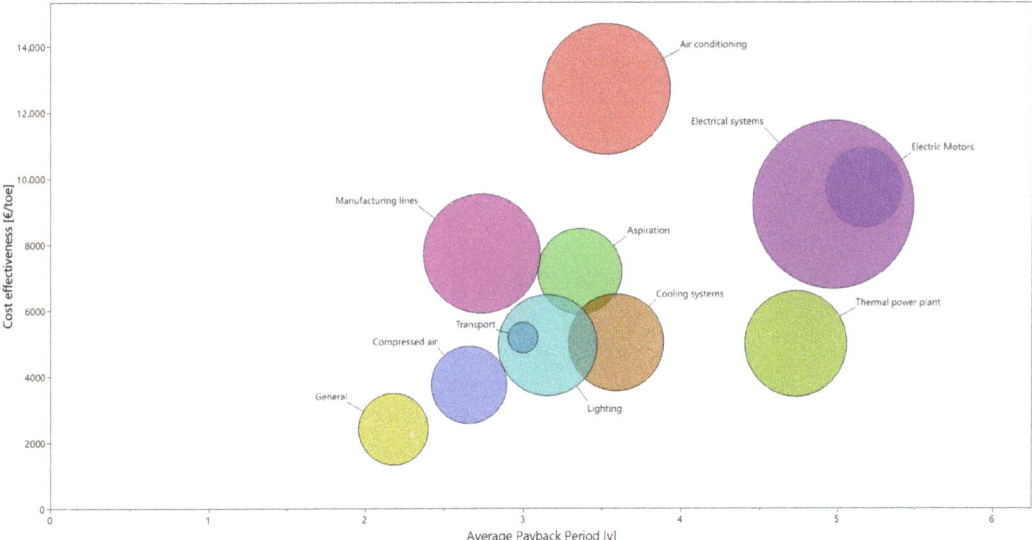

Figure 12. NACE C22.1.9. Bubble-plot for identified interventions scaled on investment. Source: Authors' elaboration with JMP statistical software V.17.

Interventions with the lowest cost-effectiveness are those related to the area 'General', consisting of management interventions such as the adoption of ISO 50001 and monitoring, followed by the areas 'Compressed Air' and 'Lighting'. Low-cost effectiveness implies a higher attitude to intervention in these areas, as shown in Figure 11a,b. In addition to low-cost effectiveness, the interventions in the areas "General", "Compressed Air", and "Lighting" present the lowest Simple Payback Period of the sector (about 2–3 years) and also a low average investment. Interventions on main activities, represented by the category "Production Lines" present a low PBP, a higher average investment and Cost Effectiveness when compared to the remaining categories.

This information, therefore, allows us to highlight how intervening on plant-critical machinery or through renewable energy production is not always convenient both from an energy and economic point of view. It is worth noting that the average investment and PBT do not include potential energy efficiency subsidies and incentive mechanisms.

Due to the low quality of the data received concerning the interventions carried out, it was not possible to perform a similar detailed analysis for this category of interventions.

3.3.2. Analysis of Energy Efficiency Interventions for the NACE C22.1.1 Sector

In the NACE C22.1.1 sector, 68 energy efficiency measures were identified in the audits, amounting to about 29 ktoe of annual energy saving, corresponding to a total investment of EUR 58 million. Moreover, 22 measures were carried out by companies, with a total of EUR 7 million invested and 2.418 toe saved each year.

Unlike the previous sector, there is a considerable discrepancy between the areas of intervention identified and carried out, as shown in Figure 13. The companies intervened mainly on the production lines (Figure 13b), making the main activities efficient according to the interventions described in Table 7. Other relevant intervention areas were pumping, compressed air production, air conditioning and lighting systems. The areas of intervention proposed in the audit (Figure 14), on the other hand, mainly refer to the streamlining of the thermal energy use by concentrating resources in the thermal power plant related to steam production. Its production is therefore made more efficient mainly through proposals to replace the boiler or through the insulation of the steam distribution plant. In any case, a good number of interventions are proposed for the main activities and on the compressed air system; this is in line with the good results obtained in the same areas with the interventions carried out. More details on the interventions carried out on production lines in NACE C22.1.1 are provided in Table 8.

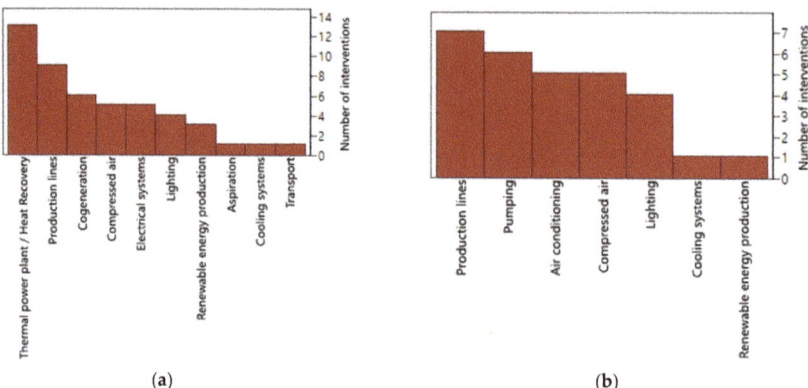

Figure 13. NACE C22.1.1. Distribution according to the main areas of the interventions identified and carried out. (**a**) Identified interventions. (**b**) Interventions carried out. Source: Authors' elaboration with JMP statistical software V. 17.

In this case, the identified interventions referred to Production lines represent, also with the "Thermal Power Plant" area, the categories with the lowest cost-effectiveness in the sector (about 2000 €/toe). This explains the high number of interventions identified (both areas) and carried out ("Production Lines" only) in these two main categories. In the compressed air production and distribution area, the average cost-effectiveness for this NACE Sector also stands at around 4000 €/toe, like the results obtained in Figure 12 for the previous sector.

In Figure 14, it is possible to distinguish two zones made up of 3 areas each; the first, consisting of "Manufacturing Lines", "Thermal Power Plant", and "Compressed Air", presents low-cost effectiveness, payback period and also average investment, determined by the size of the bubble, while the second zone, consisting of "Lighting", "Cooling Systems" and "Electrical systems", presents totally opposite characteristics. These results agree with the number of interventions proposed and carried out by the companies shown in Figure 13.

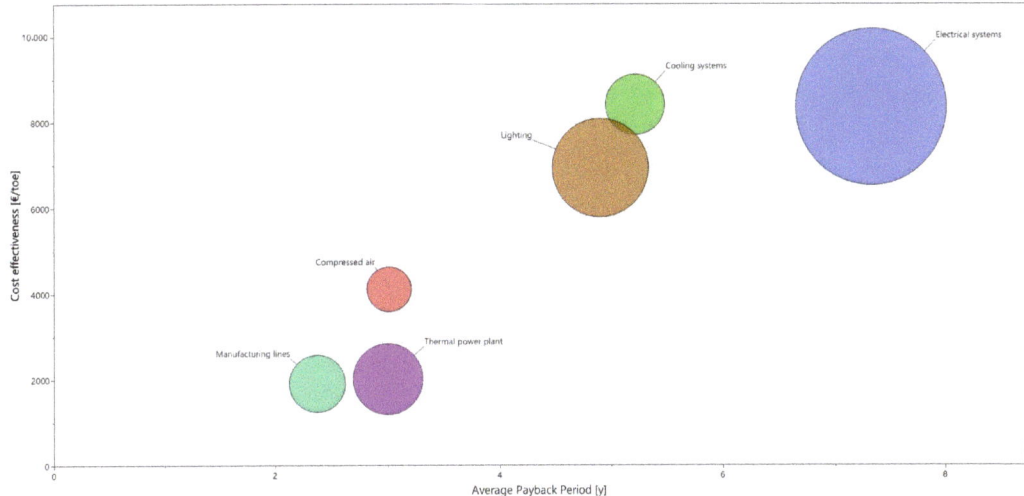

Figure 14. NACE C22.1.1. Bubble-plot for identified interventions scaled on investment. Source: Authors' elaboration with JMP statistical software V.17.

Table 8. NACE C22.1.1. Main interventions carried out on production lines.

NACE Sector C22.1.1		
Main Activities		
Production Phase	Solution	Description
Vulcanization	Firing cavity replacement	Cooking cavities that mainly use steam as an energy carrier for heating can be replaced with fewer cavities that use electrical heating elements instead, leading to an improvement in overall efficiency and a reduction in natural gas consumption.
	Efficiency of pressure level	Use a press exhaust not at atmospheric pressure but at a pressure of about 2–3 bar, increasing the proportion of heat that can be recovered from the system.
	Replacement of vulcanisers	Conversion of the water-vapour vulcanisation plant to a more efficient nitrogen plant, saving approximately 20 percent of the plant's natural gas consumption.
Mixing	Mixer replacement	For very old systems, we directly recommend replacing the mixer with new models. An example is the change from a machine working with direct current to one working with alternating current, driven by an asynchronous motor.
	Mixer efficiency	Correct metering of the amount of chilled water required to operate the machine can lead to considerable energy savings.
	Banbury Energy Efficiency	Installing pneumatic transformers for handling cylinders to limit energy losses due to rolling. They reuse the dissipated energy produced by the process to suck in and re-pressurise air from outside or from energy dense zones. They also reduce the operating pressure of the compressed air by identifying the minimum level required to move the actuator.

4. Conclusions

In this work, the analysis of the energy performance of the Italian rubber manufacturing industry based on mandatory energy audits is presented. The rubber manufacturing industry (NACE Group C22.1) is an important sector in Italy, but specific studies dedicated to the energy aspects of the sector are lacking, together with up-to-date, reliable benchmark data. The novelty of this work lies in updating, when available, or developing for the first time energy performance indicators for the sector, deepening the energy analysis of its production processes, and assessing the cost-effectiveness of energy efficiency solutions.

The analysis of about one hundred energy audits related to two main sub-sectors (NACE C22.1.9 and NACE C22.1.1) made it possible to define energy models, assess the share of different energy vectors and their contributions among the three main functional areas and among the main process activities, calculate first- and second-level energy performance indicators, and investigate the main energy efficiency interventions. Results highlighted that natural gas accounts for more than a third of the total energy consumption for NACE C22.1.9, while for NACE C22.1.1, it accounts for 50%. First-level electrical, thermal and global consumption indicators were calculated. Relative to electricity consumption average indicators, the values for C22.1.9 and C22.1.1 are 6300 kWh/t and 1800 kWh/t, respectively, both with medium/high reliability. Detailed information is available from energy audits, which also allow the computing of second-level indicators relative to specific sub-processes. For sector C22.1.9, the obtained second-level indicators have a fair level of reliability. The results show that moulding is the most energy-intensive production phase. For the NACE C22.1.1 sector, second-level indicators for vulcanisation and mixing phases have been elaborated.

The computation of energy performance indicators has been combined with the analysis of the implemented and proposed energy performance improvement actions reported in energy audits. For the NACE sector C22.1.9, a third of implemented EPIAs is associated with lighting, followed by Compressed air and Production lines interventions. In identified interventions, the most populated area is compressed air, with interventions on the search/elimination of leaks and replacement of compressors. For NACE C 22.1.1 audits, the companies intervened mainly on the production lines, and other relevant intervention areas were represented by pumping, compressed air production, air conditioning and lighting systems. On the other hand, proposed interventions are mainly related to steam production optimisation. Cost-effective indicators have also been computed. This work represents an important contribution, as detailed benchmark analyses for the rubber industry were not available in the literature. Having these up-to-date results available is very important, firstly for companies in the sector and secondly for policymakers who establish energy efficiency support measures for companies. The definition of specific energy models, in fact, can provide a useful framework for consumption accounting at the company level and for comparison over time. The analysis of the cost-effectiveness and payback time of different measures can also play a key role in planning energy efficiency support policies. Combining technical information on energy performance with economic features on interventions makes a unique data-driven and knowledge-based framework available to policymakers. The limitations of the work are related to the sample under analysis, which includes only obliged companies, i.e., large or energy-intensive enterprises, while it would be interesting to consider the entire rubber sector, namely non-energy-intensive SMEs and, with the available data, extend the analysis to other European and non-European countries. Further research could concentrate on a scenario analysis devoted to calculating the whole sector's efficiency potential. In this way, the differences among company types could also be assessed, showing the role of potential barriers to EPIA implementation.

Author Contributions: Conceptualization, M.P., C.M., F.M. and C.T.; methodology, M.P., C.M., F.M. and C.T.; investigation, M.P., C.M., F.M. and C.T.; data curation, M.P., C.M., F.M. and C.T.; writing—original draft preparation, M.P., C.M., F.M. and C.T.; writing—review and editing, M.P., C.M., F.M. and C.T.; visualisation, M.P., C.M., F.M. and C.T.; supervision, C.M., F.M. and C.T.; project administration, F.M.; funding acquisition, F.M. All authors have read and agreed to the published version of the manuscript.

Funding: This research was funded by the National Electrical System Research Program (RdS PTR 2022–2024-WP1.6), "Piano Triennale della Ricerca del Sistema Elettrico Nazionale 2022–2024", implemented under programme agreements between the Italian Ministry of Environment and Energy Security and ENEA, CNR, and RSE S.p.A.

Data Availability Statement: Data is contained within the article.

Conflicts of Interest: The authors declare no conflict of interest.

References

1. IEA 2023, Tracking Industry. Available online: https://www.iea.org/reports/industry (accessed on 1 February 2024).
2. Directive 2012/27/EU of the European Parliament and of the Council of 25 October 2012 on Energy Efficiency. 2012. Available online: https://eur-lex.europa.eu/legal-content/EN/TXT/HTML/?uri=CELEX:32012L0027&from=EN (accessed on 1 February 2024).
3. Directive 2023/1791/EU of the European Parliament and of the Council of 13 September 2023 on Energy Efficiency. 2023. Available online: https://eur-lex.europa.eu/legal-content/IT/TXT/PDF/?uri=CELEX:32023L1791 (accessed on 1 February 2024).
4. Communication from the Commission Guidelines on State Aid for Environmental Protection and Energy 2014–2020. Available online: https://eur-lex.europa.eu/legal-content/EN/TXT/HTML/?uri=CELEX:52014XC0628(01)&from=ITA (accessed on 1 February 2024).
5. Rapporto Sull'attuazione Dell'obbligo di Diagnosi 2019. 2020. Available online: https://www.efficienzaenergetica.enea.it/component/jdownloads/?task=download.send&id=556&catid=52&Itemid=101 (accessed on 1 March 2023). (In Italian)
6. Eurostat Data for 2021 Energy Consumption. Available online: https://ec.europa.eu/eurostat/databrowser/view/nrg_d_indq_n__custom_9102201/default/table?lang=en (accessed on 1 February 2024).
7. Eurostat Data for 2021 Value Added. Available online: https://ec.europa.eu/eurostat/databrowser/view/nama_10_a64_custom_9352160/default/table?lang=en (accessed on 1 February 2024).
8. Saidur, R.; Mekhilef, S. Energy use, energy savings and emission analysis in the Malaysian rubber producing industries. *Appl. Energy* **2010**, *87*, 2746–2758. [CrossRef]
9. Istat, Rapporto Sulla Competitività Dei Settori Produttivi. 2023. Available online: https://www.istat.it/it/archivio/282020 (accessed on 1 February 2024).
10. Assolombarda e Federazione Gomma Plastica. Le Industrie Della Gomma Plastica in Lombardia, Nel Contesto Europeo e Italiano. I Settori di Destinazione e la Rilevanza Della Filiera. 2023. Available online: https://www.assolombarda.it/centro-studi/le-industrie-della-gomma-plastica-in-lombardia-nel-contesto-europeo-e-italiano-report (accessed on 1 February 2024). (In Italian)
11. Herce, C.; Biele, E.; Martini, C.; Salvio, M.; Toro, C. Impact of Energy Monitoring and Management Systems on the Implementation and Planning of Energy Performance Improved Actions: An Empirical Analysis Based on Energy Audits in Italy. *Energies* **2021**, *14*, 4723. [CrossRef]
12. May, G.; Barletta, I.; Stahl, B.; Taisch, M. Energy management in production: A novel method to develop key performance indicators for improving energy efficiency. *Appl. Energy* **2015**, *149*, 46–61. [CrossRef]
13. Johnsson, S.; Andersson, E.; Thollander, P.; Karlsson, M. Energy savings and greenhouse gas mitigation potential in the Swedish wood industry. *Energy* **2019**, *187*, 115919. [CrossRef]
14. Proskuryakova, L.; Kovalev, A. Measuring energy efficiency: Is energy intensity a good evidence base? *Appl. Energy* **2015**, *138*, 450–459. [CrossRef]
15. Franco, A.; Miserocchi, L.; Testi, D. Energy Indicators for Enabling Energy Transition in Industry. *Energies* **2023**, *16*, 581. [CrossRef]
16. Morfeldt, J.; Silveira, S.; Hirsch, T.; Lindqvist, S.; Nordqvist, A.; Pettersson, J.; Pettersson, M. Improving energy and climate indicators for the steel industry—The case of Sweden. *J. Clean. Prod.* **2015**, *107*, 581–592. [CrossRef]
17. Andersson, E.; Thollander, P. Key performance indicators for energy management in the Swedish pulp and paper industry. *Energy Strategy Rev.* **2019**, *24*, 229–235. [CrossRef]
18. Haraldsson, J.; Johnsson, S.; Thollander, P.; Wallén, M. Taxonomy, saving potentials and key performance indicators for energy end-use and greenhouse gas emissions in the aluminium industry and aluminium casting foundries. *Energies* **2021**, *14*, 3571. [CrossRef]
19. Kanchiralla, F.M.; Jalo, N.; Thollander, P.; Andersson, M.; Johnsson, S. Energy use categorization with performance indicators for the food industry and a conceptual energy planning framework. *Appl. Energy* **2021**, *304*, 117788. [CrossRef]
20. Branchetti, S.; Petrovich, C.; Ciaccio, G.; De Sabbata, P.; Frascella, A.; Nigliaccio, G. Energy efficiency indicators for textile industry based on a self-analysis tool. *Commun. Comput. Inf. Sci.* **2021**, *1217*, 3–27. [CrossRef]
21. Kanchiralla, F.M.; Malik, F.; Jalo, N.; Johnsson, S.; Thollander, P.; Andersson, M. Energy end-use categorization and performance indicators for energy management in the engineering industry. *Energies* **2020**, *13*, 369. [CrossRef]
22. Phylipsen, G.J.M.; Blok, K.; Worrell, E. International comparisons of energy efficiency-Methodologies for the manufacturing industry. *Energy Policy* **1997**, *25*, 715–725. [CrossRef]
23. Bosseboeuf, D.; Chateau, B.; Lapillonne, B. Cross-country comparison on energy efficiency indicators: The on-going European effort towards a common methodology. *Energy Policy* **1997**, *25*, 673–682. [CrossRef]
24. Worrell, E.; Bernstein, L.; Roy, J.; Price, L.; Harnisch, J. Industrial energy efficiency and climate change mitigation. *Energy Effic.* **2008**, *2*, 109. [CrossRef]
25. European IPPC Bureau. BAT Reference Documents (BREFs). Available online: https://eippcb.jrc.ec.europa.eu/reference (accessed on 1 February 2024).
26. Thai Department of Alternative Energy Development and Efficiency, 2007, Project on Studying of Energy Efficiency Index in Rubber Industry. Available online: http://www2.dede.go.th/kmberc/datacenter/factory/rubber/RubberEng.pdf (accessed on 1 February 2024).

27. US Department of Energy. Bandwidth Study on Energy Use and Potential Energy Savings Opportunities in U.S. Plastics and Rubber Manufacturing. 2017. Available online: https://energy.gov/eere/amo (accessed on 1 February 2024).
28. Gudadhe, M.; Lohakare, P.; Meshram, M.; Padole, A. Energy analysis in tire manufacturing Industries. *Int. J. Mech. Eng. Technol.* **2015**, *6*, 112–119.
29. Stankevičiūtė, L. Energy Use and Energy Management in Tyre Manufacturing: The Trelleborg 1 Case. Master's Thesis, Lund University, Lund, Sweden, 2000.
30. Chikri, Y.A.; Wetzels, W.; Decarbonisation Options for the Dutch Tyre Industry. The Hague. 2019. Available online: www.pbl.nl/en (accessed on 1 February 2024).
31. Bruni, G.; De Santis, A.; Herce, C.; Leto, L.; Martini, C.; Martini, F.; Salvio, M.; Tocchetti, F.A.; Toro, C. From energy audit to energy performance indicators (Enpi): A methodology to characterize productive sectors. The Italian cement industry case study. *Energies* **2021**, *14*, 8436. [CrossRef]
32. Martini, F.; Ossidi, M.; Salvio, M.; Toro, C. Analysis of the Energy Consumption Structure and Evaluation of Energy Performance Indicators of The Italian Ceramic Industry. In Proceedings of the ECOS 2021—34th International Conference on Efficency, Cost, Optimization, Simulation and Environmental Impact of Energy Systems, Sicily, Italy, 28 June–2 July 2021; pp. 1258–1268. [CrossRef]
33. Herce, C.; Martini, C.; Salvio, M.; Toro, C. Energy Performance of Italian Oil Refineries Based on Mandatory Energy Audits. *Energies* **2022**, *15*, 532. [CrossRef]
34. Bruni, G.; Martini, C.; Martini, F.; Salvio, M. On the Energy Performance and Energy Saving Potential of the Pharmaceutical Industry: A Study Based on the Italian Energy Audits. *Processes* **2023**, *11*, 1114. [CrossRef]
35. Solnørdal, M.T.; Thyholdt, S.B. Absorptive Capacity and Energy Efficiency in Manufacturing Firms—An Empirical Analysis in Norway. *Energy Policy* **2019**, *132*, 978–990. [CrossRef]
36. Backlund, S.; Thollander, P.; Palm, J.; Ottosson, M. Extending the Energy Efficiency Gap. *Energy Policy* **2012**, *51*, 392–396. [CrossRef]
37. Cagno, E.; Worrell, E.; Trianni, A.; Pugliese, G. A Novel Approach for Barriers to Industrial Energy Efficiency. *Renew. Sustain. Energy Rev.* **2013**, *19*, 290–308. [CrossRef]
38. Santino, D.; Biele, E.; Salvio, M. *Guidelines for Energy Audits under Article 8 of the EED: Italy's Implementation Practices and Tools*; ENEA: Rome, Italy, 2019. Available online: https://www.efficienzaenergetica.enea.it/component/jdownloads/?task=download.send&id=377&catid=40&Itemid=101 (accessed on 1 February 2024).
39. European Commission. *NACE Rev. 2—Statistical Classification of Economic Activites in the European Community*; Office for Official Publications of the European Communities: Luxembourg, 2008.
40. ENEA and Federazione Gomma Plastica, Analisi dei Dati Relativi Alle Diagnosi Energetiche e Individuazione Preliminare Degli Indici di Prestazione Nei Settori Della Lavorazione Della Gomma e Della Trasformazione Delle Materie Plastiche. 2016. Available online: https://www.efficienzaenergetica.enea.it (accessed on 1 February 2024). (In Italian)
41. Mark, J.E.; Erman, B.; Roland, C.M. (Eds.) *The Science and Technology of Rubber*, 4th ed.; Academic Press: Cambridge, MA, USA, 2013.
42. *ISO 50001:2018*; Energy Management Systems—Requirements with Guidance for Use. ISO: Geneva, Switzerland, 2018.

Disclaimer/Publisher's Note: The statements, opinions and data contained in all publications are solely those of the individual author(s) and contributor(s) and not of MDPI and/or the editor(s). MDPI and/or the editor(s) disclaim responsibility for any injury to people or property resulting from any ideas, methods, instructions or products referred to in the content.

Article

An Econometric Analysis of the Energy-Saving Performance of the Italian Plastic Manufacturing Sector

Valeria Costantini [1], Mariagrazia D'Angeli [1], Martina Mancini [1], Chiara Martini [2,*] and Elena Paglialunga [1]

1. Department of Economics, Roma Tre University, 00145 Rome, Italy; valeria.costantini@uniroma3.it (V.C.); mariagrazia.dangeli@uniroma3.it (M.D.); martina.mancini2@uniroma3.it (M.M.); elena.paglialunga@uniroma3.it (E.P.)
2. Italian National Agency for New Technologies, Environment and Sustainable Economic Development (ENEA), 00196 Rome, Italy
* Correspondence: chiara.martini@enea.it

Abstract: In a scenario characterised by mitigation concerns and calls for greater resilience in the energy sector, energy audits (EAs) emerge as an essential mean for enhancing end-use energy consumption awareness and efficiency. Such a tool allows us to assess the different energy carriers consumed in a productive sector, offering insight into existing energy efficiency improvement opportunities. This opens avenues for research to devise an econometrics-based methodology that encapsulate production sites and their environmental essentials. This paper contributes to the literature by exploiting the EAs received by the Italian National agency for New technologies, Energy, and Sustainable Economic Development (ENEA) in 2019 from the Italian plastics manufacturing sector, matched with Italian firm-based data extracted from the Analisi Informatizzata delle Aziende Italiane (Italian company information and business intelligence) (AIDA) database. In particular, we investigate how the implementation of energy efficiency measures (EEMs) is influenced by a set of contextual factors, as well as features relating to the companies and EEMs themselves. The empirical investigation focuses on the EAs submitted to ENEA in 2019, which was strategically chosen due to its unique data availability and adequacy for extensive analysis. The selection of 2019 is justified as it constitutes the second mandatory reporting period for energy audits, in contrast to the 2022 data, which are currently undergoing detailed refinement. In line with the literature, the adopted empirical approach involves the use of both the OLS and logistic regression models. Empirical results confirm the relevance of economic and financial factors in guiding the decisions surrounding the sector's energy performance, alongside the analogous influence of the technical characteristics of the measures themselves and of the firms' strategies. In particular, the OLS model with no fixed effects shows that a one-percent variation in investments is associated with an increase in savings performance equal to 0.63%. As for the OLS model, including fixed effects, the elasticity among the two variables concerned reaches 0.87%, while in the logistic regression, if the investment carried out by the production sites increases, the expected percentage change in the probability that the energy-saving performance is above its average is about 187.77%. Contextual factors that prove to be equally influential include the incentive mechanism considered and the traits of the geographical area in which the companies are located. Relevant policy implications derived from this analysis include the importance of reducing informational barriers about EEMs and increasing technical assistance, which can be crucial for identifying and implementing effective energy solutions.

Keywords: energy efficiency; energy audits; industry; plastics

Citation: Costantini, V.; D'Angeli, M.; Mancini, M.; Martini, C.; Paglialunga, E. An Econometric Analysis of the Energy-Saving Performance of the Italian Plastic Manufacturing Sector. *Energies* **2024**, *17*, 811. https://doi.org/10.3390/en17040811

Academic Editor: Dimitrios Asteriou

Received: 25 December 2023
Revised: 1 February 2024
Accepted: 6 February 2024
Published: 8 February 2024

Copyright: © 2024 by the authors. Licensee MDPI, Basel, Switzerland. This article is an open access article distributed under the terms and conditions of the Creative Commons Attribution (CC BY) license (https://creativecommons.org/licenses/by/4.0/).

1. Introduction

The current policy environment in the European Union (EU) is witnessing numerous challenges affecting the stability of the macroeconomic framework of member countries, most notably the compelling need to address growing uncertainties regarding the reliability and affordability of energy flows while maintaining policy strategies in line with greenhouse gas emission reduction targets. For several years, the key institutions of the EU have shown a deep awareness of the combined benefits and urgencies associated with energy efficiency (EE), identifying energy demand moderation as a pivotal dimension of the European Energy Strategy. According to the text of Directive 2012/27/EU [1], and its 2018 [2] and 2023 [3] recasts, EE is a valuable tool whose advancements along the entire energy chain (including generation, transmission, distribution, and end-use of energy) would generate benefits directly attributable to the wellbeing of the EU population, such as those related to the environment, air quality and public health, and even positive impacts related to the energy security of member countries, by reducing their dependence on energy imports from outside the Union. As a result, energy costs for households and businesses would decrease, thus alleviating energy poverty, fostering business competitiveness and the dynamics of labour markets. In 2012, the European Energy Efficiency Directive 2012/27/EU (EED) [1] was therefore enacted, with the aim of regulating the matter and introducing additional stimulus mechanisms for the adoption of energy efficiency measures. The integrated approach used is reflected in the request to Member States to establish "national energy efficiency targets expressed through national action plans to be published every three years", and the complementary request to define and promote financing instruments or fiscal incentives that lead to the implementation of energy management practices and energy-efficient technologies or techniques. The legislation on the subject is, in fact, constantly evolving. On 10 October 2023, a new Directive on energy efficiency entered into force [3], containing updated norms on the matter to reduce final energy consumption at an EU level of 11.7% by 2030, which translates into an upper limit on the EU final energy consumption of 763 million toe and 993 million toe for primary consumption. The 2023 revision of the Directive is part of the EU Green Deal package, and the ambition of the 2021 initial proposal was strengthened in the REPowerEU plan. For the first time, the "energy efficiency first" principle takes on legal standing, meaning that EU countries must consider EE in all relevant policy and investment decisions. The present work aims to offer a method to analyse the impact of the existing mechanism of mandatory energy audits on the adoption of EEMs, applying it to a specific manufacturing sector, prior to the introduction of these new obligations.

In line with European commitments, Italy pursues a path to improve the country's energy security, environmental protection, and accessibility of energy services, recognising the benefits associated with energy efficiency and therefore identifying its strategic role through the Legislative Decree No. 102 of 2014 on energy efficiency [4]. The Decree establishes a framework of measures for the promotion and improvement of energy efficiency that contribute to the achievement of the national energy savings target indicated in Article 3 and to the implementation of the European principle that puts energy efficiency first. In detail, it sets a national obligation to save 20 Mtoe per year of primary energy, equivalent to 15.5 Mtoe per annum of final energy through mechanisms and tools outlined by the Legislative Decree itself. For what concerns these latter, as stressed by the Italian Action Plan for EE, Italy mainly refers to the White Certificates obligation scheme, accompanied by other support tools for energy efficiency enhancement measures such as tax deductions for the energy requalification of buildings and the renewable energy for heating and cooling support scheme (*Conto termico*), which incentivises interventions to increase energy efficiency and thermal energy production from renewable sources for small-scale plants. EE Directive 2012/27/EU [1] introduced the obligation for large enterprises to carry out an energy audit on their production sites, starting from December 2015 and subsequently every 4 years. Among the horizontal policies mentioned in the Decree, in line with the requirements of the Directive, EAs are recognised as a key tool capable of stimulating a conscious management of energy consumption within production activities, while identifying feasible efficiency measures and mechanisms for their implementation, breaking down some of the barriers associated with the adoption of EEMs.

In this study, we exploit the large potential offered by data collected through mandatory EAs, a crucial source of detailed and reliable information on energy-saving opportunities. The focus on the plastics manufacturing sector is motivated by its economic significance and the abundance of EEMs identified in the EAs. We adopt an econometric approach which involves the use of both OLS and logistic regression models to assess the contributions of economic, financial, and contextual factors to a firm's energy performance. Such econometric approach represents a significant starting point to explore firm dynamics in relation to EEMs. However, the unique nature of the EAs' data limits comparability to other studies.

This analysis aims to understand the broader influence of financial and managerial practices on EE. The main focus of this paper is the examination of how certain aspects of corporate productivity, including broader management practices not directly related to energy consumption behaviours, influence industrial energy performances. For instance, a specific area of interest is the role of debt-financed funding in implementing energy efficiency measures, as this could reveal key business strategies for long-term cost management and the impact of these financial decisions on energy efficiency of the sector.

Simultaneously, our investigation delves into the technical characteristics of the energy efficiency interventions. We aim to identify which specific measures and characteristics might lead to enhanced energy performance in the plastics manufacturing sector. This entails examining discrepancies in energy performance impact, attributable to the efficacy of interventions, cost–benefit ratios, adaptability to corporate needs, and technological innovations.

Furthermore, the study critically assesses existing incentives, with a particular focus on the Italian White Certificate scheme, and their effectiveness in supporting enterprises overcome economic barriers to implementing EEMs. Considering the several challenges that companies face in implementing EEMs, it should be considered how the specific characteristics of production realities influence their energy performance and their propensity to implement EEMs. In fact, a uniform approach to incentives may not be sufficient in addressing these challenges. Market-based incentives based solely on the level of energy savings achieved (such as White Certificates) may not be sufficient to achieve sector-specific EE targets, especially if there are discrepancies compounded by differences in energy behaviour related to the geographical location of companies. In acknowledging the inherent limitations of this study, including the nature of the EA information, the specific focus on the plastics manufacturing sector, and the heterogeneity in the energy audits, we emphasise the need for a cautious interpretation of our findings. These factors highlight the potential for further research, particularly in areas where data and methodological limitations currently constrain our understanding of regional disparities and sector-specific challenges in energy efficiency.

2. Literature Review

2.1. Energy Efficiency in Productive Activities

2.1.1. Barriers and Drivers of Energy Efficiency Interventions

The emphasis on incentive mechanisms and energy audits, as part of the European approach, is based on the imperative for Member States to reduce energy intensity of their production processes at an increasing pace. However, businesses, leading players in the clean energy transition, still face several obstacles in pursuing this path.

A distinctive branch of the literature attempts to gain an in-depth understanding of the forces (external and behavioural) that inhibit investment in energy-efficient technologies, i.e., barriers to adoption [5], with the aim of expanding the insufficient adoption of EEMs notwithstanding the benefits in terms of enhanced competitiveness and decreased energy costs. Multiple studies emphasise the crucial importance of financial elements.

In particular, access to capital appears to be one of the main impediments, strongly felt by small and medium enterprises (SMEs) [6–9]. Newell and Anderson [10] substantiates how the initial expenditure incurred by firms and the relative payback period are negatively correlated with the rate of EEM adoption. Therefore, economic incentives are particularly effective in the short term as they abate resistance stemming from implementation costs. However, there may be many benefits and risks that these financial measures do not capture. The specific features of an EEM can be an equally relevant factor for the purpose of implementation, shaping the difficulties faced by the company. In this regard, the authors of [11] identify macro-areas for defining these characteristics; the first deals with the relative benefit of the measure itself (including financial aspects and noneconomic benefits such as those related to the reduction in environmental damage), while the second area deals with the technical context of the intervention and thus with all the elements that characterise the type of EEM, such as its magnitude and duration.

An additional problem arises considering that these specificities can in turn acquire different degrees of relevance depending on the characteristics of the company in consideration and the market in which it operates; in fact, different sizes, market competitiveness, the organisational and knowledge management model adopted, etc., are all traits that can interact with the specificities of the intervention, thus widening or reducing existing barriers [8]. Finally, among the most frequently encountered reasons for insufficient adoption is the perception of the intervention itself, i.e., the priority assigned to it in relation to the corporate strategy adopted. Zhang et al. [12] suggests that whether an EEM is perceived as strategic depends not only on the characteristics of the intervention itself but to an even greater extent on the culture and priorities of the company.

These discussions highlight the importance of the regulatory instruments adopted in the EU to create an economic, legal, and social context capable of transforming corporate culture and strategies, while financially supporting companies in implementing EEMs. To this end, the role of financial, governmental, and non-governmental institutions in promoting energy efficiency in the various production sectors and/or end-uses and guiding financial markets is evident. Trianni and Cagno [8] demonstrates this role through an analysis of the factors capable of stimulating the adoption of energy conservation measures (i.e., energy efficiency drivers). Indeed, along with economic factors, the authors emphasise the influence of regulatory instruments. These can be distinguished between internal, such as voluntary agreements made by companies, and external, such as obligations and constraints imposed on companies, including those related to the performance of energy consumption audits. These obligations, together with the use of information tools and the necessary professional training, increase the efforts made by the production activities and the ease with which they are carried out, increasing awareness both on energy consumption and energy efficiency solutions for the production site, thus breaking down informational barriers.

In addition to internal business organisational factors, which can potentially hinder EEM implementation, it is crucial to address the influences of specific regulatory and policy environments on energy efficiency trajectories. By analysing the broader interconnections and systemic links between renewable energy consumption, financial development, and public health [11], and in particular by exploring the two-way influences and long-term linkages between these variables in different nations, the authors clarify how renewable energy consumption not only supports the reduction in environmental pollution, but also enhances public health. This tripartite relationship, explored through robust econometric methods, reinforces the need for policies that balance immediate financial incentives with long-term health and sustainability goals. From a business perspective, this causality relationship suggests that investment in renewable energy and prioritising financial growth may have more far-reaching implications. This could affect not only immediate operational costs, but also public welfare, and could potentially shape both market reception and competitive edge. When examining the effects of energy policies and their influence on corporate health, we can gain some fascinating insights. Through examining how China has intertwined different policy instruments in the field of energy conservation and emission reduction, Zhang et al. [12] enlightens us on the role of policy coordination in the broader framework of economic growth. When administrative and financial measures are coordinated, there is a positive knock-on effect on economic growth. It is as if these two forces, once aligned, not only serve the goal of energy conservation, but also give the economy a boost.

Zhang et al. [12] also presents a contrasting scenario where administrative measures are implemented alongside fiscal and taxation policies, or when they are aligned with broader policy orientations. In this case, the impact on economic growth goes in the opposite direction. The results suggest the need for a delicate balancing act when introducing policies; policy instruments must be rooted in the specific implementation context and work in harmony, simultaneously supporting sustainable energy goals and economic growth.

In conclusion, these studies shed light on the complexity and multidimensionality of energy efficiency in manufacturing sectors. They emphasise the central role of internal corporate culture and external policy frameworks, along with governmental coordination strategies, in influencing the adoption and effectiveness of EEMs. This interplay of factors accentuates the need for a holistic approach, complemented by an in-depth understanding of the unique green transition needs of the productive sectors, for the creation of customised policy instruments that meet sector-specific needs.

2.1.2. Empirical Analysis of Barriers and Drivers of Energy Efficiency Interventions

This section focuses on the literature on which this work is based, which relates to the use of econometric methods for the sectoral analysis of observed energy efficiency pathways. The use of such methodologies allows for the identification of the foundational traits of the company and the interaction between productive activities and the socioeconomic context capable of influencing the ability of production sites to implement EEMs. Sectoral studies in the field of energy efficiency mostly focus on the analysis of the benefits associated with the implementation of specific technology systems and behavioural measures [13,14]. Few empirical contributions, on the other hand, take a generalised approach aimed at estimating how various elements, inherent in multiple aspects of the company and the context, enable industries to move towards better energy-saving performances. A relevant strand of the literature consists of the studies conducted by Cagno and Trianni; the two authors, in collaboration with several researchers, have carried out multiple empirical works regarding the drivers and barriers that can influence the adoption of EEMs [15–18]. Although these contributions mostly adopt statistical approaches (correlation analysis) methodologically different from econometric models, there are important exceptions, such as [19,20]. The latter studies—in the context of SMEs in the Slovenian manufacturing sector—how salient the barriers and drivers are for EEM adoption in comparison to other contextual influences. Trianni and Cagno [8] constructed a linear probability and logit model to investigate the likelihood of EEM implementation by separately examining elements influencing past and future EE investment decisions. The authors found that companies' internal culture and their innovative capabilities raise the emphasis on energy conservation. Considering past investment decisions, they observed how greater leverage negatively affects the probability of introducing EEMs. They also show that businesses that have implemented awareness programs and performed an energy audit are much more likely to implement EEMs. As for the results related to future investment decisions, the firm's ability to innovate becomes central; research and development activities to achieve an energy-conscious corporate culture are an essential stimulus for EEM deployment. The deployment appears to be highly influenced by organisational and skill-related barriers. Finally, behavioural and technological barriers reduce the likelihood of future EEM introduction, showing once again how building internal awareness, while leveraging external linkages, can influence the energy conservation pattern of a manufacturing sector. In this regard, Hrovatin et al. [20] exploits panel data pertaining to actual investment decisions through two probit models and evaluate the probability for firms to carry out an environmental investment and then, separately, the probability associated with carrying out investments in clean technologies and in energy efficiency measures. Economic factors related to the characteristics of the enterprises as profitability (ROA), indebtedness (debt ratio), innovative capabilities (R&D activity), and energy costs were used in these approaches as predictors, together with market-driven elements and regulations. Regarding the results inherent in investments in EEMs, the authors suggest that both firm and sectoral characteristics can shape them; in particular, firm-specific and market-related economic factors are of crucial importance. For instance, firm size affects the level of EE investments due to the greater exposure to international competition and higher energy costs; consequently, these become necessary means for large enterprises to lower their manufacturing costs and enhance their competitive edge.

Finally, this type of investment is influenced by favourable expectations about future demand and to be relatively more resilient, during and after the 2008 financial crisis. The reason for this probably lies in the fact that EE investments are relatively cost-effective, which makes it easier "to obtain the corresponding financing despite the stringent banking regulations and scarce financial resources" [20]. In their groundbreaking 2014 study, Boyd and Curtis [21] employed a two-phase approach (which at first establishes a benchmark for plant-level energy intensity and subsequently analyses the impact of management practices on this relative energy intensity using a linear regression model) to delve into the impact of corporate management practices on energy performance in the U.S. manufacturing sector. The study incorporated plant-level fixed effects, allowing for the control of unique, invariant characteristics of each plant that could sway energy efficiency. The findings reveal a complex relationship between management practices and energy efficiency. Certain practices, such as effective monitoring and well-structured incentive systems, are linked to a decrease in energy usage. However, a focus on aggressive production targets can paradoxically lead to an increase in energy consumption. Following Boyd and Curtis's approach to examining the role of management in operational efficiency, Blass et al. [22] delves into how top operations managers within small and medium-sized manufacturing firms influence the adoption of energy-saving practices. By employing logistic and OLS regression models with fixed effects for multisite firms, Blass et al. [22] reveals that the engagement of operationally focused senior managers significantly boosts the implementation of recommended practices by 13.4% and increases the adoption of substantive process changes. These insights highlight the importance of managerial influence in enhancing EE for SMEs, aligning with, and expanding upon, the foundational findings of [21]. In parallel with the examination of policy and managerial influences on energy efficiency, the 2017 research conducted by Nicola Cantore [23] presents an in-depth analysis of 214 companies across diverse industries and countries. Cantore's study, utilising logistic regression models complemented by Principal Component Analysis, sheds light on the determinants of energy efficiency technology adoption within the manufacturing sector. Contrary to expectations, the study reveals that geographical location is not a significant factor in investment decisions related to energy efficiency. Instead, it emphasises the decisive roles of internal management, prior experience with EE projects, and microeconomic barriers such as capital constraints. These insights emphasise the importance of a company's internal organisational structure and management practices in the evolution of industrial energy trajectories, advocating for a tailored approach to energy conservation promotion that caters to the specificities of different manufacturing sectors and developmental contexts. Notably, the study uncovers a counterintuitive yet significant relationship where the adoption of certified management systems may deter further energy efficiency investments. From a management perspective, achieving certification might be perceived as an end-goal, potentially reducing the motivation for ongoing investment in EEMs. Further, another significant empirical contribution is [10] on the relevance of information programs for technology adoption. Specifically, the authors considered a set of potential energy-saving projects related to the U.S. manufacturing sector, recommended by auditors of the U.S. Department of Energy's Industrial Assessment Centre (IAC) program, therefore using data resulting from EAs. The utilisation of EA information in this study significantly enhances its contribution to the existing literature, offering a foundation upon which our article draws considerable inspiration. The authors exploited a logit model with plant-level fixed effects [24,25] to control for unobserved differences between plants in the propensity to adopt EE technologies. Newell and Anderson [10] found that adoption rates are higher for projects with shorter payback periods, lower costs, higher resulting annual savings, higher energy prices, and greater energy conservation; results also emphasised how installations are 40% more sensitive to upfront costs than to annual savings, identifying important policy implications, as subsidies appear to be more effective in promoting energy-efficient technologies than increases in energy prices. The study highlights that although numerous cost-effective energy-efficient projects are implemented by manufacturing plants, a significant number remain unadopted. This is mainly due to high implicit discount rates and overlooked factors such as project risks and internal barriers within firms. These findings indicate that decision-making within firms involves more than just financial calculations, encompassing a range of unquantified elements that influence the adoption of energy-efficient technologies.

The use of data from energy audits appears to be an under-used methodological choice, especially considering the econometric formulations. However, there is the extensive opportunity to extract relevant information from them, as demonstrated by [26]. This work is an effort to compute energy performance indices in the cement manufacturing sector, clearly framing the type of information that can be extracted and the methodology used in the analysis. An empirical analysis aimed at assessing the impact of energy monitoring and energy management systems (EnMSs) for implementing and planning actions to improve energy performance was also developed [27]. In particular, the studies emphasised the richness of the data retrieved from audits, which allow for a delineation of the characteristics of the analysed sectors (plastic and ceramic manufacturing), as well as the relevance of promoting energy monitoring and EnMSs as a pivotal part of energy efficiency policies.

The possibilities offered by the data contained in the audits, together with the wide range of applicable econometric approaches, provide the foundations for a broader research path aimed at identifying a standard methodology for analysing the energy performance of productive sectors.

2.2. The Italian Energy Audit System

For the purposes of this analysis, prior to proceeding with the discussion related to the data employed, it is crucial to provide a framework for a proper understanding of the tool used as a source of information, i.e., the EA, and ultimately how this is regulated in Italy.

Article 8 on EAs and EnMSs of the EE Directive 2012/27/EU [1] imposes two major obligations on Member States: to promote the availability of EAs among final customers in all sectors, and to ensure that enterprises that are not SMEs conduct energy audits at least every four years.

Consequently, EAs become the most qualified tool in the EU for analysing the energy flows of production activities and their management, represented by a "systematic evaluation of energy uses from their point of entry up to their final uses, able consequently to track and quantify energy saving opportunities" [1]. On the one hand, energy audits constitute a tool to understand and represent the structure of energy consumption in the various functional areas of production activities; on the other hand, they support companies in identifying the possible reductions in energy consumption, economic costs, and carbon emissions.

In line with the EU requirements, the Italian Legislative Decree 102/2014 [4] identifies large enterprises and energy-intensive enterprises as the subjects obligated to carry out an EA at their production sites, excluding, however, large enterprises with a total annual energy consumption of less than 50 toe. In addition, the Decree entrusts ENEA as the entity in charge of establishing and managing a database of enterprises subject to audit, with the task of carrying out controls to verify the conformity of the audits with legal requirements.

The standard EN 16247.1 "Energy Audit–General Requirements" [28] and subsequent regulatory tools define the methodology for obtaining a good-quality energy audit of a production site. In this standard, the Italian approach to energy audit has been referred to; indeed, the guidelines for energy audit contents adopted in Italy have been considered to be the best practice in the EU. The audit process involves defining the professional prerequisites of the energy auditor and assigning the responsibility of gathering all information necessary and useful for gaining an understanding of the various areas of energy consumption, along with its supply sources. The role of the auditor is central due to the specific knowledge in managing energy and implementing potential interventions, and the potential to reduce resistance to adopting energy management practices by conducting analyses in the field.

Companies are required to send an audit report to ENEA every four years; the report should gather all the essential information to draw a complete picture of the company's funding characteristics, while also providing a summary of energy consumption and its deployment in the different functional areas. The section of the report aimed at describing the energy efficiency interventions already introduced and those implementable (as identified by the auditor) is particularly relevant. Table 1 lists all the information contained in an audit report.

Table 1. Information contained in the energy audits.

Audit Report Content
1. Note on who drew the energy audit.
2. General data of the company.
3. Data of the production site for which the audit is made.
4. The reference period of the audit.
5. The units of measurement adopted within the audit.
6. Energy consumption.
7. Raw materials used in production processes.
8. Production process description.
9. Description of finished products.
10. Reference energy indicators.
11. Information on data collection method.
12. Description of implementation of monitoring strategy.
13. Energy models.
14. Calculation of energy indicators.
15. Interventions carried out in the past.
16. Individuation of possible interventions.
17. Summary of identified interventions:
a. Net Present Value (NPV);
b. Investment (I);
c. Cash flow (FC);
d. Energy savings;
e. Payback period (TR);
f. IRR.

Source: ENEA, The Energy Audit Guidelines and Operating Manual.

3. Data and Methodology

3.1. Data

By receiving the information resulting from the energy audit obligation from a large number of companies every four years, ENEA manages a unique database, offering the opportunity for an in-depth analysis of the characteristics of the companies and their impact on the past and future evolution of EEMs in different sectors.

This work exploits a cross-sectional database, which consists of the information extracted from the 684 energy audits sent to ENEA in 2019 by over 600 obligated manufacturing enterprises pertaining to the Italian production of plastic articles. The year 2019 represents the starting year of the second obligation period for energy audit policy (after its introduction in 2015) and a significantly higher number of audits has been collected by ENEA relative to following obligation years (energy audits collected in December 2019 are relative to 2018; more information on the obligation years can be found in the report "Rapporto sull'attuazione dell'obbligo di diagnosi" drafted for the Ministry of Environment and Energy Security, available here https://www.efficienzaenergetica.enea.it/servizi-per/imprese/diagnosi-energetiche/pubblicazioni-e-atti.html (accessed on 23 December 2023)). In particular, the sample includes audits received from firms operating in the plastics sector, classified as ATECO 22.21.00, 22.22.00, and 22.29.09, equivalent to NACE C22.1.1, C22.2.2, and C22.2.9. (The ATECO and NACE codes mentioned here refer to specific classifications within the European system for categorising various types of economic activities; ATECO (Attività Economiche) is the Italian version, while NACE (Nomenclature statistique des Activités économiques dans la Communauté Européenne) is the broader European standard.) These codes predominantly include small and medium-sized energy-intensive enterprises, with a high share of multisite enterprises. An in-depth discussion of the business area under consideration will be provided in the next section. The uniqueness of the database is the informative scope of the energy audit, which allows us to capture data inherent to implementable EEMs as well as intrinsic to production sites.

With regard to the former, these cover the description, type, and energy saving of the identified energy efficiency measures, allowing us to consider the technical specificity of the interventions themselves. Accordingly, information is also disclosed on the different types of energy savings that could be attained: annual electric energy savings (toe), annual thermal energy savings (toe), other annual savings (toe), total annual savings (toe), primary electric savings (toe), total primary savings (toe), and annual fuel savings (toe). The database also incorporates details concerning the investment required for EEMs' implementation and the associated net present value, and payback period, calculated without incentives. Cost-effectiveness can thus be derived from the information in the database. In detail, the technical traits of the intervention are considered by classifying the energy efficiency measures into different areas of categorisation: HVAC (heating, ventilation, and air conditioning); building envelope; compressed air systems; engines, inverters, and other electrical installations; electric systems; general/managerial (monitoring, training, energy management system, ISO 50001 [29]); distribution networks; lighting; power factor correction; cold production unit; production lines; intake systems; thermal power plant/heat recovery; transport; production from renewable sources and cogeneration/trigeneration; and a miscellaneous category of measures not elsewhere classified.

With regard to the variables tied to enterprises, ENEA collects information concerning the company name and the ID code for the identification of production sites associated with the same VAT (multi-site companies); the region and province in which the production sites are located; the presence of energy consumption monitoring systems; and the presence of ISO 50001 certification. Finally, it is possible to make a distinction between the obligated parties, i.e., to assess whether the company considered is a large company or an energy-intensive company. Table 2 summarises the different information included in the database.

Table 2. Information obtained from the energy audits.

Information Related to the Firm	Information Related to EEMs
Company name	Investment level
ID code	Payback period (PBP)
VAT	Cost-effectiveness
Region and province	Area of categorisation
Energy monitoring mechanism	Realizable energy savings
ISO 50001 [29]	Net present value
Energy intensity	
Dimension	
Energy consumption level	

Source: Authors' own elaboration based on ENEA data.

Furthermore, for the purposes of this work, energy consumption levels of the production sites considered were included as well (primary electric consumption (toe), total primary consumption (toe), total final consumption (toe), thermal energy consumption (toe), and gas consumption (toe)). This information can be derived from the ENEA energy audit database, as described in [27], in terms of electrical, thermal, and gas consumption levels.

Given that this information for three NACE C22 sectors has never been processed before, considerable effort was required to validate the data. The adjustments that were made to the source database are listed in detail in Table 3.

Table 3. Adjustments applied to the database.

Data Validation Steps
Energy efficiency interventions with missing total energy savings were removed as they were not relevant for the purposes of the analysis.
Interventions resulting from audits not relevant to the investigated obligation period were removed from the database.
Energy savings and investments values, associated with interventions, which were particularly distant from their average (with reference to both tails of the distribution) were checked by directly verifying the contents of the energy audits.
Adjustments were made, where necessary, to the values of energy savings by applying due conversions.
Where investment and cost-effectiveness levels were missing, the energy audits were inspected and, if necessary, the database was integrated on the basis of the content of the audits.
Payback periods were included where missing if related information was present in the energy audits.

Source: Authors' own elaboration.

Finally, to broaden the lens on business attributes that are capable of influencing energy-saving performance, ENEA EA data were merged with data extracted from AIDA (Informed Analysis of Italian Companies) database. More specifically, through the identification code of the companies, information on productivity and sector characteristics was extracted. These turned out to be available for about 400 of the companies subject to the obligation. The merge, in combination with the previous data preparation phases, resulted in a narrower number of observations: 1304 implementable EEMs constituted the sample analysed.

3.1.1. Variable Description

The selection of suitable econometric methods reflected the recognition that the database includes variables connected to the company rather than only to individual EEMs, such as energy consumption or those obtained from the AIDA database. As a result, if more than one intervention is proposed for a company based on the energy audit, then the values of the company-level parameters are reiterated for each intervention. This peculiarity of the data was evaluated along with the possibility that multiple production sites (therefore multiple interventions) are tied to the same enterprise, thus generating the presence of specific uncontrolled elements common to different production sites, able to affect the energy-saving performance.

Based on the previous description of the information that could be extracted from an EA, the predictors used in the models are summarised in Table 4.

Table 4. Summary of the variables included in the logistic and linear regressions.

Predictor Names $(X_1, X_2 \ldots X_k)$	Variable Description	Logistic Regression	Pooled OLS	Pooled OLS (FE)
INVESTMENT	Investment level (EUR) associated with the EEM.	✓	✓	✓
NPV	Net present value of the investment (EUR).	✓	✓	✓
PBP	Payback period associated with the investment (Years).	✓	✓	✓
MONITORING	Dichotomous variable = 1 if the EEM considered comes from a production site subjected to monitoring.	✓	✓	

Table 4. *Cont.*

Predictor Names ($X_1, X_2 \ldots X_k$)	Variable Description	Logistic Regression	Pooled OLS	Pooled OLS (FE)
ISO 50001	Dichotomous variable = 1 if the EEM considered comes from a production site owning a system for managing energy vectors.	✓	✓	
FIXED EFFECTS	N-1 dummy for each production site having the same ID code, so pertaining to the same firm.			✓
NORTH–EAST	Dichotomous variable = 1 if the intervention considered comes from a production site located in the northeastern area of Italy.	✓	✓	✓
NORTH–WEST	Dichotomous variable = 1 if the intervention considered comes from a production site located in the northwestern area of Italy.	✓	✓	✓
SOUTH	Dichotomous variable = 1 if the intervention considered comes from a production site located in the southern area of Italy.	✓	✓	✓
AIDA variables related to firms' productivity				
REVENUES	Company revenues from sales in 2018 (EUR).			
EMPLOYEE	Company number of employees in 2018.			
DEBT_RATIO	Debt ratio indicating the company's total debt, as a percentage of its total assets in 2018.	✓	✓	✓
CAPITAL_TURNOVER	Denotes the number of times the invested capital has returned in the form of sales in 2018.	✓	✓	✓
REVENUES_PROCAP	Company revenues per employee in 2018 (EUR).	✓	✓	✓

Source: Authors' own elaboration based on ENEA and AIDA data.

With the intent of mirroring in the models the impact undertaken by the technical characteristics of the interventions, dichotomous variables indicating the area of categorisation of the energy efficiency measures were introduced (see Table 5). It is thereby possible to verify whether investing in certain areas negatively affects energy-saving performance.

Table 5. Summary of the constructed dummies for intervention areas.

Dummy Variables for Intervention Areas	
HVAC	Dichotomous variable = 1 if the intervention belongs to the Air conditioning area.
Cogeneration/Trigeneration	Dichotomous variable = 1 if the intervention belongs to the Cogeneration/Trigeneration area.
Compressed air systems	Dichotomous variable = 1 if the intervention belongs to the Compressed air area.
Engines/Inverters	Dichotomous variable = 1 if the intervention belongs to the Engines/Inverters area.
Electric systems	Dichotomous variable = 1 if the intervention belongs to the Electric systems area.
General/Managerial	Dichotomous variable = 1 if the intervention belongs to the General/Managerial area.
Other	Dichotomous variable = 1 if the intervention belongs to the Not elsewhere classified area.
Cold production units	Dichotomous variable = 1 if the intervention belongs to the Cold production units area.
Production from renewables sources	Dichotomous variable = 1 if the intervention belongs to the Production from renewable source area.
Production lines	Dichotomous variable = 1 if the intervention belongs to the Production lines area.
Intake systems	Dichotomous variable = 1 if the intervention belongs to the Intake systems area.
Thermal power plant/Heat recovery	Dichotomous variable = 1 if the intervention belongs to the Thermal power plant/Heat recovery area.
Transport	Dichotomous variable = 1 if the intervention belongs to the Transport area.

Source: Authors' own elaboration based on ENEA data.

3.1.2. The Energy Performance

The methodology adopted in this study encompasses various econometric techniques; hence, this section will scrutinise the response variable and its transformations as employed across the distinct methodologies implemented. The response variable investigated here is measured as the ratio between the potential energy savings related to the identified EEMs and the total consumption of energy carriers in the production site considered. In particular, for each individual intervention identified (potential), the total achievable energy savings—expressed as an aggregate of electric, thermal, and other savings—is measured in tons of oil equivalent (toe). This conversion allows us to jointly evaluate the different types of interventions. These primary savings in toe are then divided by the overall level of energy consumption (again considered in terms of primary energy consumption) of the enterprise. Hence, we obtain, overall, the energy performance of the sector that would occur if these interventions were implemented (information on actual implementation will be available through the refinement of data from the 2022 energy audits). Therefore, factors capable of increasing or reducing the sector's potential energy performance are explored. The dichotomous version of the variable, investigated trough the logistic regression, reports a value of 1 if the observed energy performance for a given production site exceeds the average of its distribution. On the other hand, in the OLS model, the response variable is not exhibited as dichotomous, but as a continuous variable; and by examining its distribution graphically, we observed that there is a concentration of observations for values lower than 0.25. This means that the energy performance of companies in the sector is predominantly below this threshold. Consequently, to reduce the incidence of extreme values related to large-scale energy self-production interventions (cogeneration/trigeneration), a logarithmic transformation of the variable has been introduced for the OLS regressions.

3.1.3. The White Certificate Variable

In order to verify the extent to which the presence of policies that encourage the adoption of EE interventions can boost energy-saving performances, a special dichotomous variable was constructed, related to the White Certificate scheme. As highlighted in the introduction, this incentive tool is one of the main policy instruments employed to foster the implementation of energy efficiency measures in the Italian industry. Thanks to the White Certificate guidelines issued by the Gestore Servizi Energetici (GSE) [30], it was possible to detect which EEMs were eligible for this support policy. According to the guidance offered by the GSE and the reported background information, the White Certificate variable was built as detailed in Table 6. NA has been attributed to Building Envelope since no intervention in this area can be associated with the White Certificate policy. Other energy efficiency incentive mechanisms are applicable, namely tax deductions.

For the cogeneration/trigeneration and the production from renewables areas, there are some clarifications to be made. Regarding the former, high-efficiency cogeneration (CAR) under the White Certificate scheme are regulated by the Directive 2004/8/EC, transposed in Italy by Legislative Decree 20/2007, which stipulated that, as of 2011, the condition for which the combined production of electricity and heat can qualify as "High-Efficiency Cogeneration" is based on the primary energy saving (PES) parameter. White Certificates for cogeneration are therefore assigned annually by the GSE, if the PES value is at least 10% or, in the case of micro-cogeneration units (<50 kWe) or small cogeneration units (<1 MWe), when it takes any positive value. In this study, the set of cogeneration interventions benefiting from the White Certificates is thus underestimated, as the information available has not always made it possible to evaluate access to the incentive. Regarding incentives for the installation of energy production plants from renewable sources in Italy, in continuity with Ministerial Decree 06/07/2012 and Ministerial Decree 23/06/2016, from which part of the structure is inherited, the Ministerial Decree 04/07/2019 promotes, through economic support, their diffusion. The plants eligible for incentives under the Decree are described in different categories, including newly built photovoltaic ones. Eligible to apply for incentives will be only those plants that were placed in a useful position in the rankings of one of seven competitive registry or downward auction procedures on the value of the incentive, drawn up by the GSE based on specific priority criteria.

Table 6. Description of the construction of the White Certificate variable.

Intervention Areas	White Certificate Variable
HVAC	1 if the intervention concerns a heat recovery and/or refrigeration system; 0 otherwise.
Building envelope	NA
Cogeneration/Trigeneration	1 if the audit reveals PES greater than 10% and/or if the plant capacity < 50 kWe or <1 Mwe; 0 otherwise.
Compressed air	1 for intervention on compressed air/compressor replacement; 0 otherwise.
Engines/Inverters	1 for motor/inverter replacement if savings > 5 toe; 0 otherwise.
Electric systems	1 only for power quality interventions 0 otherwise.
General/Managerial	1 per each introduced management and monitoring system; 0 otherwise.
Lighting	1 for LED introduction if energy savings > 5 toe; 0 otherwise.
Cold production units	1 per replacement/introduction of refrigerant system; 0 otherwise.
Production from renewables	1 if the intervention entails the installation of a new plant; 0 otherwise.
Production lines	1 only in case of press replacement excluding hydraulic presses; 0 otherwise.
Intake systems	0 for all the interventions.
Thermal power plant/Heat recovery	1 for any registered heat recovery and for EEMs referring to process chiller; 0 otherwise.
Transport	0 for all the interventions in light of the fact that the interventions in question are not eligible for a White Certificate.

Source: Authors' own elaboration based on ENEA data.

3.2. Methodology

Econometric Approach and the Energy-Saving Performance

The literature reviewed in the previous section has revealed a prevalence of the use of nonlinear regression models that exploit a binary response variable that is equal to 1 in the case of the implementation of the EEM, and 0 otherwise [8,10,17,18]. In parallel to these perspectives, the present paper involves the use of a logistic regression model. In the current study, the primary variable of interest is quantified as the proportion of potential energy savings attributable to the proposed EEMs relative to the aggregate energy carrier consumption at the production facility in question. Thus, the main purpose of this modelling choice is to assess the presence and impact of relevant factors (inherent to both the interventions and the companies themselves) capable of stimulating above-average energy-saving performance.

To carry out the analysis, the RStudio software (version 2023.09.1) was utilised. RStudio is a widely used statistical software used to perform various statistical tasks. The sensitivity of the logit models to the choice of threshold for the dichotomous response variable was tested relative to the alternative option of a binary variable, which assumes a value of one if energy performance is greater than the median of its own distribution. Although the obtained significances are in line, the latter have a much lower goodness of fit.

The logistic regression is compared with an ordinary least squares (OLS) linear regression model, an embryonic approach chosen for its capacity to shed light on future analytical directions, especially in the absence of benchmark studies using similar datasets. In adopting this approach, we align with the methodologies utilised by the authors of [21], who demonstrated the efficacy of linear regression in elucidating the influences on energy-saving performance within manufacturing sectors. Indeed, we introduce fixed effects for the site identification code in our linear regression formulation to account for the presence of multisite enterprises, a decision that echoes Curtis and Boyd's considerations of uncontrolled common characteristics among production activities that can potentially influence energy-saving performance. To ensure the robustness of the OLS model, as fixed-effects control for common factors of production sites belonging to the same company, the variables related to monitoring and ISO 50001 are removed. Finally, for the latter model, logarithmic transformation is applied to the response variable (the performance) in order to reduce the incidence of extreme values related to large-scale energy self-production facilities. Based on the results in terms of goodness of the models, the fixed effects were removed from the logistic regression and kept instead in the OLS model.

In detail, the logistic regression model is formulated as shown in Equation (1):

$$P(y_i = 1|X_i) = \frac{e^{\beta_0 + \beta_1 X_1 + \beta_2 X_2 \ldots B_k X_k}}{(1 + e^{\beta_0 + \beta_1 X_1 + \beta_2 X_2 \ldots B_k X_k})} \quad (1)$$

where $P(y_i = 1|X_i)$ is the odds that, given the set of explanatory variables $(X_1, \ldots, X_k)$ described in the previous section (investment, PBP, White Certificates, etc.), the intervention i will result in an above-average energy-saving performance (y), which can be written as follows:

$$Performance_{i,f \ (x = \frac{Total\ savings\ (toe)_i}{Total\ consumption\ (toe)_i})} = \begin{cases} 1; \ x > mean(x) \\ 0; \ x \leq mean(x) \end{cases} \quad (2)$$

In which i represents the index of the i-th intervention and f represents the f-th firm.

While the OLS model is depicted by Equation (3), Equation (4) illustrates how the regression differs when we introduce fixed effects for the enterprise ID code, where the X stands for the chosen predictors, β for the associated coefficients, ϵ is the residual error term, while γ denotes the fixed effects, a set of unobserved variables that are constant over time for each observed unit, but vary between different companies f.

$$Log(Y_{i,f}) = \beta_0 + \beta_1 X_1 + \beta_2 X_2 \ldots + \beta_k X_k + \varepsilon \quad (3)$$

$$Log(Y_{i,f}) = \beta_1 X_1 + \ldots + \beta_k X_k + \gamma_f + \varepsilon \quad (4)$$

In the OLS models, the response variable is not exhibited as dichotomous but as a continuous variable to which we applied the logarithm. The reference period for each variable was not detailed in the formulation of the model since the data only refer to the year 2018. All the models are estimated using maximum likelihood (ML) estimation method.

4. Empirical Results

4.1. Descriptive Analysis of the Italian Plastic Production Sector

As for the economic magnitude of this industrial branch, according to the Italian National Statistic Institution [31], in Italy in 2019, the entire plastic processing sector involved almost 10,000 enterprises with more than 175,000 employees. The companies also show increasing levels of profitability with a dependence on imports higher than their contribution on exports. There is a clear concentration of production activities in the northern regions of the peninsula, accounting for more than half of the total. The sampled businesses account for about 5 percent of the total number of production sites belonging to NACE code 22.

For the 453 companies included in the final sample obtained after the merge, AIDA data highlighted that in 2018, they employed 35,527 individuals for an average revenue per employee that reached EUR 370.664. Total turnover exceeded EUR 11 million. The companies also displayed a turnover rate of invested capital of 1.14, interpretable as an efficient ability to achieve high levels of revenue by rationally deploying available resources, confirming the dynamism of the sector itself.

The energy consumption levels of these enterprises reiterate the importance of the sector in terms of energy efficiency potential. Electricity constitutes most of the energy uses of the Italian plastic manufacturing sector, followed by thermal and gas consumption, indicating a strong electrification of the entire production system.

The energy savings flows are equally relevant: 70,231.97 toe are the primary savings that can be achieved through EEMs, which are inherent in the implementation of energy self-production interventions (renewable energy production and cogeneration/trigeneration plants) and 13,003.85 toe of final energy savings achievable by deploying all the other EEMs. These stylised facts relate to a period prior to the economic contractions caused by the COVID-19 pandemic, which, together with the current energy crisis hampering production, increased the uncertainty associated with production activities. Uncertainty pertains also to the potential future progression of taxation on plastics (currently applied only to single-use products). More information is included in Tables 7 and 8. In the first table, potential savings are calculated in final energy for all areas except for two of them; for these two areas, involving self-production of energy, potential savings are shown in the second table in primary energy.

Table 7. Summary of identified interventions pertaining to final energy savings.

	Interventions	Interventions, %	Energy Savings (Final Energy Savings, Toe)	Savings, %	Average Cost-Effectiveness	Investment (EUR)
HVAC	46.0	1.76	1247.0	1.77	4043.66	5,751,737.00
Building envelope	7.0	0.27	177.2	0.25	10,634.08	946,274.00
Compressed air	248.0	9.51	8414.8	11.96	5345.02	22,365,921.38
Electric system	127.0	4.87	3229.0	4.59	6673.73	10,723,177.41
General/Managerial	166.0	6.37	16,086.0	22.86	4861.37	22,510,741.31
Lighting	298.0	11.43	16,533.7	23.50	6276.83	35,104,974.62
Other	6.0	0.23	127.7	0.18	4468.71	557,059.00
Cold production units	44.0	1.69	2695.4	3.83	8215.31	5,643,930.00
Productive lines	117.0	4.49	8150.3	11.58	6693.79	21,466,966.53
Intake systems	16.0	0.61	116.2	0.17	6476.49	610,861.00
Thermal power plant/Heat recovery	37.0	1.42	4068.2	5.78	4609.74	8,095,459.00
Transport	30.0	1.15	100.9	0.14	5959.97	490,400.00

Source: Authors' own elaboration based on ENEA data.

Table 8. Summary of identified interventions pertaining to primary energy savings.

	Interventions	Interventions, %	Energy Savings (Primary Energy Savings, Toe)	Average Cost-Effectiveness	Investment (EUR)
Cogeneration/Trigeneration	68.0	2.61	3465.3	6162.90	12,219,494.44
Production from renewable sources	171.0	6.56	10,215.2	6973.03	25,159,500.59

Source: Authors' own elaboration based on ENEA data.

Focusing on the different areas of energy efficiency interventions, the values reported in Table 7 show the high numerosity of interventions concerning the lighting system, which constitute 11.4% of the total, followed by those aimed at improving compressed air (9.5% of the total) and those concerning the general/managerial area (6.4% of the total). Slightly less numerous are instead the EEMs associated with production lines (4.5%), signalling opportunity for companies to review production processes to improve the efficiency in the use of resources, including energy vectors. The terminology of the above areas should not mislead readers, as EEMs could refer to extensive and complex operations as well as to simpler ones. For example, energy efficiency in production lines could be attained through the replacement of a melting furnace with another one with a higher energy efficiency. In contrast, energy efficiency in the compressed air area is often associated with leak detection, which is a relatively simpler form of intervention. The production of energy from renewable sources, on the other hand, includes 171 interventions (6.6% of the total), many of which refer to installing photovoltaic panels, which increases the company's independence from the purchase of energy flows. Many efficiency measures have adaptable characteristics. For example, heat recovery can also be implemented by referring to compressed air or production lines equipment, making it relevant not only to the thermal power plant/heat recovery area, but also for the others. Therefore, the absence of such measures does not necessarily imply that there are no identified interventions that could affect the above areas.

The replacement of lighting equipment with LED is a widely identifiable and feasible intervention that would lead to the saving of 16,533.7 toe. However, it is necessary to underline how an excessive emphasis on such interventions would prevent the delineation of more efficient technological trajectories pertaining to the production processes; such emphasis is already supported by the relatively low cost of such interventions.

Concerning the extent to which the concept of innovation is extended to the introduction of new organisational forms, the general/managerial area interventions provide a better understanding of the current management of energy flows; by broadening the knowledge base, the search for new rationalisation mechanisms is stimulated. The 166 interventions identified in relation to the introduction of new organisational practices, energy management tools, and the adoption of ISO 50001 certification (general/managerial area) would allow for a reduction in consumption by 16,086.0 toe, confirming that there is large room for improvement when a more organised and careful management of energy sources is implemented within the company.

The greatest source of savings is identified instead in the area pertaining to the installation of cogenerators and trigenerators. In recent years, these systems have been the subject of in-depth studies within the European Union with the objective to identify new ways of achieving energy efficiency in buildings and industry, as they allow for the combined and simultaneous production of electrical and thermal energy, thus increasing the efficiency of installations and optimising energy self-sufficiency. This type of intervention is characterised by its criticality, relatively large scale except for micro-cogenerators, and long technical life and payback period. As a result, such measures may be implemented with resistance, which is exacerbated by the current economic conditions.

However, the implementation of the above measures would lead to a reduction in consumption of 3465.3 tons of oil equivalent for an equally important investment of EUR 12,219,494.44. Similarly, production from renewable energy sources involves a high level of initial investment, equivalent to EUR 25,159,500.59, and relatively long payback periods, which could be shortened using existing incentives. Accordingly, the associated energy savings reach 10,215.2 toe. These EEMs offer significant benefits, not only in terms of reduced consumption, cost reductions, and productivity improvements, but also in terms of competitiveness and the ability of the companies to meet their social and environmental commitments. As highlighted in Figure 1, electricity flows not only account for most of the consumption in the three NACE sectors considered, but also have the most significant savings potential; in fact, the total energy savings is given by the sum of electrical savings, which accounts for more than half of the total (75%), thermal savings, fuel savings, and other savings.

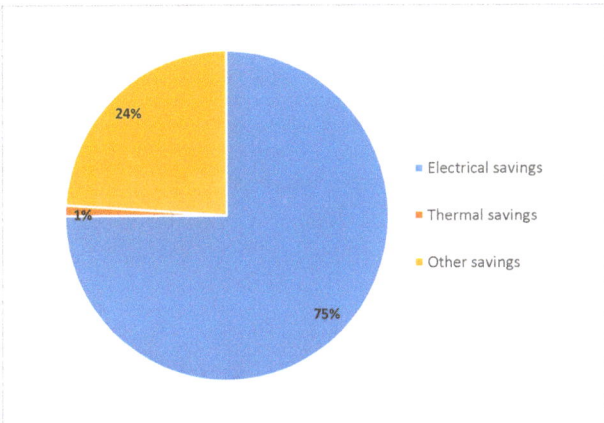

Figure 1. Breakdown of energy savings (%, calculated on savings expressed in toe). Source: Authors' own elaboration based on ENEA data.

The observations obtained from the audits also display a concentration of identified interventions related to low levels of investment and savings; overall, the companies subjected to energy audits with an investment of EUR 193,525,069.46 would be able to obtain 13,680.5 toe of primary energy savings, associated with the areas cogeneration/trigeneration and production from renewables, and 70,361.2 toe of final energy savings, relative to all other intervention areas. These savings represent an upper threshold, since not all the EEMs identified in the energy audits will be chosen for implementation.

Furthermore, given the relevance of the cost-effectiveness to business decisions regarding the adoption of EEMs, it is possible to see, as shown in Figure 2, how most interventions display levels of such indicators around EUR 5001 and 15,000, signalling a positive economic performance for many of the interventions (this comparison excludes those sectors where the financial characteristics of the investments are not comparable on a large scale, such as production of renewable energy and CHP/trigeneration.

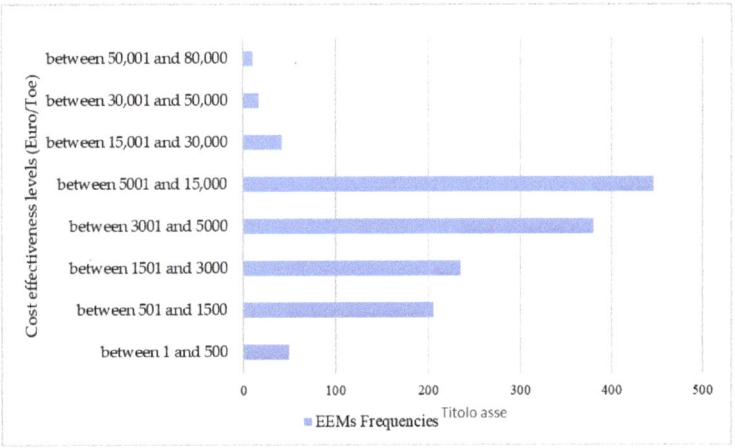

Figure 2. Cost-effectiveness levels (EUR/toe) associated with identified EEMs. Source: Authors' own elaboration based on ENEA data.

Turning our attention to the relationship between energy savings and the payback period of the investments necessary for their realisation, the weight assumed for the total savings by cogeneration/trigeneration measures and those related to the self-production of energy through production from renewables prompted us to exclude them from Figure 3. In Figure 3, we can observe a clear concentration of the achievable savings on investments characterised by medium–long payback periods (up to 5 years), which may constitute a potential constraint for the propensity to adopt the efficiency measure and for the realisation of the savings themselves. At the same time, about 26.84% of the potential savings can be achieved through less risky investment options from the point of view of business decision makers and thus characterised by shorter payback periods (up to 3 years).

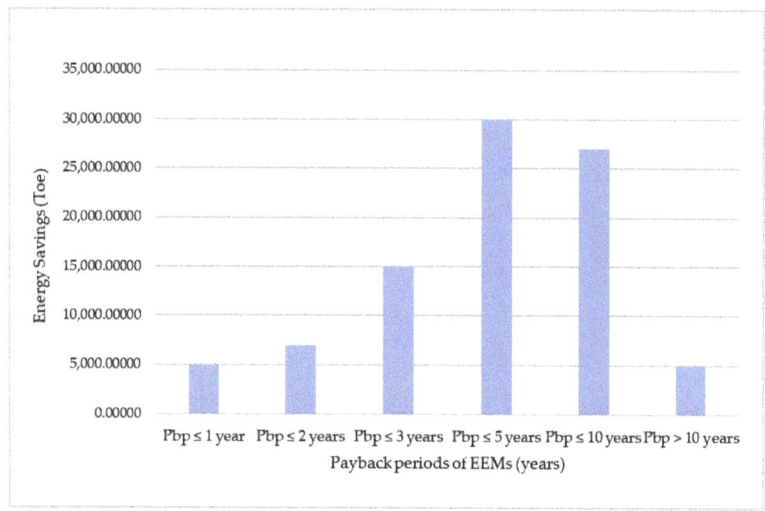

Figure 3. Payback periods associated with identified EEMs. Source: Authors' own elaboration based on ENEA data.

The regional distribution of the identified EEMs (the regional distribution refers to all energy efficiency interventions identified by the energy audits received by the plastics sector in 2019) (Figure 4) reflects the location of examined companies in the most industrialised regions of the Peninsula, namely in Northern Italy.

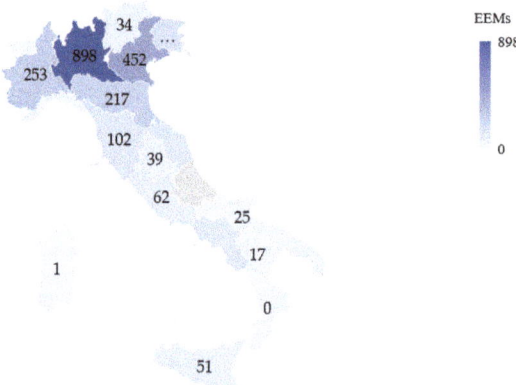

Figure 4. Regional distribution of identified EEMs. Source: Authors' own elaboration based on ENEA data.

Indeed, the analysis shows a clear concentration of identified interventions in Lombardy, Piedmont, and Veneto. Such EEMs could be implemented or not depending also on the commitment of regional governments to promote the ecological and energy transition and the consequent reduction of environmental impacts. No interventions and therefore savings are found instead for the regions of Calabria and Val D'Aosta, from which no energy audits belonging to the examined sectors were received.

4.2. Econometric Results

This section is dedicated to the discussion of empirical findings obtained through the application of the methodologies described in Section 2. The approaches employed have confirmed the great potential of the analysis of EA information, combining elements associated with the intrinsic characteristics of companies and information on measures to achieve energy savings, capable of driving the energy performance of the plastics manufacturing sector. Results are shown in Table 9. In the field of econometrics, the Breusch–Pagan (BP) test is utilised to examine the presence of heteroskedasticity, which refers to the situation when the variance of the residuals is not constant, within a linear regression model. The test is conducted in order to evaluate the null hypothesis of homoskedasticity. In this specific case, the BP test is applied to the pooled OLS model, resulting in a BP value of 42.532, accompanied by a p-value of 0.01124. Consequently, we reject the null hypothesis of homoskedasticity and reach the conclusion that heteroskedasticity does exist in our pooled OLS model. Considering the potential limitations of the OLS model identified through the BP test, we opt for the adoption of a nonlinear logistic regression model, incorporating a binary response variable. The results also report the mean values of the variance inflation factor (VIF) for the ordinary least squares (OLS) model. These metrics are used to gauge the degree of multicollinearity among the predictor variables. In this particular case, the low VIF values observed (1.460, 1.398, and 1.390) suggest that the variables included in the model exhibit a significant degree of independence from one another. In all models, the potential savings are calculated as primary energy savings for all interventions areas.

Table 9. Summary of econometric results.

	Pooled OLS	Pooled OLS Fixed-Effects	Logistic Regression
n (Investment)	0.626 *** (0.025)	0.870 *** (0.018)	1.057 *** (0.118)
White Certificate	0.161 ** (0.077)	0.053 (0.048)	0.224 (0.307)
PBP	−0.081 *** (0.012)	−0.103 *** (0.009)	−0.101 ** (0.046)
Cost-effectiveness	−0.00003 *** (0.00000)	−0.00004 *** (0.00000)	−0.0001 *** (0.00004)
MONITORING	−0.437 *** (0.069)		−1.070 *** (0.247)
ISO 50001	0.098 (0.169)		0.147 (0.685)
NORTH–EAST	0.036 (0.092)	−4.829 ** (2.239)	0.331 (0.329)
NORTH–WEST	0.159 ** (0.075)	−9.790 *** (1.342)	0.125 (0.271)
SOUTH	−0.143 (0.131)	−7.698 *** (2.640)	0.554 (0.407)
DEBT_RATIO	0.008 *** (0.002)	−0.086 ** (0.034)	−0.003 (0.009)

Table 9. Cont.

	Pooled OLS	Pooled OLS Fixed-Effects	Logistic Regression
CAPITAL_TURNOVER	−0.027 (0.082)	−4.782 *** (1.494)	−0.146 (0.287)
HVAC	0.350 * (0.179)	0.264 ** (0.111)	0.757 (0.752)
Compressed Air	0.559 *** (0.096)	0.558 *** (0.059)	0.417 (0.660)
Cogeneration/Trigeneration	1.568 *** (0.179)	1.057 *** (0.118)	3.653 *** (0.818)
Thermal Power Plant/Heat Recovery	1.108 *** (0.203)	1.456 *** (0.127)	2.957 *** (0.650)
Engines/Inverters	−0.130 (0.114)	0.073 (0.071)	1.039 * (0.569)
Electric Systems	0.460 *** (0.165)	0.084 (0.104)	1.778 *** (0.527)
General/Managerial	0.747 *** (0.111)	0.630 *** (0.069)	1.294 ** (0.554)
Production from Renewables	0.675 *** (0.132)	0.033 (0.087)	2.167 *** (0.502)
Cold Production Units	0.337 * (0.178)	0.252 ** (0.110)	1.492 ** (0.582)
Production Lines	0.486 *** (0.121)	0.326 *** (0.076)	2.000 *** (0.512)
Intake Systems	−0.284 (0.267)	−0.064 (0.173)	−11.192 (577.260)
Transport	−0.241 (0.204)	−0.187 (0.119)	2.544 ** (1.030)
CONSTANT	−10.129 *** (0.309)		−12.283 *** (1.272)
Observations	1320	1320	1320
R2	0.627	0.992	
Adjusted R2	0.620	0.988	
Akaike Inf. Crit			52.000
Residual Std. Error		1.070 (df = 1294)	0.541 (df = 891)
F-Statistic		87.110 *** (df = 25; 1294)	248.279 *** (df = 429; 891)
Studentised Breusch–Pagan Test	BP = 42.532, (df = 24, p-value = 0.01124)		
Average VIF Value	1.460	1.398	1.390

Notes: *, **, *** represent, respectively, statistically significance levels of 0.05, 0.01, and 0.001; a dot (.) represents a significance level of 0.1. These markers provide a quick visual cue to assess the statistical robustness of the findings. Source: Authors' own elaboration.

First, the coefficients related to the economic dimension of the energy efficiency measures show very encouraging results. All the specifications used led us to observe a significant and positive role of the investment level in influencing the future level of energy performances in the sector of plastic production: the greater the initial expenditure made for the implementation of the measures, the higher the likelihood of businesses experiencing a higher energy performance. In the pooled OLS model without fixed effects, the impact of the investment level on the dependent variable can be interpreted in terms of elasticity, as the logarithm is applied to this predictor (to reduce the impact of observations particularly distant from the mean). Therefore, a one-percent variation in investments is associated with an increase in savings performance equal to 0.63%. As for the OLS model including fixed effects, the elasticity among the two variables concerned reaches 0.87%, while in the logistic regression, if the investment carried out by the production sites increases, the expected percentage change in the probability that the energy-saving performance is above its average is about 187.77%.

Moreover, if the payback period (PBP) increases, the financial concern for the company decision makers increases as well, discouraging the implementation of EEMs. In the OLS without fixed effects, we observe a 0.08% reduction in the dependent variable related to a unit increase in the predictor, while once we control for firms' characteristics, this percentage reduction is equal to 0.10%. Considering the logistic formulation, the same alteration of PBP leads to a higher expected percentual change of the odds ratio, consisting of 10.62%. Overall, the outcomes relating to PBP are in line with the empirical findings of [10], confirming a role of the risk level of EEMs, therefore inherent to their different technical nature and in some cases to their economic dimension. Thus, the control effectively acts as a barrier to the adoption of EEMs.

The existing relationship between costs faced by companies and the attainable benefits proves to be a widely explored relation in the literature on energy efficiency, with the objective of reducing the resistance to the adoption of EEMs. This study, using the information extracted from the audits, contributes to empirically demonstrate the beneficial impact, albeit small, that increases in cost-effectiveness produce on energy-saving performances. All the econometric models considered confirm the significance of this parameter.

Moving on to the variables inherent to the firm's traits, in the OLS without fixed effects and in the logistic regression, it is possible to recognise the role played by the parameter monitoring, which brings a reduction in energy-saving performance (0.44% in the OLS, 191.53% in the odds ratio), although minor with respect to the other company attributes examined. Nevertheless, this does not directly imply that the monitoring of production activities is detrimental to the development of energy efficiency pathways. On the contrary, this could be related to the fact that companies already having a monitoring system have more accurate estimations of the potential savings associated with identified EEMs. For this reason, the saving potential and the performance could be adversely impacted, resulting in a means that is more "realistic".

The dummy variables representing the geographical location of the firms from which the audits, and so the relative interventions, have been received, assume distinct roles in the models considered. In the linear regression formulation with fixed effects, we clearly see the significance of all the three geographical macro areas. In the northeast, compared to all other locations, including those referring to the central part of Italy used as the reference area, a 4.83% decrease in energy performance is observed. The drop is even higher when considering the southern regions of the peninsula instead. In fact, the south parameter resulted in a 7.7% reduction in the level of energy-saving performance, compared to the other macro regions. Finally, for the north–west variable, the highest relative decrease in the dependent variable was found to be around 9.8%. Alternatively, when examining the model without fixed effects, only one macro area assumes significance, namely the northwest area, while no significance is observed for the geographic areas in the logistic modelling choice.

It is interesting to note that all these variables become significant only when we include a component that allows us to account the fact that the production sites belong to the same company, and thus the existence of common strategies and elements. Nonetheless, the contradictory results obtained by the various econometric models suggest the possibility of refining the approach to better reflect the role of the geographical location of firms and the consequences of their geographical proximity, by relying on spatial econometric methods.

Moving to the variables obtained from the AIDA database, the related parameters assume relevance exclusively for the linear regression models. Regarding the outcomes achieved, including the firm's fixed effects, we observe that the coefficients associated with the variables debt ratio and capital turnover are statistically significant; in particular, this result indicates that the greater the number of times in which the company can recover the capital invested in management through sales and revenues, the lower the rate of achievable energy-saving performance. Despite the general expectation that firms with higher profitability—and in this case, a higher aptitude for dynamic cash flows—face fewer financial constraints and thus are more inclined to adopt EEMs, the empirical evidence in several contributions, especially for manufacturing SMEs, did not confirm this hypothesis. The analysis detects a clear negative role of capital turnover, indicating how an efficient management is not necessarily linked to increases in energy efficiency: in fact, a percentage reduction of the performance equal to 4.78% has been found for the capital turnover unitarian variation.

Shifting our attention to the impact of the financial liabilities of firms (debt ratio) in the literature, this impact is generally assessed in terms of whether the availability of financing affects energy efficiency pathways, alternatively measured by considering the company's profitability. Companies encounter numerous challenges related to debt servicing due to a relatively higher share of liabilities, which discourages the adoption of the measures. Regardless, the contributions on the issue once again offer opposing propositions; for instance, the authors of [32] found that profitability has a deleterious influence on energy efficiency investments, while the authors of [33] did not find that the debt-to-equity ratio behaves as a real barrier. Concerning the finding obtained here for indebtedness, the variable debt ratio assumes relevance in both OLS approaches, but once again, conflicting results are found. In the modelling choice without fixed effects, a unit increase in liability leads to a slight percentage increment in the dependent variable (0.008%). In the alternative modelling choice, controlling for factors common to the production sites examined, the coefficient of this variable becomes negative, although it remains particularly low (-0.086%). These results suggest that the influence of an increase in the amount of debt-financed assets has a different impact on energy-saving performance depending on the characteristics of the firm.

As for the results inherent to fixed effects themselves (OLS model), since these dummies are constructed by exploiting the site ID code, they cannot be explicitly listed in this analysis as particularly sensitive information would be displayed; therefore, here, we simply confirm that they are significant for the majority of the sites considered, validating the existence of specific features, not controlled by the other predictors, common to sites belonging to the same VAT. These findings open the way for further investigation of the characteristics capable of generating better energy performances.

The intervention areas show different behavioural patterns, although most areas are significant in all the models exhibited. Beginning with the OLS model without fixed effects, we observe how, relative to all the other intervention areas and to the baseline area (represented by lighting, not incorporated in the modelling), cogeneration/trigeneration, thermal power plant/heat recovery and general/managerial assume the highest positive impact on the response parameter. Regarding the first area, such a result is not surprising given the magnitude of the measures themselves. However, it is interesting to observe the role played by the thermal power plant/heat recovery area; in fact, despite gathering 1.4% of the total identified EEMs and corresponding to 5.78% of the potential energy savings, the performance variation associated with its own unit variation is relatively high and equal to 1.11%. With regard to the general/managerial area, to quantify the associated 0.71% increase in energy performance, it is necessary to consider that the total number of interventions in this area would allow us to achieve 22.86% of potential energy savings. Production from renewable sources, similarly to the other areas, also positively affects the response variable. In the case of the fixed-effects OLS model, the findings are reasonably close to the previous ones, while in the logit model, the intervention areas where very interesting behaviours can be observed are production lines, transport, and cold production units. The last two areas account only for 1.15% and 1.69% of the total interventions individuated and are associated with an expected change in odds ratio of about 1173.05% and 344.59%, respectively.

No significance was found, in any of the model specifications considered, for the NPV associated with the energy efficiency interventions and for the per capita revenue of the production sites.

Finally, regarding the incentive policy tool included (White Certificate), its significance is detected only in the OLS formulation (without fixed effects); in particular, the coefficient suggests a performance change of 0.16%. It is, however, necessary to say that these outcomes are influenced by the assumptions underlying the construction of the White Certificate variable, which leads to an underestimation of the interventions incentivised. Moreover, the information content of the variable itself is relatively poor: more information on the economic dimension of these incentives would enrich the analysis. In addition, the literature on the subject still seems far from including instruments reflecting national and/or regional policies, but the results obtained confirm the need to broaden the prospects for the adoption of EEMs.

5. Discussion

This paper leverages a comprehensive and unique database, comprising information gathered from energy audits, a pivotal tool currently utilised for assessing energy consumption across productive sectors. The database specifically focuses on energy consumption and potential EEMs based on audits received by ENEA from the plastic production sector in 2019. The empirical methodology is anchored in econometric techniques, aimed at identifying the characteristics of companies, the socioeconomic context, and EEMs that are likely to yield above-average energy savings. The study critically examines the role of incentive policies, albeit limited to the introduction of a singular instrument, the White Certificates, at a national level. This presents an opportunity to enhance the approach by broadening the spectrum of incentive mechanisms in the model and acknowledging the influence of regional policies common across various production sites, potentially resulting in varying energy-saving performances. Moreover, the information encapsulated in the White Certificate variable could be enriched by considering the economic magnitude of such political support, rather than its mere presence or absence. This, however, necessitates comprehensive information and a data management process; indeed, it should be verified for each EEM if the White Certificate amount is included in the business plans developed in the energy audits, and then a new variable should be added to the database.

The research investigates the existence of barriers and driving factors for the adoption of measures aimed at reducing energy consumption, employing both logistic regression and linear models, resonating with the findings highlighted in the literature review. The latter model is distinguished by the inclusion of fixed effects for site identification codes, exploring potential shared characteristics among production sites within the same corporate entity. Despite testing different methodologies, the empirical results can be synthesised by acknowledging that characteristics at both the company and EEM levels significantly influence the energy efficiency trajectories of production sites. The presence of energy monitoring mechanisms, interestingly, exhibits a negative impact on the response variable. These results, however, must be contextualised considering that our dependent variable is a ratio, with the denominator representing the companies' consumption flows. This variable has greater weight than the potential savings constituting the numerator, particularly since a considerable portion of the EEMs in the database, individually, do not project exceedingly high savings in tons of oil equivalent. This indicates that monitoring mechanisms enable more precise energy savings estimates, thus enhancing the diagnostic information's quality.

The analysis also incorporates parameters pertaining to the economic productivity of the companies from which the audits were sourced, utilising an additional database on Italian companies, AIDA. A specific focus was placed on the role of companies' indebtedness;

an increase in a company's financial liabilities impacts energy performance positively in a general context but negatively when the unique characteristics of the company are considered. The interaction among companies' traits and substantial indebtedness may alter their business strategies, diverting focus from energy efficiency investments to other areas, driven by the need to manage the financial burden of debt, including interest payments and debt repayments. This shift potentially diminishes their capacity to invest in energy efficiency projects, with companies possibly prioritising debt management and creditor repayments over EE investments, adversely affecting their overall energy performance. These findings suggest the importance of technical assistance to companies, for example, in the form of energy consulting or industrial expertise, and the useful role of one-stop shops in supporting the identification and implementation of effective energy solutions, mitigating reliance on indebtedness for competencies or resources. Concurrently, encouraging the sharing of best practices among companies enhances learning and may improve EE without necessitating significant investments. In this context, the significance of Article 14 of the Decree-Law n. 17/2022 becomes apparent, promoting EEM investments in Southern Italy under specific conditions. Although a modest allocation of EUR 145 million has been earmarked for 2022 and 2023 in the form of a tax credit, the impact assessment of these policies opens avenues for future research. Since 2016, several regional calls have also been issued to finance energy audits in SMEs and sometimes also in large enterprises; in some cases, the financing was also granted for implementing EEMs identified in the energy audit. The access to such calls has been relatively limited as well as their effective results [34]. Both types of policies could not be included in the examined database.

Regarding capital turnover, an indicator of managerial efficiency, it paradoxically exerts a negative influence on our response variable. This is counterintuitive to the expectation that more dynamic companies, with fluid cash flows, would be predisposed to invest more substantially in energy efficiency measures. This anomaly could stem from prioritisation in investment choices, with companies having higher capital turnover possibly favouring investments in non-energy efficiency areas.

Another consideration relates to the investment payback period; companies with high capital turnover might gravitate towards investments with shorter recovery durations, seeking swift financial returns. Investments in energy efficiency, conversely, might necessitate longer recovery times due to higher initial costs and the gradual accumulation of energy savings. Simultaneously, per capita revenues seemingly do not influence the energy behaviour of companies, underscoring that the decision to implement measures is derived from evaluating multiple aspects, not necessarily correlated to the company's profitability.

Contrary to the findings of [23], the presence of certification related to the implementation of an energy management system (ISO 50001) did not yield conclusive results in this study. The paper also aligns with the findings of [10,22], emphasising the economic/financial dimension linked to EEMs in determining energy trajectories: higher investment levels lead to enhanced energy-saving performances. Generally, an extended period required for companies to recoup the initial expenditure acts as a deterrent to implementing EEMs. The associated risk with a longer payback period thus emerges as a significant barrier, its impact varying based on the technical nature of the measure. These findings resonate with [35]: "The probability of an investment being made decreases as the payback time in years increases, while sensitivity to these increases".

Regarding the categorisation of intervention areas, the results seem particularly noteworthy. The general categorisation area reveals that modifying energy consumption behaviours is not only broadly implementable in the sector, but also capable of affecting substantial improvements in energy performances. This mitigates concerns about the sector's future energy trajectory, particularly considering the significance of economic barriers. Furthermore, the challenge in estimating benefits associated with this intervention type affects the analysis, necessitating a nuanced evaluation of payback periods information, considering these limitations.

Geographical location, delineated by administrative divisions, exhibits distinct "genetic" traits, such as local policies, existing industrial clusters, and perceptions of necessary actions to combat climate change. These factors either directly or indirectly foster or impede future energy efficiency trajectories. Consequently, geographical variables were integrated into the models developed in this study. These variables were derived by assessing the origin of the identified measures, controlling for the macro-regional area of the companies using dichotomous variables. In summary, based on the coefficients obtained, firms in the northwestern region exhibit the lowest average energy performance, followed sequentially by those in the south and the northeast. These results validate our anticipation of regional disparities, pinpointing relatively inferior performance in the northwest. Such discrepancies are not explained by the quantity and scope of energy efficiency interventions or the number of companies, suggesting the presence of unaccounted factors in this context that might hinder the adoption of energy efficiency measures in northwestern companies. The significance of these indicators stresses the potential for further refinement of the methodology to capture the role of these spatial components more accurately, i.e., the attributes that define the genetic makeup of a specific geographical location in determining the energy performance of companies.

Furthermore, the White Certificate variable is observed to have a positive, albeit small, impact on performance. This finding necessitates an interpretation mindful of the inherent limitations of the variable. Our available data allowed only for an assessment of the obtainability of the White Certificate, not its economic value. Acquiring comprehensive information on the incentives received by companies, especially those influenced by regional policies, is a difficult task; data are scarce and elusive to ascertain. While this analysis underscores the need for more detailed and refined incentive data, it also tentatively supports the hypothesis that an incentive model based on achieved savings might not realise the anticipated impact. When the unique characteristics of companies are factored in through fixed effects, the incentive ceases to significantly influence performance. This highlights how pathways to energy efficiency can differ markedly across companies in this sector, an insight that this differentiation should inform the design of targeted policies. The heterogeneity observed at the sectoral level, as underscored by the authors of [23], reinforces these considerations, although it lies outside the current analysis's scope.

The literature published on EE improvements encompasses a diverse array of insights and conclusions. Yet, few studies venture into the realms of spatial and political contexts, integrating diverse contextual elements beyond market conditions and competitive dynamics into their analyses. The application of econometric models to evaluate viable pathways for energy efficiency is an area ripe with potential, particularly regarding the incorporation of models capable of recognising spatial correlations. Such approaches could broaden the contextual elements typically considered in models, providing enhanced opportunities to leverage robust data maintained by agencies tasked with managing energy audit databases. This study not only showcases some of the feasible analyses that can be conducted using these rich data, but also introduces a broad spectrum for future research endeavours.

6. Conclusions

Our research employs a range of econometric approaches, including three distinct models, to thoroughly examine the influence of economic, technical, and contextual factors on energy performance in the Italian plastics sector. This paper is rooted in data extracted from the 2019 energy audits of a representative sample of companies in the Italian plastics industry, encompassing over 600 enterprises. These data were further combined with management-related information extracted from the AIDA database. The analysis highlighted in multiple phases how the quality and homogeneity of EAs are crucial for the future use of these data, which possess significant analytical potential.

However, we faced significant challenges in accessing detailed information on policy tools, particularly concerning the companies that benefit from such incentives. The clarity and completeness of this information are critical for effectively assessing the efficacy of energy policies. Additionally, despite the presence of regional incentives and calls for proposals, we identified a lack of comprehensive data on these policies and their economic scope, suggesting the need for a more uniform and transparent approach to energy policies at the regional level.

Regarding company characteristics, our investigation shed light on how corporate indebtedness significantly affects their energy performance. An increase in debt can negatively impact a company's energy behaviour in relation to its unique characteristics. The presence of certain specific traits might lead highly indebted companies to reprioritise investments, shifting from energy efficiency to other areas due to financial burden. In this context, reducing informational barriers about efficiency measures could lead to a revaluation of the associated benefits and a revision of corporate strategies; technical assistance, such as energy consultancy and access to industrial expertise, can be crucial for identifying and implementing effective energy solutions, reducing dependence on indebtedness. Our study also highlighted that per capita revenues do not seem to significantly influence the energy behaviour of companies, suggesting that the decision to implement energy efficiency measures is based on the interaction of multiple factors, not necessarily linked to a company's profitability. Regarding capital turnover, we observed a negative impact on energy performance, which might also be linked to the priority assigned to investments in energy efficiency and the pursuit of rapid financial returns.

Considering these results, the authors emphasise the importance of behavioural modifications for reducing energy consumption. Although this topic has been addressed on several occasions, it tends to be underestimated, even though the implementation of such measures is generally more feasible than others. Greater promotion of efficient energy practices would lead to extensive energy conservation, especially considering the relatively lower level of investment and shorter payback periods, although difficulties in estimating the associated benefits might lead to an underestimation of the same. Furthermore, we confirmed the importance of the economic/financial dimension related to EE interventions in determining energy trajectories: higher levels of investment lead to better energy-saving performances. A prolonged payback period acts as a deterrent to the implementation of EEMs, a risk that varies depending on the technical nature of the measures. In summary, these results suggest that economic incentives aimed at reducing the resistance associated with initial investment, technical assistance, and the sharing of best practices are fundamental to stimulate the adoption of energy efficiency measures.

Turning our gaze onto the technical characteristics of the identified EEMs, our analysis displays that specific intervention areas, such as cogeneration/trigeneration and thermal power plant/heat recovery, have a more pronounced impact on the energy performance. This result underscores the potential benefits of focusing on the promotion of specific types of interventions to achieve significant energy savings in the plastics manufacturing sector. Additionally, given the relevance of energy self-production, namely cogenerators/trigenerators and plants using renewable sources, there is a need to develop ad hoc models for these intervention areas to understand the most appropriate strategy to stimulate their implementation.

From a policy perspective, our work underscores the need for tailored incentives and support mechanisms that address the specificities of companies and achievable energy savings to effectively promote energy efficiency in the plastics production sector. The current approach, which mainly design incentives based on achievable energy savings, may not be sufficiently effective in reducing the resistance associated with the implementation of measures, as evidenced by the variable impact of different EEMs and the influence of company-specific factors. Therefore, policy measures should consider the unique characteristics and needs of the specific industrial sector to facilitate the adoption of energy-efficient technologies.

Looking to the future, the analysis conducted in the plastics sector can be extended to other industrial sectors, providing insights into the specific needs of different industries and tools to compare them. The robustness of the models and results also implies the possibility of expanding the characteristics of the companies and interventions considered to obtain a comprehensive and detailed picture of the barriers and drivers to energy efficiency present in the analysed sector.

In addition, a future path for research opens, considering the possibility of exploring localised energy dynamics and the interaction between companies, foreseeing the implementation of econometric approaches capable of capturing the spatial impact of corporate behaviours; mechanisms of interaction and spillover that enhance or inhibit industrial energy performance could be identified and promoted. The process of refining data related to the third cycle of EAs (2023) will also allow us to assess which interventions have been implemented, analysing the factors that influenced the process. This approach will enable us to reach even more precise conclusions on the path to follow to stimulate the adoption of energy efficiency interventions in the considered sector.

Further research opportunities would also be possible thanks to continuous efforts of the ENEA for the promotion of high-quality energy audits and the future changes in the subjects obligated to perform EAs, an evolution that the recent legislation in this matter leaves us to suppose. In conclusion, our study provides a significant contribution to the understanding of energy efficiency in the Italian plastics manufacturing sector. Our findings offer insights for formulating effective policies and indicate promising directions for future research in the field of energy efficiency and sustainable development. By highlighting key drivers and barriers to energy efficiency, we aim to inform more effective policy decisions and encourage continued research in this vital area, emphasising the importance of tailored incentives and the need for greater transparency in information on business access to existing incentives.

Author Contributions: Conceptualisation, V.C., M.D., M.M., C.M. and E.P.; methodology, V.C., M.D., M.M., C.M. and E.P.; investigation, V.C., M.D., M.M., C.M. and E.P.; data curation, V.C., M.D., M.M., C.M. and E.P.; writing—original draft preparation, V.C., M.D., M.M., C.M. and E.P.; writing—review and editing, V.C., M.D., M.M., C.M. and E.P.; supervision, V.C., C.M. and E.P.; project administration, V.C. and C.M.; funding acquisition, V.C. and C.M. All authors have read and agreed to the published version of the manuscript.

Funding: This research was funded by the National Electrical System Research Programme (RdS PTR 2022–2024), "Piano Triennale della Ricerca del Sistema Elettrico Nazionale 2022–2024" implemented under programme agreements between the Italian Ministry of Environment and Energy Security and ENEA, CNR, and RSE S.p.A.

Data Availability Statement: Data are contained within the article.

Acknowledgments: Comments, discussions, and suggestions from several persons enriched the current development of the research.

Conflicts of Interest: The authors declare no conflicts of interest.

References

1. European Parliament; Council of the European Union. Energy Efficiency Directive. Directive 2012/27/EU of the European Parliament and of the Council of 25 October 2012 on Energy Efficiency, Amending Directives 2009/125/EC and 2010/30/EU and Repealing Directives 2004/8/EC and 2006/32/EC. *Off. J. Eur. Union* **2012**, L315/1. Available online: https://eur-lex.europa.eu/eli/dir/2012/27/oj (accessed on 23 December 2023).
2. European Parliament; Council of the European Union. Directive (EU) 2018/2002 of the European Parliament and of the Council of 11 December 2018 Amending Directive 2012/27/EU on Energy Efficiency. *Off. J. Eur. Union* **2018**. Available online: https://eur-lex.europa.eu/eli/dir/2018/2002/oj (accessed on 23 December 2023).
3. European Parliament; Council of the European Union. Revised Energy Efficiency Directive. Directive 2023/1791 of the European 2012/27/EU of the European Parliament and of the Council on Energy Efficiency and Amending Regulation (EU) 2023/955 (Recast). *Off. J. Eur. Union* **2023**. Available online: https://eur-lex.europa.eu/eli/dir/2023/1791/oj (accessed on 23 December 2023).
4. Italian Government. Legislative Decree No 102 of 4 July 2014: Implementation of Directive 2012/27/EU on Energy Efficiency, Amending Directives 2009/125/EC and 2010/30/EU and Repealing Directives 2004/8/EC and 2006/32/EC; Gazzetta Ufficiale 165. 2014. Available online: https://www.gazzettaufficiale.it/eli/id/2014/07/18/14G00113/sg (accessed on 23 December 2023).

5. Reddy, B.S. Barriers and Drivers to Energy Efficiency—A New Taxonomical Approach. *Energy Convers. Manag.* **2013**, *74*, 403–416. [CrossRef]
6. de Groot, H.L.F.; Verhoef, E.T.; Nijkamp, P. Energy Saving by Firms: Decision-Making, Barriers and Policies. *Energy Econ.* **2001**, *23*, 171–174. [CrossRef]
7. Thollander, P.; Danestig, M.; Rohdin, P. Energy Policies for Increased Industrial Energy Efficiency: Evaluation of a Local Energy Programme for Manufacturing SMEs. *Energy Policy* **2007**, *35*, 5774–5783. [CrossRef]
8. Trianni, A.; Cagno, E. Dealing with Barriers to Energy Efficiency and SMEs: Some Empirical Evidence. *Energy* **2012**, *37*, 494–504. [CrossRef]
9. Fleiter, T.; Schleich, J.; Ravivanpong, P. Adoption of Energy-Efficiency Measures in SMEs—An Empirical Analysis Based on Energy Audit Data from Germany. *Energy Policy* **2012**, *51*, 863–875. [CrossRef]
10. Newell, R.G.; Anderson, S. Information Programs for Technology Adoption: The Case of Energy-Efficiency Audits. *Resour. Energy Econ.* **2004**, *26*, 27–50.
11. Zhang, Z.; Zhang, Y.; Zhao, M.; Muttarak, R.; Feng, Y. What is the global causality among renewable energy consumption, financial development, and public health? New perspective of mineral energy substitution. *Resour. Policy* **2023**, *85*, 104036. [CrossRef]
12. Zhang, G.; Zhang, Z.; Gao, X.; Yu, L.; Wang, S.; Wang, Y. Impact of Energy Conservation and Emissions Reduction Policy Means Coordination on Economic Growth: Quantitative Evidence from China. *Sustainability* **2017**, *9*, 686. [CrossRef]
13. Fleiter, T.; Hirzel, S.; Worrell, E. The Characteristics of Energy-Efficiency Measures—A Neglected Dimension. *Energy Policy* **2012**, *51*, 502–513. [CrossRef]
14. Cooremans, C.; Schönenberger, A. Energy Management: A Key Driver of Energy-Efficiency Investment? *J. Clean. Prod.* **2019**, *203*, 264–275. [CrossRef]
15. Trianni, A.; Cagno, E.; Farnè, S. An Empirical Investigation of Barriers, Drivers and Practices for Energy Efficiency in Primary Metals Manufacturing SMEs. *Energy Procedia* **2014**, *61*, 1252–1255. [CrossRef]
16. Trianni, A.; Cagno, E. Diffusion of Motor Systems Energy Efficiency Measures: An Empirical Study within Italian Manufacturing SMEs. *Energy Procedia* **2015**, *75*, 2569–2574. [CrossRef]
17. Cagno, E.; Accordini, D.; Trianni, A. A Framework to Characterize Factors Affecting the Adoption of Energy Efficiency Measures Within Electric Motors Systems. *Energy Procedia* **2019**, *158*, 3352–3357. [CrossRef]
18. Neri, A.; Cagno, E.; Trianni, A. Barriers and Drivers for the Adoption of Industrial Sustainability Measures in European SMEs: Empirical Evidence from Chemical and Metalworking Sectors. *Sustain. Prod. Consum.* **2021**, *28*, 1433–1464. [CrossRef]
19. Hrovatin, N.; Cagno, E.; Dolšak, J.; Zorić, J. How Important Are Perceived Barriers and Drivers Versus Other Contextual Factors for the Adoption of Energy Efficiency Measures: An Empirical Investigation in Manufacturing SMEs. *J. Clean. Prod.* **2021**, *323*, 129123. [CrossRef]
20. Hrovatin, N.; Dolsak, N.; Zorić, J. Factors Impacting Investments in Energy Efficiency and Clean Technologies: Empirical Evidence from Slovenian Manufacturing Firms. *J. Clean. Prod.* **2016**, *127*, 475–486. [CrossRef]
21. Boyd, G.A.; Curtis, E.M. Evidence of an 'Energy-Management Gap' in U.S. Manufacturing: Spillovers from Firm Management Practices to Energy Efficiency. *J. Environ. Econ. Manag.* **2014**, *68*, 463–479. [CrossRef]
22. Blass, V.; Corbett, C.J.; Delmas, M.A.; Muthulingam, S. Top management and the adoption of energy efficiency practices: Evidence from small and medium-sized manufacturing firms in the US. *Energy* **2014**, *65*, 560–571. [CrossRef]
23. Cantore, N. Factors Affecting the Adoption of Energy Efficiency in the Manufacturing Sector of Developing Countries. *Energy Effic.* **2017**, *10*, 743–752. [CrossRef]
24. Chamberlain, G. Analysis of Covariance with Qualitative Data. *Rev. Econ. Stud.* **1980**, *47*, 225–238. [CrossRef]
25. Hamerle, A.; Ronning, G. Panel Analysis for Qualitative Variables. In *Handbook of Statistical Modeling for the Social and Behavioral Sciences*; Springer: New York, NY, USA, 1995; pp. 401–451.
26. Bruni, G.; De Santis, A.; Herce, C.; Leto, L.; Martini, C.; Martini, F.; Salvio, M.; Tocchetti, F.A.; Toro, C. From Energy Audit to Energy Performance Indicators: A Methodology to Characterize Productive Sectors. The Italian Cement Industry Case Study. *Energies* **2021**, *14*, 8436. [CrossRef]
27. Herce, C.; Biele, E.; Martini, C.; Salvio, M.; Toro, C. Impact of Energy Monitoring and Management Systems on the Implementation and Planning of Energy Performance Improved Actions: An Empirical Analysis Based on Energy Audits in Italy. *Energies* **2021**, *14*, 4723. [CrossRef]
28. *EN 16247.1:2022*; Energy Audit—General Requirements. European Union: Brussels, Belgium, 2022.
29. *ISO 50001*; Energy Management. ISO: Geneva, Switzerland, 2018.
30. Gestore Servizi Energetici. Types of Projects to Be Submitted to the Final Report, 1–4. Available online: https://www.mercatoelettrico.org/en/mercati/tee/CosaSonoTee.aspx (accessed on 23 December 2023).
31. Istituto Nazionale di Statistica. Competitiveness of Productive Sectors. Sector 22–Manufacture of Rubber and Plastic Articles. 2018.
32. Kounetas, K.; Skuras, D.; Tsekouras, K. Promoting Energy Efficiency Policies Over the Information Barrier. *Inf. Econ. Policy* **2011**, *23*, 72–84. [CrossRef]
33. DeCanio, S.; Watkins, W. Investment in Energy Efficiency: Do the Characteristics of Firms Matter? *Rev. Econ. Stat.* **1998**, *80*, 95–107. [CrossRef]

34. Toro, C.; Biele, E.; Herce, C.; Martini, C.; Salvio, M.; Threpsiadi, A.; Wilkinson-Dix, J. Overview of Energy Efficiency Policies and Programmes for SMEs in Italy. In Proceedings of the 2022 Energy Evaluation Europe Conference, Palaiseau, France, 28–30 September 2022.
35. Abadie, L.M.; Ortiz, R.A.; Galarraga, I. Determinants of energy efficiency investments in the US. *Energy Policy* **2012**, *45*, 551–566. [CrossRef]

Disclaimer/Publisher's Note: The statements, opinions and data contained in all publications are solely those of the individual author(s) and contributor(s) and not of MDPI and/or the editor(s). MDPI and/or the editor(s) disclaim responsibility for any injury to people or property resulting from any ideas, methods, instructions or products referred to in the content.

Article

Energy Efficiency of AGV-Drone Joint In-Plant Supply of Production Lines

Tamás Bányai

Institute of Logistics, University of Miskolc, 3515 Miskolc, Hungary; tamas.banyai@uni-miskolc.hu

Abstract: Energy efficiency plays an increasingly important role not only in supply chains, but also in in-plant supply systems. Manufacturing companies are increasingly using energy-efficient material handling equipment to solve their in-plant material handling tasks. A new example of this effort is the use of drones for in-plant transportation of small components. Within the frame of this article, a new AGV-drone joint in-plant supply model is described. The joint service of AGV-based milkrun trolleys and drones makes it possible to optimize the in-plant supply in production lines. This article discusses the mathematical description of AGV-drone joint in-plant supply solutions. The numerical analysis of the different AGV-drone joint in-plant supply solutions shows that this new approach can lead to an energy consumption reduction of about 30%, which also has a significant impact on GHG emission.

Keywords: emission reduction; energy efficiency; logistics; optimization; scheduling; service level; vehicle routing

Citation: Bányai, T. Energy Efficiency of AGV-Drone Joint In-Plant Supply of Production Lines. *Energies* 2023, 16, 4109. https://doi.org/10.3390/en16104109

Academic Editors: Chiara Martini and Claudia Toro

Received: 14 April 2023
Revised: 30 April 2023
Accepted: 6 May 2023
Published: 16 May 2023

Copyright: © 2023 by the author. Licensee MDPI, Basel, Switzerland. This article is an open access article distributed under the terms and conditions of the Creative Commons Attribution (CC BY) license (https://creativecommons.org/licenses/by/4.0/).

1. Introduction

Milkrun supply is an extremely popular in-plant supply solution, where a wide range of components has to be picked up and delivered among warehouses, supermarkets, and production resources. The design of milkrun routes represents a problem of dynamic systems; therefore, its processes cannot be planned in advance, but dynamic design and operation methodologies have to be used instead of conventional in-advance design [1].

Milkrun in-plant supply is defined as a supply system including periodically moving vehicles which perform the material supply of manufacturing and assembly cells in different predefined routes. The milkrun supply generally can be taken into consideration as a lean distribution system which standardizes the in-plant logistics processes, the logistics resources, and the strategies. Although there is a wide range of studies focusing on milkrun supply, most of them discuss supply chain-related milkrun solutions and only a few of them discuss in-plant milkrun solutions. The studies generally focus on the optimization of the required resources, and the transportation distances and the potential of using mixed resources (e.g., linked AGV-drone services) are not analyzed [2].

Based on the results of the Fortune Business Insight Study, the unmanned aircraft market has been significantly increased and a Compound Annual Growth Rate of about 43% is expected, while the market size value is growing from USD 13.48 billion in 2022 to USD 232.8 billion by 2029 [3]. The application of drones includes a wide range of services, such as pandemic vaccine distribution [4], as well as local and global supply chain distribution [5]. The application of drones focuses on both single unmanned vehicles and cloud-based supply chain solutions, where not only individual drones, but also drone clusters can be taken into consideration as potential supply solutions [6]. The control of complex drone-based supply solutions is generally based on artificial intelligence methods, which enables the optimization of single echelon and multi-echelon supply chains [7,8]. Most of the literature has discussed the application of drones only in the case of global and local supply chains, and only a few of the studies discuss the in-plant applications of drones.

The design and operation of in-plant milkrun supply solutions can be defined as an NP-hard optimization problem [2]; therefore, most of the optimization-related approaches are using heuristics or metaheuristics. The objective functions of different approaches of the design of milkrun-based in-plant supply integrate the following aspects: minimization of required AGVs, minimization of milkrun trolleys' capacities, minimization of routes of milkruns, minimization of work in process inventories (WIP), minimization of material handling operations, or minimization of inventory cost, number, and capacity of required supermarkets [9].

The impact of milkrun-based in-plant supply can be analyzed using discrete event simulation, where both the resources and the traveling cycles among storages, supermarkets, manufacturing, and assembly cells can be taken into consideration. The simulation makes it possible to analyze the impact of different disturbances on the efficiency of logistics and manufacturing [10].

After this introduction, the remaining part of the paper is divided into three sections. Section 2 presents a literature review to summarize the research background regarding design and optimization of in-plant supply focusing on milkrun solutions. Section 3 presents a new approach, which makes it possible to model and analyze conventional AGV-based and AGV-drone joint in-plant supply solutions from an energy efficiency point of view. Section 4 presents the results of the numerical analysis. Conclusions and future research directions are discussed in the last section.

2. Literature Review

Production processes can be generally described as stochastic flow lines, where storage capacities and limited material supply operations can significantly influence and increase the uncertainties. In the case of an uncertain production environment, the performance analysis and the optimization of stochastic in-plant supply processes play an important role in the efficiency improvement [11]. Holistic approaches can also be used to optimize the in-plant milkrun systems. The planning and dimensioning tasks of in-plant milkrun services can integrate a wide range of logistics operations and logistics-related topics focusing on consignment, storage, time management, and ergonomics [12].

Milkrun solutions are used in a wide range of industries: agricultural machinery [13], food industry [14], automotive supplier [15,16], cable manufacturing [17], or washing machine manufacturing [18].

In the context of Industry 4.0, the application of dynamic simulation models in the optimization of intralogistics processes becomes more and more important. The interactive layout design is an important tool which can have a great impact on the efficiency of milkrun design [19]. As a case study focusing on the transportation of part supply improvement shows, poorly managed in-plant supply processes can lead to increased costs and decreased product quality, while the ability to fulfill customers' demands can also be decreased. Lean tools can support the improvement of in-plant supply processes. As a case study shows, the most important problem of the operation of milkrun supply solutions is the asynchronicity between the demands of production or assembly lines and in-plant supply processes [13].

The design process of in-plant supply solutions is a complex engineering task, where a wide range of influencing factors has to be taken into consideration. The calculation of the intensity of dimensioning parameters can significantly improve the efficiency of milkrun design [20], and it is especially important in the case of AGV-drone joint supply, where the integration of different technologies is a core problem. Milkrun supply can be used both for external logistics and for intralogistics. The design problems of milkrun solutions represent in both cases complex optimization problems. In the case of external logistics, manufacturers, retailers, and suppliers are integrated into a value chain, while in the case of intralogistics, the production and assembly resources are integrated by milkrun solutions. As research by [14] shows, the Analytical Hierarchy Process is a suitable design methodology to optimize milkrun processes. The design tasks of in-plant supply solutions can be solved using analytical methods [21], heuristics [22], and simulation [23,24], but empirical studies also discuss important design aspects [25]. As a 3D micro-simulation of milkruns and pickers in warehouses shows, discrete event simulation and agent-based simulation make it possible to take a wide range of parameters into consideration including storage strategy, quantity of demands, structure of shifts, volume of components, arrival rate of requests, and the available number of milkrun trolleys [26]. As this research shows, milkrun design requires a holistic approach. A new approach focuses on the structural optimization of milkrun-based in-plant supply, where time- and capacity-based constraints are taken into consideration [27]. Merging the payload cycles can also lead to optimized milkrun solutions, as a case study shows in the case of an automotive supplier, where the optimization of loading capacities in the restructuring and rerouting of milkrun routes resulted in a more efficient milkrun-based supply [16].

The design of an intralogistics system integrates different aspects of material handling design including layout planning, vehicle routing, and scheduling. As a black hole heuristics-based optimization approach shows, the integrated solution of milkrun services in production processes can lead to efficient intralogistics operations [28]. One of the first real-time milkrun design approaches was published by [29]. In this research, an in-plant milkrun control methodology is shown focusing on the morphological classification of static and dynamic approaches. Value stream mapping and lean metrics can also support the design of milkrun supply processes. The flexible routing of milkrun trolleys can lead to reduced stocks (inventory in warehouse and work in process inventory), while the lean rate increases [30]. In the case of complex in-plant supply solutions, milkrun routing and scheduling are subject to a trade-off between vehicle fleet size and storage capacity. In this case, periodic distribution policies can support the optimal process, which is based on the identification of the relationship between tact time and the size of replenishment batches [31]. In the case of stochastic processes, probabilities can be added to the predefined schedules and requests, and this model can minimize the order cycle time and the average picking effort [32].

A wide range of methods and tools can be used for the optimization of milkrun supply solutions, such as linear programming [11], Analytical Hierarchy Process [14], Saving Matrix Methods [15], black hole heuristics [28], Linear Temporal Logic [33], and simulation [34]. The application of product allocation in a polling-based milkrun picking system is a complex design problem. As the research results in the case of exhaustive, locally-gated, and globally-gated picking strategies show, the cyclic polling system with simultaneous batch arrivals is a suitable solution for milkrun services [35,36]. Drones can also be used in polling-based milkrun picking systems.

The mentioned mathematical methods, algorithms, and tools are suitable to solve the design and operation problems of various milkrun-based in-plant supply solutions, depending on the complexity of the optimization problem. Linear programming, Analytical Hierarchy Process, and Saving Matrix Methods are generally used for analytical problems, while heuristic methods are suitable for NP-hard optimization problems. Uncertainties can be taken into consideration by using different simulation tools.

The above-mentioned research results indicate the scientific potential of the optimization of in-plant supply solutions. The articles that addressed the design and control problems of in-plant supply solutions focus on AGV-based milkrun services, and only a few of them describe the potential of the application of drones in intralogistics, especially in the field of in-plant supply [37]. According to that, the focus of this research is on the analyses of the impact of AGV-drone joint in-plant supply of production lines on energy efficiency.

As a consequence, the main contributions of this article are the following: (1) model framework of AGV-drone joint in-plant supply solutions; (2) mathematical description of AGV-drone joint in-plant supply solutions; (3) numerical analysis of the impact of different AGV-drone joint in-plant supply solutions on energy efficiency, focusing on both energy consumption and virtual GHG emission.

3. Materials and Methods

Within the frame of this chapter, the mathematical models of different AGV-drone joint in-plant material supply systems will be described. These mathematical models are based on the previous work of Bányai focusing on the evaluation of energy efficiency of integrated first-mile and last-mile drone-based delivery operations [38]. The chapter focuses on the following topics: (i) definition of the input parameters of AGV-drone joint in-plant material supply systems; (ii) definition of specific supply models from the general input parameters depending on the cooperation level of AGV-based milkrun trolley and drone; (iii) mathematical models of typical AGV-drone joint in-plant material supply systems. The structure of the proposed models is demonstrated in Figure 1.

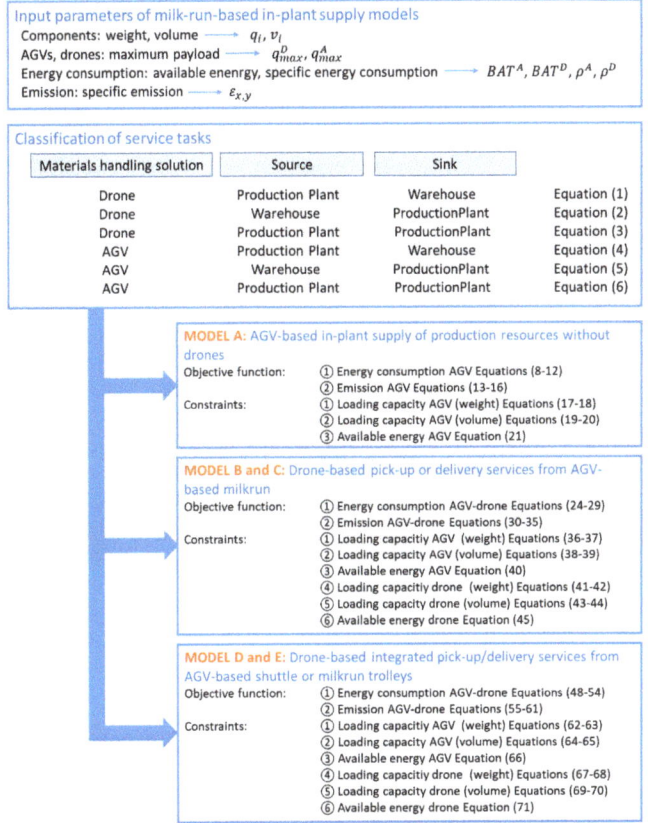

Figure 1. The structure of the proposed models (input parameters, classification of service tasks, typical service models, objective functions, and constraints).

The following typical models are discussed:
- Model A: conventional AGV-based in-plant supply of production resources without drones;
- Model B: drone-based pick-up services from AGV-based milkrun trolleys;
- Model C: drone-based delivery services from AGV-based milkrun trolleys;
- Model D: drone-based integrated pick-up and delivery shuttle services from AGV-based milkrun trolleys;
- Model E: drone-based integrated pick-up and delivery milkrun services from AGV-based milkrun trolleys.

These five service solutions can be integrated into three different mathematical models:
- Modeling of AGV-based in-plant supply of production resources without drones (see Section 3.1);
- Modeling of drone-based pick-up/delivery services from AGV-based milkrun (see Section 3.2);
- Modeling of the drone-based integrated pick-up/delivery shuttle and milkrun services from AGV-based milkrun trolleys (see Section 3.3).

The input parameters for the analysis and optimization of AGV-drone joint in-plant material supply systems are the following:
- q_i: weight of components of pick-up and delivery service i in kg;
- v_i: volume of components of pick-up and delivery service in L;
- z_i: type of material supply service i;
- q_{max}^D: maximum payload of drones in kg;
- q_{max}^A: maximum payload of AGV-based milkrun trolleys in kg;
- BAT^A: available power capacity stored in the battery of AGV-based milkrun trolleys in kWh;
- BAT^D: available power capacity stored in the battery of drones in kWh;
- ρ^A: specific energy consumption of AGV-based milkrun trolleys in kWh/km, which depends on the payload;
- ρ^D: specific energy consumption of drones in kWh/km, which depends on the payload;
- $\varepsilon_{x,y}$: specific emission of CO_2, SO_2, CO, HC, NO_X, and PM, depending on the generation source x of electricity in g/kWh in the case of GHG y.

Depending on the constraints regarding weight and volume of components of service tasks, it is possible to define typical relations in the production plant and typical in-plant service tasks.

If the constraints regarding weight and volume of components to be supplied in the production plant make it possible to fulfill the service for the requested demand by drone, while the location of service location is on the route of the AGV-based milkrun trolley and the departure is the warehouse, where AGV and drone pools are also located, then the service can be assigned to the set of in-plant service operations including pick-up service from production resource to warehouse to be performed by drone:

$$q_i \leq q_{max}^D \wedge v_i \leq v_{max}^D \wedge z_i = PPW \rightarrow q_i \in Q^{DPW}, \quad (1)$$

where q_i is the weight of component i to be picked up, q_{max}^D is the loading capacity of the drone, v_i is the volume of component i to be picked up, v_{max}^D is the upper limit of volume of components suitable for transportation by drone, Q^{DPW} is the set of pick-up service tasks from the production plant to the warehouse suitable for drone-based pick-up supply, and z_i is the type of the delivery tasks, where $z_i \in [PPW, DWP, PP, DP]$ and PPW is for pick-up service from production plant to warehouse, DWP is for delivery from the warehouse to the production plant, PP is for pick-up service in the production plant, and DP is for delivery service in the production plant.

If the constraints regarding weight and volume of components to be supplied in the production plant make it possible to fulfill the service demand by drone, while the location of pick-up is in the warehouse and the departure location is a production resource on the route of the AGV-based milkrun trolley, then the service can be assigned to the set of in-plant service operations including delivery service from warehouse to production resource to be performed by drone:

$$q_i \leq q_{max}^D \wedge v_i \leq v_{max}^D \wedge z_i = DWP \rightarrow q_i \in Q^{DDW}, \quad (2)$$

where Q^{DDW} is the set of delivery service tasks from the warehouse to the production plant suitable for drone-based delivery service.

If the constraints regarding weight and volume of components to be supplied in the production plant make it possible to fulfill the service demand by drone, while the location of pick-up and delivery is a production resource on the route of the AGV-based milkrun trolley, then the service can be assigned to the set of in-plant service operations including delivery service between two production resources to be performed by drone:

$$q_i \leq q_{max}^D \wedge v_i \leq v_{max}^D \wedge z_i \in [PP, DP] \rightarrow q_i \in Q^{DP}, \tag{3}$$

where Q^{DP} is the set of delivery tasks inside the production plant between production resources suitable for drone-based delivery. If $Q^{DP} > 0$, then the service task is a pick-up operation; otherwise, it is a delivery task between two production resources of the production plant.

If the constraints regarding weight and volume of components to be supplied in the production plant do not make it possible to fulfill the service demand by drone but by AGV-based milkrun trolley, while the location of pick-up is on the route of the AGV-based milkrun trolley and the departure is the warehouse, then the service can be assigned to the set of in-plant service operations including pick-up service from production resource to warehouse to be performed by AGV-based milkrun trolley:

$$\left(q_i > q_{max}^D \vee v_i > v_{max}^D\right) \wedge z_i = PPW \rightarrow q_i \in Q^{APW}, \tag{4}$$

where Q^{APW} is the set of pick-up service tasks from the production plant to the warehouse suitable for AGV-based milkrun trolley pick-up supply.

If the constraints regarding weight and volume of components to be supplied in the production plant do not make it possible to fulfill the service demand by drone but by AGV-based milkrun trolley, while the location of pick-up is in the warehouse and the departure location is a production resource on the route of the AGV-based milkrun trolley, then the service can be assigned to the set of in-plant service operations including delivery service from warehouse to production resource to be performed by AGV-based milkrun trolley:

$$\left(q_i > q_{max}^D \vee v_i > v_{max}^D\right) \wedge z_i = DWP \rightarrow q_i \in Q^{ADW}, \tag{5}$$

where Q^{ADW} is the set of delivery service tasks from the warehouse to the production plant suitable for AGV-based milkrun trolley delivery supply.

If the constraints regarding weight and volume of components to be supplied in the production plant do not make it possible to fulfill the service demand by drone but by AGV-based milkrun trolley, while the location of pick-up and delivery is a production resource on the route of the AGV-based milkrun trolley, then the service can be assigned to the set of in-plant service operations including delivery service between two production resources to be performed by AGV-based milkrun trolley:

$$\left(q_i > q_{max}^D \vee v_i > v_{max}^D\right) \wedge z_i \in [PP, DP] \rightarrow q_i \in Q^{AP}, \tag{6}$$

where Q^{AP} is the set of delivery tasks inside the production plant between production resources suitable for AGV-based milkrun trolley delivery. If $Q^{DP} > 0$, then the service task is a pick-up operation; otherwise, it is a delivery task between two production resources of the production plant. These sets make it possible to define five different models of AGV-drone joint in-plant supply of production lines.

3.1. Modeling of AGV-Based In-Plant Supply of Production Resources without Drones

In this case, all pick-up and delivery services are performed by AGV-based milkrun trolleys. Using the Q^{DPW}, Q^{DDW}, Q^{DP}, Q^{APW}, Q^{ADW}, and Q^{AP} matrices, it is possible to define the input parameters of the AGV-based in-plant supply model. The potential pick-up and delivery tasks of the in-plant supply model are as follows:

$$q_i^A = \begin{cases} \forall i \in [1,\ldots,\vartheta_1] : q_i^A = q_i^{DPW} \\ \forall i \in [1+\vartheta_1,\ldots,\vartheta_2] : q_i^A = q_{i-\vartheta_1}^{DDW} \\ \forall i \in [1+\vartheta_2,\ldots,\vartheta_3] : q_i^A = q_{i-\vartheta_2}^{DP} \\ \forall i \in [1+\vartheta_3,\ldots,\vartheta_4] : q_i^A = q_{i-\vartheta_3}^{APW} \\ \forall i \in [1+\vartheta_4,\ldots,\vartheta_5] : q_i^A = q_{i-\vartheta_4}^{ADW} \\ \forall i \in [1+\vartheta_5,\ldots,\vartheta_6] : q_i^A = q_{i-\vartheta_5}^{AP} \end{cases}, \quad (7)$$

where $\vartheta_s = \sum_{j=1}^{s} \beta_j$. The objective function of the optimization is the minimization of energy consumption, which also has a significant impact on the emission of CO_2, SO_2, CO, HC, NO_X, and PM. In this article, automated guided vehicles and drones are performing in-plant supply services; therefore, the emission can be calculated as a virtual emission taking the emission rates of the generation source of electricity, which can be lignite, coal, oil, natural gas, photovoltaic, biomass, nuclear, water, wind, and their mix.

3.1.1. The Objective Function

Within the frame of this model, both the energy consumption of the drone and the AGV-based milkrun trolley, and their virtual GHG emission can be defined.

The minimization of the energy consumption can be defined depending on the different services performed by the AGV-based milkrun trolley:

$$C^1 = C^{1AWP} + C^{1AP} + C^{1APW} \rightarrow min., \quad (8)$$

where C^1 is the energy consumption of AGVs and drones, C^{aAWP} is the energy consumption of the AGV-based milkrun trolley within the first section of the delivery route from the warehouse or AGV pool to the first production resource in model a, C^{aAP} is the energy consumption of the AGV-based milkrun trolley between two production resources of the production plant in model a, and C^{aAPW} is the energy consumption of the AGV-based milkrun trolley within the closing section of the delivery route from the last production resource to the warehouse or AGV depot in the case of model a (e.g., C^{1APW} is the energy consumption of the AGV-based milkrun trolley from the production resources to the warehouse for the first model).

The energy consumption for the first section of the in-plant supply route from the warehouse to the first production plant can be described depending on the route length of the section between the warehouse and the first production plant of the service route, the loading of the AGV-based milkrun trolley, and the specific energy consumption of the AGV-based milkrun trolley:

$$C^{1AWP} = \left(q^{ANET} + \sum_{i=1, z_i^* = DWP}^{\vartheta_6} q_{p_i}^{*A} \right) \cdot l_{D,p_1} \cdot \rho^A, \quad (9)$$

where q^{ANET} is the net weight of the AGV-based milkrun trolley, l_{D,p_1} is the length of the milkrun section between the warehouse and the first production plant, ρ^T is the specific energy consumption of the AGV-based milkrun trolley [kWh/(kg·km)], $\bar{p} = [p_i]$ is the assignment matrix representing the optimal solution as a permutation matrix. This matrix makes it possible to define the weight of components assigned to supply tasks for each scheduled delivery as:

$$\forall i \in \left[1, \ldots, \sum_{j=1}^{6} \beta_j \right] : q_{p_i}^{*A} = q_i^A. \quad (10)$$

The energy consumption among production resources can be defined depending on the length of the milkrun sections between the pick-up and delivery locations, the loading of the AGV-based milkrun trolley, and the specific energy consumption:

$$C^{1AP} = \sum_{k=1}^{\vartheta_6 - 1} \left(q^{ANET} + \sum_{i=k, z_i^* = DWP}^{\vartheta_6} q_{p_i}^{*A} + \sum_{i=1, z_i^* \in [PP, DP]}^{k} q_{p_i}^{*A} \right) \cdot l_{p_k, p_{k+1}} \cdot \rho^A. \quad (11)$$

where $l_{p_k,p_{k+1}}$ is the length of the milkrun section between the scheduled supply operations assigned to production resources k and $k + 1$.

The energy consumption of the transportation in the last section of the milkrun route from the last production resource to the warehouse can be defined depending on the length of the milkrun sections from the last production resource to the warehouse, the loading of the AGV-based milkrun trolley, and the specific energy consumption:

$$C^{1APW} = \left(q^{ANET} + \sum_{i=1, z_i^* = PPW}^{\sum_{j=1}^{6} \beta_j} q_{p_i}^{*A} \right) \cdot l_{D,p_1} \cdot \rho^A. \tag{12}$$

The weight of the components assigned to pick-up and delivery tasks between the first and last sections of the milkrun route has no impact on the final payload of the AGV-based milkrun trolley, because all of these operations are finished before the last section of the milkrun.

The minimization of the CO_2, SO_2, CO, HC, NO_X, and PM emission can be expressed as:

$$E_{x,y}^1 = E_{x,y}^{1AWP} + E_{x,y}^{1AP} + E_{x,y}^{1APW} \to min., \tag{13}$$

where $E_{x,y}^1$ is the energy consumption of the conventional model based on operations performed by the AGV-based milkrun trolley, $E_{x,y}^{aAWP}$ is the virtual GHG emission generated within the first section of the milkrun from the warehouse to the first production resource in model a, $E_{x,y}^{aAP}$ is the virtual GHG emission generated by the milkrun route between the first and last production plant in a, $E_{x,y}^{aAPW}$ is the virtual GHG emission generated within the last section of the milkrun from the last production plant to the warehouse in model a, x defines the type of electricity generation sources, and y defines the type of GHG (CO_2, SO_2, CO, HC, NO_X, PM).

The virtual GHG emission generated in the first section of the milkrun from the warehouse to the first production plant can be written as:

$$E_{x,y}^{1AWP} = \left(q^{ANET} + \sum_{i=1, z_i^* = DWP}^{\vartheta_6} q_{p_i}^{*A} \right) \cdot l_{D,p_1} \cdot \rho^A \cdot \varepsilon_{x,y}. \tag{14}$$

The virtual GHG emission of the milkrun among the first and last production resources of the production plant can be defined as:

$$E_{x,y}^{1AP} = \sum_{k=1}^{\vartheta_6 - 1} \left(q^{ANET} + \sum_{i=k, z_i^* = DWP}^{\vartheta_6} q_{p_i}^{*T} + \sum_{i=1, z_i^* \in [PP,DP]}^{k} q_{p_i}^{*A} \right) \cdot l_{p_k, p_{k+1}} \cdot \rho^A \cdot \varepsilon_{x,y}. \tag{15}$$

The virtual GHG emission of the last section of the milkrun from the last production resource to the warehouse is as follows:

$$E_{x,y}^{1APW} = \left(q^{ANET} + \sum_{i=1, z_i^* = PPW}^{\vartheta_6} q_{p_{\vartheta_6}}^{*A} \right) \cdot l_{\vartheta_6, D} \cdot \rho^A \cdot \varepsilon_{x,y}. \tag{16}$$

3.1.2. The Constraints

This model takes three constraints regarding available capacities and energies into consideration.

Constraint 1 for AGV-based milkrun trolley: the vehicle routing problem of the AGV-based milkrun trolley must be solved so that it is not allowed to exceed this maximum payload of the AGV-based milkrun trolley:

$$q_{p_i}^L \le q^{Amax}, \tag{17}$$

where q^{Amax} is the maximum payload of the AGV-based milkrun trolley and

$$q_{p_i}^L = q^{ANET} + \sum_{i=k, z_i^* = DWP}^{\vartheta_6} q_{p_i}^{*A} + \sum_{i=1, z_i^* \in [PP,DP]}^{k} q_{p_i}^{*A}, \tag{18}$$

where $q_{p_i}^L$ is the weight of the load of the AGV-based milkrun trolley at production resource p_i. In the case of this constraint, the weight of the AGV can be calculated as the sum of the net weight of the AGV-based milkrun trolley, the weight of components delivered from the warehouse to the production resources, and the weight of components picked up and delivered between production resources.

Constraint 2 for AGV-based milkrun trolley: it is not allowed to exceed the available loading volume of the AGV-based milkrun trolley:

$$v_{p_i}^L \le v^{Amax}, \tag{19}$$

where

$$v_{p_i}^L = \sum_{i=k, z_i^* = DWP}^{\vartheta_6} v_{p_i}^{*A} + \sum_{i=1, z_i^* \in [PP, DP]}^{k} v_{p_i}^{*A}. \qquad (20)$$

where $v_{p_i}^{*A}$ is the volume of component p_i, and $v_{p_i}^L$ is the volume of components on the AGV-based milkrun trolley at production resource p_i and v^{Amax} is the maximum loading volume of the AGV-based milkrun trolley.

In the case of this constraint, the current volume of the loading can be calculated as the sum of the volume of components transported from the warehouse to the production resources, and the volume of components picked up or delivered between production resources.

Constraint 3 for AGV-based milkrun trolley: the third constraint defines the upper limit of available power capacity stored in the battery of the AGV-based milkrun trolley:

$$C^1 \leq BAT^A, \qquad (21)$$

where BAT^A is the capacity of the AGV-based milkrun trolley's battery.

3.1.3. Decision Variables

The decision variable of this milkrun supply problem is the $\overline{p} = [p_i]$ permutation matrix, which defines the optimal sequence of production resources to be supplied by required components, where the value of p_i defines the ID of in-plant supply demand of a production resource to be scheduled.

3.2. Modeling of Drone-Based Pick-Up/Delivery Services from AGV-Based Milkrun

In this milkrun-based in-plant supply solution, the pick-up operations can be assigned to a drone depending on their suitability defined by the weight and volume of the component, while delivery tasks are performed by AGV-based milkrun trolley. The basic operations of this type of AGV-drone joint in-plant supply are the following:

- The pick-up service tasks from the first production resource to the warehouse are performed in the relation (production resource—AGV-based milkrun trolley—warehouse), if the capacity-related constraints focusing on weight and volume of the components to be supplied make it possible. Between the production resource and the AGV-based milkrun trolley, the transportation is assigned to a drone, while in the case of the relation (AGV-based milkrun trolley—warehouse), the component is transported by the AGV-based milkrun trolley.
- The delivery services from the warehouse to the production resource are performed by the AGV-based milkrun trolley, because in this model the joint solution means that drones are not performing direct supply from/to the warehouse.
- The pick-up services between two production resources are performed in the following way: the pick-up service is performed by the drone if the capacity-related constraints focusing on weight and volume of the components to be supplied make it possible in the relation (production resource—AGV-based milkrun trolley) and the delivery to the next production resource is performed by the AGV-based milkrun trolley.

In this case, using the Q^{DPW}, Q^{DDW}, Q^{DP}, Q^{APW}, Q^{ADW}, and Q^{AP} matrices, we can define the basic parameters of this model as follows. The potential pick-up and delivery tasks of the in-plant supply model can be written as follows both for the drone and the AGV-based milkrun trolley:

$$q_i^A = \begin{cases} \forall i \in [1, \ldots, \beta_2] : q_i^A = q_i^{DDW} \\ \forall i \in [1 + \beta_2, \ldots, \beta_2 + \beta_3] : q_i^A = q_{i-\beta_2}^{DP} \\ \forall i \in [1 + \beta_2 + \beta_3, \ldots, \vartheta_4 - \beta_1] : q_i^A = q_{i-\beta_2+\beta_3}^{APW} \\ \forall i \in [\vartheta_4 - \beta_1 + 1, \ldots, \vartheta_5 - \beta_1] : q_i^A = q_{i-\vartheta_4-\beta_1}^{ADW} \\ \forall i \in [\vartheta_5 - \beta_1 + 1, \ldots, \vartheta_6 - \beta_1] : q_i^A = q_{i-\vartheta_5-\beta_1}^{AP} \end{cases}, \qquad (22)$$

and

$$q_g^D = \begin{cases} \forall g \in [1,\ldots,\vartheta_1] : q_g^D = q_g^{DPW} \\ \forall g \in [1+\vartheta_1,\ldots,\vartheta_1+\vartheta_3-\vartheta_2] : q_g^D = q_{g-\vartheta_3}^{DP} \end{cases}. \quad (23)$$

3.2.1. The Objective Function

Within the frame of this model, both the energy consumption of the drone and the AGV-based milkrun trolley, and their virtual GHG emission can be defined. The minimization of the energy consumption performed by the drone and the AGV-based milkrun trolley is as follows:

$$C^2 = C^{2AWP} + C^{2AP} + C^{2APW} + C^{2DP} + C^{2DPW} \to min., \quad (24)$$

where C^2 is the energy consumption of the milkrun supply including energy consumption of the drone and the AGV-based milkrun trolley, C^{aDP} is the energy consumption of the drone between a production resource and the AGV-based milkrun trolley in model a, and C^{aDPW} is the energy consumption of the drone from the last production resource to the warehouse in model a.

The energy consumption of the first section of the milkrun from the warehouse to the first production resource can be defined depending on the length of the milkrun sections between the warehouse and the first production resource, the loading of the AGV-based milkrun trolley, and the specific energy consumption:

$$C^{2AWP} = \left(q^{ANET} + \sum_{i=1+\vartheta_1, z_i^* = DWP}^{\vartheta_6} q_{p_i}^{*A}\right) \cdot l_{D,p_1} \cdot \rho^A. \quad (25)$$

The energy consumption of the AGV-based milkrun trolley generated by the transportation services between the first and last production resource can be defined depending on the length of the milkrun sections between the production resources, the loading of the AGV-based milkrun trolley, and the specific energy consumption:

$$C^{2AP} = \sum_{k=1+\vartheta_1}^{\vartheta_6-1} \left(q^{ANET} + \sum_{i=k, z_i^* = DWP}^{\vartheta_6} q_{p_i}^{*T} + \sum_{i=1+\vartheta_1, z_i^* \in [PP,DP]}^{k} q_{p_i}^{*A}\right) \cdot l_{p_k, p_{k+1}} \cdot \rho^A. \quad (26)$$

The energy consumption of the last section of the milkrun from the last production resource of the route to the warehouse can be defined depending on the length of the milkrun sections between the last production resource and the warehouse, the loading of the AGV-based milkrun trolley, and the specific energy consumption:

$$C^{2APW} = \left(q^{ANET} + \sum_{i=1+\vartheta_1, z_i^* = PPW}^{\vartheta_6} q_{p_1}^{*A}\right) \cdot l_{D,p_1} \cdot \rho^A. \quad (27)$$

The energy consumption of the drone within the in-plant supply route between the first and last production resource which results from pick-up services can be formulated as:

$$C^{2DP} = \sum_{k=1+\vartheta_1, z_k^* \in [PP]}^{\vartheta_1+\vartheta_3-\vartheta_2} \left(q^{DNET} + q_{p_k}^{*D}\right) \cdot 2 \cdot l_{p_k, p_{TR}} \cdot \rho^D. \quad (28)$$

where $l_{p_k, p_{TR}}$ is the travelling distance between the production resource p_k and the current position of the AGV-based milkrun trolley, and ρ^D is the specific energy consumption of the drone.

The energy consumption of the drone between the production resource and the warehouse can be calculated as:

$$C^{2DPW} = \sum_{k=1, z_k^* \in [PP]}^{\vartheta_1} \left(q^{DNET} + q_{p_k}^{*D}\right) \cdot 2 \cdot l_{p_k, D} \cdot \rho^D. \quad (29)$$

The minimization of the CO_2, SO_2, CO, HC, NO_X, and PM emission can be written in the following form:

$$E_{x,y}^2 = E_{x,y}^{2AWP} + E_{x,y}^{2AP} + E_{x,y}^{2APW} + E_{x,y}^{2DP} + E_{x,y}^{2DPW} \to min. \quad (30)$$

where $E_{x,y}^{aDP}$ is the virtual GHG emission of the drone within the in-plant supply route between the first and last production resource in model a, and $E_{x,y}^{aDPW}$ is the virtual GHG emission of the drone between the production resource and the warehouse in model a.

The virtual GHG emission of the first section of the milkrun from the warehouse to the first production resource in the case of the AGV-based milkrun trolley can be expressed as:

$$E_{x,y}^{2AWP} = \left(q^{ANET} + \sum_{i=1+\vartheta_1, z_i^* = DWP}^{\vartheta_6} q_{p_i}^{*A} \right) \cdot l_{D,p_1} \cdot \rho^A \cdot \varepsilon_{x,y}. \tag{31}$$

The virtual GHG emission of the milkrun between the first and last production resource in the case of the AGV-based milkrun trolley is as follows:

$$E_{x,y}^{2AP} = \sum_{k=1+\vartheta_1}^{\vartheta_6-1} \left(q^{ANET} + \sum_{i=k, z_i^* = DWP}^{\vartheta_6} q_{p_i}^{*A} + \sum_{i=1+\vartheta_1, z_i^* \in [PP,DP]}^{k} q_{p_i}^{*A} \right) \cdot l_{p_k, p_{k+1}} \cdot \varepsilon_{x,y}. \tag{32}$$

The virtual GHG emission of the last section of the in-plant supply route from the last production resource to the warehouse in the case of the AGV-based milkrun trolley can be formulated as:

$$E_{x,y}^{2APW} = \left(q^{ANET} + \sum_{i=1+\vartheta_1, z_i^* = PPW}^{\vartheta_6} q_{p_1}^{*T} \right) \cdot l_{D,p_1} \cdot \rho^a \cdot \varepsilon_{x,y}. \tag{33}$$

The drone's virtual GHG emission generated between the first and last production resource is as follows:

$$E_{x,y}^{2DP} = \sum_{k=1+\vartheta_1, z_k^* \in [PP]}^{\vartheta_1+\vartheta_3-\vartheta_2} \left(q^{DNET} + q_{p_k}^{*D} \right) \cdot 2 \cdot l_{p_k, p_{TR}} \cdot \rho^D \cdot \varepsilon_{x,y}. \tag{34}$$

The drone's virtual GHG emission between production resources and warehouse can be expressed as:

$$E_{x,y}^{2DPW} = \sum_{k=1, z_k^* \in [PP]}^{\vartheta_1} \left(q^{DNET} + q_{p_k}^{*D} \right) \cdot 2 \cdot l_{p_k, D} \cdot \rho^D \cdot \varepsilon_{x,y}. \tag{35}$$

3.2.2. The Constraints

We can define constraints for in-plant supply services performed by both the AGV-based milkrun trolley and the drone.

Constraint 1 for AGV-based milkrun trolley: the in-plant supply route of the AGV-based milkrun trolley must be designed so that maximum payload must be taken into consideration:

$$\forall i \in [\vartheta_6 - \beta_1] : q_{p_i}^L \leq q^{Amax}, \tag{36}$$

where

$$\forall i \in [\vartheta_6 - \beta_1] : q_{p_i}^L = q^{ANET} + \sum_{i=k, z_i^* = DWP}^{\vartheta_6 - \beta_1} q_{p_i}^{*T} + \sum_{i=1, z_i^* \in [PP,DP]}^{k} q_{p_i}^{*A}. \tag{37}$$

In the case of this constraint, the weight of the AGV can be calculated as the sum of the net weight of the AGV-based milkrun trolley, the weight of components delivered from the warehouse to the production resources, and the weight of components picked up and delivered between production resources. The difference between constraints (17–18) and (36–37) is that in the case of the conventional, AGV-based model, the payload-related constraint is calculated for all production resources, while in the case of this AGV-drone joint service, only production resources, where no drone is performing a service task, are taken into consideration.

Constraint 2 for AGV-based milkrun trolley: the in-plant supply route must be designed so that maximum loading volume of the AGV-based milkrun trolley must be taken into consideration, and it is not allowed to exceed this value:

$$\forall i \in [\vartheta_6 - \beta_1] : v_{p_i}^L \leq v^{Amax}, \tag{38}$$

where

$$\forall i \in [\vartheta_6 - \beta_1] : v_{p_i}^L = q^{ANET} + \sum_{i=k, z_i^* = DWP}^{\vartheta_6 - \beta_1} v_{p_i}^{*A} + \sum_{i=1, z_i^* \in [PP,DP]}^{k} v_{p_i}^{*A}. \tag{39}$$

In the case of this constraint, the current volume of the loading can be calculated as the sum of the volume of components transported from the warehouse to the production resources, and the volume of components picked up or delivered between production resources. The difference between constraints (19–20) and (38–39) is that in the case of the conventional, AGV-based model, the volume-related constraint is calculated for all production resources, while in the case of this AGV-drone joint service, only production resources, where no drone is performing a service task, are taken into consideration.

Constraint 3 for AGV-based milkrun trolley: it is not allowed to consume more energy than the available power capacity stored in the battery of the AGV-based milkrun trolley.

$$C^{2AWP} + C^{2AP} + C^{2APW} \leq CAP^T. \tag{40}$$

In the case of the drone-based in-plant supply, we can also define three constraints.

Constraint 1 for drone: this constraint defines the maximum payload of the drone. In the case of shuttle service (no milkrun is performed by the drone), this maximum payload of the drone can be written as follows:

$$\forall i \in [\vartheta_1 + \vartheta_3 - \vartheta_2] : q_{p_i} \leq q^{Dmax}, \tag{41}$$

where q^{Dmax} is the maximum payload of the drone. In the case of milkrun, the weight of the components picked up by the drone has an upper limit, which can be expressed as:

$$\forall i : \sum_{i \in \theta} q_{p_i} \leq q^{Dmax}, \tag{42}$$

where θ is the set of milkrun routes of the drone.

In the case of this constraint, the current payload of the drone can be calculated as the weight of the current payload.

Constraint 2 for drone: it is not allowed to exceed the maximum available loading volume of the drone. In the case of shuttle service (no milkrun is performed by the drone), this constraint can be written as follows:

$$\forall i \in [\vartheta_1 + \vartheta_3 - \vartheta_2] : v_{p_i} \leq v^{Dmax}, \tag{43}$$

where v^{Dmax} is the maximum loading volume of the drone. In the case of shuttle service (no milkrun is performed by the drone), this constraint can be written as follows:

$$\forall i : \sum_{i \in \theta} v_{p_i} \leq v^{Dmax}. \tag{44}$$

In the case of this constraint, the current volume of the load of the drone can be calculated as the volume of the current payload, because the drone performs only shuttle services.

Constraint 3 for drone: it defines the upper limit of available power capacity stored in the battery of the drone.

$$\forall i : \sum_{i \in \theta} C_i^{*2DP} \leq BAT^D \wedge \forall i : \sum_{i \in \theta} C_i^{*2DPW} \leq BAT^D, \tag{45}$$

where BAT^D is the capacity of the drone's battery.

3.2.3. Decision Variables

The decision variable of the above-mentioned AGV-drone joint in-plant supply service with drone-based pick-up operations problem is the $\overline{p} = [p_i]$ matrix.

3.3. Modeling of the Drone-Based Integrated Pick-Up/Delivery Shuttle and Milkrun Services from AGV-Based Milkrun Trolleys

In this model, the drone can perform both the pick-up and the delivery operations suitable for drone-based services.

Using the Q^{DPW}, Q^{DDW}, Q^{DP}, Q^{APW}, Q^{ADW}, and Q^{AP} matrices, we can formulate the input parameters of the AGV-drone joint in-plant supply model as follows. The matrices of the available delivery tasks are as follows:

$$q_g^D = \begin{cases} \forall i \in [1, \ldots, \vartheta_1] : q_g^D = q_g^{DPW} \\ \forall i \in [1+\vartheta_1, \ldots, \vartheta_2] : q_g^D = q_{g-\vartheta_1}^{DDW}, \\ \forall i \in [1+\vartheta_2, \ldots, \vartheta_3] : q_g^D = q_{g-\vartheta_2}^{DP} \end{cases} \tag{46}$$

and

$$q_i^T = \begin{cases} \forall i \in [1,\ldots,\vartheta_4 - \vartheta_3] : q_i^A = q_i^{APW} \\ \forall i \in [\beta_4+1,\ldots,\beta_4+\beta_5] : q_i^A = q_{i-\beta_4}^{ADW} \\ \forall i \in [\beta_4+\beta_5+1,\ldots,\vartheta_6-\vartheta_3] : q_i^A = q_{i-\beta_4-\beta_5}^{AP} \end{cases} . \quad (47)$$

3.3.1. The Objective Function

In this milkrun supply model, both the energy consumption of the drone and the AGV-based milkrun trolley, and their virtual GHG emission can be defined. The minimization of the energy consumption performed by the drone and the AGV-based milkrun trolley is as follows:

$$C^3 = C^{3AWP} + C^{3AP} + C^{3APW} + C^{3DPW} + C^{3DP} + C^{3DWP} \to min., \quad (48)$$

where C^3 is the energy consumption of the whole drone-based integrated pick-up/delivery shuttle and milkrun services from AGV-based milkrun trolleys.

The energy consumption generated in the first milkrun section from the warehouse to the first production resource can be formulated depending on the length of milkrun sections between the warehouse and the first production resource, the loading of the AGV-based milkrun trolley, and the specific energy consumption:

$$C^{3AWP} = \left(q^{ANET} + \sum_{i=\beta_4+1, z_i^* = DWP}^{\beta_4+\beta_5} q_{p_i}^{*A} \right) \cdot l_{D,p_1} \cdot \rho^A. \quad (49)$$

The energy consumption of the AGV-based milkrun trolley within the in-plant supply route between the first and last production resources can be expressed depending on the length of the milkrun sections between the production resources, the loading of the AGV-based milkrun trolley, and the specific energy consumption:

$$C^{3AP} = \sum_{k=\beta_4+\beta_5+1}^{\vartheta_6-\vartheta_3} \left(q^{ANET} + \sum_{i=k, z_i^* = DWP}^{\vartheta_6-\vartheta_3} q_{p_i}^{*T} + \sum_{i=\beta_4+\beta_5+1, z_i^* \in [PP, DP]}^{k} q_{p_i}^{*A} \right) \cdot l_{p_k, p_{k+1}} \cdot \rho^A. \quad (50)$$

The energy consumption of the last milkrun section from the last production resource to the warehouse is a function of the length of the transportation between the last production resource and the warehouse, the loading of the AGV-based milkrun trolley, and the specific energy consumption:

$$C^{3APW} = \left(q^{ANET} + \sum_{i=\beta_4+1, z_i^* = PPW}^{\beta_4+\beta_5} q_{p_1}^{*A} \right) \cdot l_{D,p_1} \cdot \rho^A. \quad (51)$$

The drone's energy consumption generated between the first and last production resource performing pick-up supply services can be written as:

$$C^{3DP} = \sum_{k=\beta_4+\beta_5+1, z_k^* \in [PP]}^{\vartheta_6-\vartheta_3} \left(q^{DNET} + q_{p_k}^{*D} \right) \cdot 2 \cdot l_{p_k, p_{TR}} \cdot \rho^D. \quad (52)$$

where $l_{p_k, p_{TR}}$ is the travelling distance between the pick-up operation assigned to production resource p_k and the current position of the AGV-based milkrun trolley.

The drone's energy consumption between production resources and the warehouse is as follows:

$$C^{3DWP} = \sum_{k=1, z_k^* \in [PP]}^{\vartheta_4-\vartheta_3} \left(q^{DNET} + q_{p_k}^{*D} \right) \cdot 2 \cdot l_{p_k, D} \cdot \rho^D. \quad (53)$$

The energy consumption in the first section of the in-plant supply route from the warehouse to the first production resource for the drone is as follows:

$$C^{3DPW} = \left(q^{DNET} + \sum_{i=\beta_4+1, z_i^* = DWP}^{\beta_4+\beta_5} q_{p_i}^{*A} \right) \cdot l_{D,p_1} \cdot \rho^A. \quad (54)$$

The emission reduction, as objective function, can be expressed as a function of the source of energy generation and GHG's type:

$$E_{x,y}^3 = E_{x,y}^{3AWP} + E_{x,y}^{3AP} + E_{x,y}^{3APW} + E_{x,y}^{3DPW} + E_{x,y}^{3DP} + E_{x,y}^{3DWP} \to min. \quad (55)$$

The virtual GHG emission of the first section of milkrun from the warehouse to the first production resource in the case of the AGV-based milkrun trolley is as follows:

$$E_{x,y}^{3AWP} = \left(q^{ANET} + \sum_{i=\beta_4+1, z_i^* = DWP}^{\beta_4+\beta_5} q_{p_i}^{*A} \right) \cdot l_{D,p_1} \cdot \rho^A \cdot \varepsilon_{x,y}. \quad (56)$$

The virtual GHG emission between the first and last production resource in the case of the AGV-based milkrun trolley is as follows:

$$E^{3AP}_{x,y} = \sum_{k=\beta_4+\beta_5+1}^{\vartheta_6-\vartheta_3} \left(q^{ANET} + \sum_{i=k,z_i^*=DWP}^{\vartheta_6-\vartheta_3} q_{p_i}^{*T} + \sum_{i=\beta_4+\beta_5+1,z_i^* \in [PP,DP]}^{k} q_{p_i}^{*A} \right) \cdot l_{p_k,p_{k+1}} \cdot \rho^A \cdot \varepsilon_{x,y}. \tag{57}$$

The virtual GHG emission of the last section of the milkrun from the last production resource to the warehouse in the case of the AGV-based milkrun trolley is expressed as:

$$E^{3APW}_{x,y} = \left(q^{ANET} + \sum_{i=\beta_4+1,z_i^*=PPW}^{\beta_4+\beta_5} q_{p_i}^{*A} \right) \cdot l_{D,p_1} \cdot \rho^A \cdot \varepsilon_{x,y}. \tag{58}$$

The drone's virtual GHG emission between the first and last production resource is as follows:

$$E^{3DP}_{x,y} = \sum_{k=\beta_4+\beta_5+1, z_k^* \in [PP]}^{\vartheta_6-\vartheta_3} \left(q^{DNET} + q_{p_k}^{*D} \right) \cdot 2 \cdot l_{p_k,p_{TR}} \cdot \rho^D \cdot \varepsilon_{x,y}. \tag{59}$$

The drone's virtual GHG emission between production resources and warehouse is as follows:

$$E^{3DWP}_{x,y} = \sum_{k=1, z_k^* \in [PP]}^{\vartheta_4-\vartheta_3} \left(q^{DNET} + q_{p_k}^{*D} \right) \cdot 2 \cdot l_{p_k,D} \cdot \rho^D \cdot \varepsilon_{x,y}. \tag{60}$$

The drone's virtual GHG emission in the first milkrun section from the warehouse to the first production resource is as follows:

$$E^{3DPW}_{x,y} = \left(q^{DNET} + \sum_{i=\beta_4+1, z_i^*=DWP}^{\beta_4+\beta_5} q_{p_i}^{*A} \right) \cdot l_{D,p_1} \cdot \rho^A \cdot \varepsilon_{x,y}. \tag{61}$$

3.3.2. The Constraints

We can define constraints for in-plant supply services performed by both the AGV-based milkrun trolley and the drone.

Constraint 1 for AGV-based milkrun trolley: the in-plant supply route of the AGV-based milkrun trolley must be designed so that the upper limit of allowed payload must be taken into consideration:

$$\forall i \in [1,\ldots,\vartheta_6-\vartheta_3] : q_{p_i}^L \leq q^{Amax}, \tag{62}$$

where

$$\forall i \in [1\ldots\vartheta_6-\vartheta_3] : q_{p_i}^L = q^{ANET} + \sum_{i=k,z_i^*=DWP}^{\vartheta_6-\vartheta_3} q_{p_i}^{*A} + \sum_{i=1,z_i^* \in [PP,DP]}^{k} q_{p_i}^{*A}. \tag{63}$$

In the case of this constraint, the weight of the AGV can be calculated as the sum of the net weight of the AGV-based milkrun trolley, the weight of components delivered from the warehouse to the production resources, and the weight of components picked up and delivered between production resources. The difference between constraints (17–18), (36–37), and (62–63) is that the set of production resources to be taken into consideration is not the same in the case of the proposed models.

Constraint 2 for AGV-based milkrun trolley: the in-plant supply route must be designed so that the maximum loading volume of the AGV-based milkrun trolley must be taken into consideration:

$$\forall i \in [1,\ldots,\vartheta_6-\vartheta_3] : v_{p_i}^L \leq v^{Amax}, \tag{64}$$

where

$$\forall i \in [1,\ldots,\vartheta_6-\vartheta_3] : v_{p_i}^L = q^{ANET} + \sum_{i=k,z_i^*=DWP}^{\vartheta_6-\vartheta_3} v_{p_i}^{*A} + \sum_{i=1,z_i^* \in [PP,DP]}^{k} v_{p_i}^{*A}. \tag{65}$$

In the case of this constraint, the volume of the AGV can be calculated as the sum of the net weight of the AGV-based milkrun trolley, the weight of components delivered from the warehouse to the production resources, and the weight of components picked up and delivered between production resources. The difference between constraints (19–20), (38–39), and (64–65) is that the set of production resources to be taken into consideration is not the same in the case of the proposed models.

Constraint 3 for AGV-based milkrun trolley: the upper limit of available power capacity stored in the battery of the AGV-based milkrun trolley must be taken into consideration.

$$C^{3AWP} + C^{3AP} + C^{3APW} \leq CAP^A. \tag{66}$$

In the case of the drone, we can also define three constraints.

Constraint 1 for drone: in the case of shuttle service (no milkrun is performed by the drone), this constraint can be written as follows:

$$\forall i \in [1, \ldots, \vartheta_3] : q_{p_i} \leq q^{Dmax}, \tag{67}$$

where q^{Dmax} is the maximum payload of the drone. If the drone performs milkrun routes, the weight of the components picked up cannot be higher than the maximum payload:

$$\forall i : \sum_{i \in \theta} q_{p_i} \leq q^{Dmax}, \tag{68}$$

where θ is the set of milkrun routes performed by the drone.

In the case of this constraint, the current weight of the payload of the drone can be calculated as the weight of the current payload. The difference between Constraints (41), (42), (67) and (68) is that the set of production resources to be taken into consideration is not the same in the case of the proposed models.

Constraint 2 for drone: it is not allowed to exceed the maximum available loading volume of the drone. In the case of shuttle service (no milkrun is performed by the drone), this constraint can be written as follows:

$$\forall i \in [1, \ldots, \vartheta_3] : v_{p_i} \leq v^{Dmax}, \tag{69}$$

where v^{Dmax} is the upper limit of the loading. In this case, the weight of the components picked up cannot exceed the maximum payload of the drone:

$$\forall i : \sum_{i \in \theta} v_{p_i} \leq v^{Dmax}. \tag{70}$$

In the case of this constraint, the current volume of the payload of the drone can be calculated as the volume of the current payload. The difference between constraints (43–44) and (69–70) is that the set of production resources to be taken into consideration is not the same in the case of the proposed models.

Constraint 3 for drone: this constraint expresses the upper limit of power capacity stored in the battery of the drone's battery.

$$\forall i : \sum_{i \in \theta} C_i^{*2DP} \leq BAT^D \wedge \forall i : \sum_{i \in \theta} C_i^{*2DWP} \leq BAT^D, \tag{71}$$

where BAT^D is the capacity of the drone's battery.

3.3.3. Decision Variables

The decision variable of this drone-based integrated pick-up/delivery shuttle and milkrun services from AGV-based milkrun trolleys problem is the $\overline{p} = [p_i]$ permutation matrix. Within the frame of the next chapter, the above-mentioned models will be analyzed through five scenarios.

4. Results of Numerical Analyses

In this chapter, the above-discussed types of AGV-drone joint in-plant supply solutions are analyzed. The above-mentioned models are solved using the heuristic option of the Excel Solver. The scenario analysis focuses on the following three models:

- AGV-based in-plant supply of production resources without drones: in this in-plant supply model, pick-up and delivery operations are performed by AGV-based milkrun.
- Drone-based pick-up services from AGV-based milkrun: in this model, the suitable pick-up services are performed by the drone from AGV-based milkrun. The suitability depends on the weight constraints of the delivery drone.
- Drone-based delivery services from AGV-based milkrun: in this model, the suitable delivery services are performed by the drone from AGV-based milkrun.

- Integrated pick-up/delivery shuttle services from AGV-based milkrun: in this model, all suitable pick-up and delivery services are performed as shuttle services by the drone from AGV-based milkrun.
- Integrated pick-up/delivery milkrun services from AGV-based milkrun: in this model, all suitable pick-up and delivery services are performed as milkrun services by the drone from AGV-based milkrun.

The input parameters of the AGV-drone joint in-plant supply models are the following: location of production resources (see Table A1), weight and volume of components to be transported from/to production resources (see Table A2), maximum payload of AGV-based milkrun trolleys and drones, available maximum capacity of battery in AGV-based milkrun trolleys and drones, specific energy consumption of AGV-based milkrun trolleys and drones, specific emission of CO_2, SO_2, CO, HC, NO_X, and PM depending on the electricity generation source mix.

The maximum payload of AGV-based milkrun trolley is 80 kg, while the carrying capability of the drone is 5 kg. The average speed of the AGV-based milkrun trolley is about 0.4 m/s; the average loading and unloading time of pick-up and delivery services is 32 s per component. The scenarios take the impact of the weight of components on the loading time into consideration. The standard energy consumption of the AGV-based milkrun trolley is about 120 Wh/km, for the drone it is about 25 Wh/km, but this energy consumption can be influenced by the load of the AGV-based milkrun trolley and the drone. The specific virtual CO_2, SO_2, CO, HC, NO_X, and PM emission is shown in Table A3 [39]. In the analyzed scenarios, there are 25 milkruns per shift, and the analyzed production plan has six production lines. The scenarios show one milkrun in the case of a production line, and the energy consumption and the virtual CO_2, SO_2, CO, HC, NO_X, and PM emission is calculated for a whole shift including 25 milkruns and six production lines.

4.1. Scenario 1: AGV-Based In-Plant Supply of Production Resources without Drones

The total length of the optimized AGV-based in-plant supply (see Table A4 and Figure 2) performed by the AGV-based milkrun trolley is 434.56 m, the required transportation time is 1165 s, and the required loading and unloading time is 754 s.

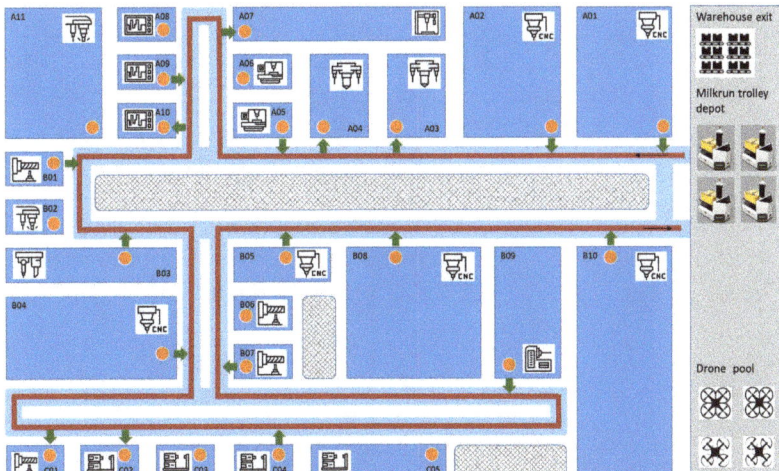

Figure 2. The scheduled and performed in-plant supply operations by AGV-based milkrun trolley in the case of Scenario 1 (green arrow is for pick-up and delivery operations of AGV-based milkrun trolley, black arrow shows the direction of the milkrun route, orange dots are for pick-up and delivery locations of machines).

The virtual CO_2, SO_2, CO, HC, NO_X, and PM emission of the AGV-based in-plant supply can be calculated based on the total length of the milkrun routes and the loading of trolleys. The energy consumption of the AGV-based in-plant supply is 156.9 Wh in the case of the analyzed route; the total energy consumption for the shift including 25 milkruns and six production lines is 23.56 kWh. Table 1 shows the virtual CO_2, SO_2, CO, HC, NO_X, and PM emission.

Table 1. The virtual CO_2, SO_2, CO, HC, NO_X, and PM emission of AGV-based milkrun trolleys in the case of Scenario 1.

Electricity Generation Source	Emission (g)					
	CO_2	SO_2	CO	HC	NO_X	PM
Lignite	24,799.485	0.753	20.705	11.294	111.998	0.941
Coal	20,893.684	0.659	17.247	9.412	93.175	0.706
Oil	17,246.701	0.518	14.470	7.882	78.210	0.659
Natural gas	11,740.933	0.376	9.835	5.365	52.375	0.447
Photovoltaic	1999.958	0.047	1.718	0.941	9.317	0.071
Biomass	1058.802	0.024	0.894	0.494	4.823	0.047
Nuclear	682.339	<0.001	0.565	0.306	3.106	0.024
Water	611.752	<0.001	0.518	0.282	2.800	0.024
Wind	611.752	<0.001	0.518	0.282	2.800	0.024
Mix 1: 40% Oil–60%Biomass	7533.961	0.221	6.324	3.449	34.178	0.291
Mix 2: 60% Coal–40%Water	12,780.91	0.395	10.555	5.759	57.024	0.432

4.2. Scenario 2: AGV-Drone Joint In-Plant Supply: Pick-Up Service by Drones

The length of the AGV-based milkrun trolley per milkrun route is 383.04 m, the required transportation time is 957.6 s, and the required loading and unloading time is 669 s (see Table A5 and Figure 3). The length of the drone per milkrun route is 423.4 m, the required transportation time is 201.6 s, and the required loading and unloading time is 84 s (see Table A6 and Figure 3).

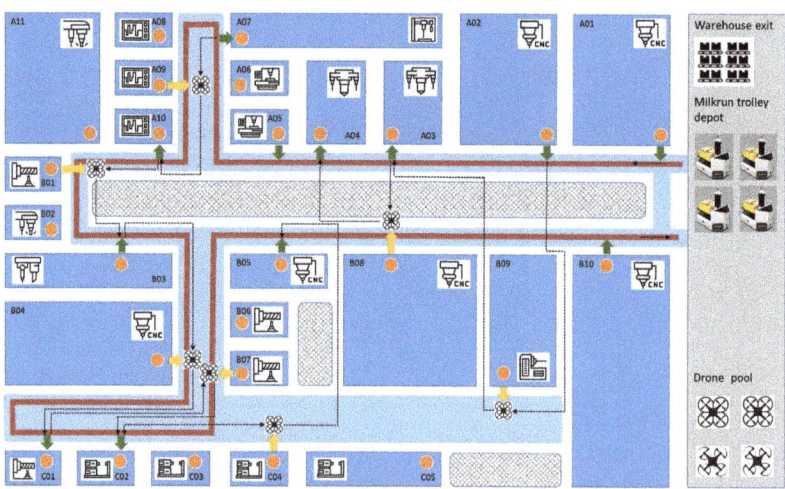

Figure 3. The scheduled and performed in-plant supply operations, where suitable pick-up services are performed by the drone (green arrow is for pick-up and delivery operations of AGV-based milkrun trolley, yellow arrow is for pick-up and delivery operations of drone, black arrow shows the direction of the milkrun route, orange dots are for pick-up and delivery locations of machines).

The virtual CO_2, SO_2, CO, HC, NO_X, and PM emission of the AGV-based in-plant supply can be calculated based on the total length of the milkrun routes and the loading of trolleys. The energy consumption of the AGV-based milkrun trolley is 133.9 Wh in the case of the analyzed route; the total energy consumption for the shift including 25 milkruns and six production lines is 20.09 kWh. The energy consumption of the drone is 11.18 Wh in the case of the analyzed route; the total energy consumption for the shift including 25 milkruns and six production lines is 1.67 kWh. The total energy consumption of the AGV-drone joint in-plant supply with pick-up service by drones for the shift including 25 milkruns and six production lines is 21.76 kWh, which means an 8% savings in energy cost. Table 2 shows the virtual CO_2, SO_2, CO, HC, NO_X, and PM emission.

Table 2. The virtual CO_2, SO_2, CO, HC, NO_X, and PM emission of AGV-based milkrun trolley and drone in the case of Scenario 2.

Electricity Generation Source	Emission (g)					
	CO_2	SO_2	CO	HC	NO_X	PM
Lignite	22,944.168	0.697	19.156	10.449	103.619	0.871
Coal	19,330.570	0.610	15.956	8.707	86.204	0.653
Oil	15,956.428	0.479	13.388	7.293	72.359	0.610
Natural gas	10,862.561	0.348	9.099	4.963	48.457	0.414
Photovoltaic	1850.336	0.044	1.589	0.871	8.620	0.065
Biomass	979.590	0.022	0.827	0.457	4.463	0.044
Nuclear	631.291	<0.001	0.522	0.283	2.873	0.022
Water	565.985	<0.001	0.479	0.261	2.590	0.022
Wind	565.985	<0.001	0.479	0.261	2.590	0.022
Mix 1: 40% Oil–60% Biomass	6970.324	0.204	5.851	3.191	31.621	0.269
Mix 2: 60% Coal–40% Water	11,824.736	0.365	9.765	5.328	52.758	0.400

4.3. Scenario 3: AGV-Drone Joint In-Plant Supply: Delivery Service by Drones

The length of the AGV-based milkrun trolley per milkrun route is 414.4 m, the required transportation time is 1036 s, and the required loading and unloading time is 599.4 s (see Table A7 and Figure 4). The length of the drone per milkrun route is 213.9 m, the required transportation time is 101.9 s, and the required loading and unloading time is 48 s (see Table A8 and Figure 4).

The virtual CO_2, SO_2, CO, HC, NO_X, and PM emission of the AGV-based in-plant supply can be calculated based on the total length of the milkrun routes and the loading of trolleys. The energy consumption of the AGV-based milkrun trolley is 132 Wh in the case of the analyzed route; the total energy consumption for the shift including 25 milkruns and six production lines is 19.79 kWh. The energy consumption of the drone is 5.53 Wh in the case of the analyzed route; the total energy consumption for the shift including 25 milkruns and six production lines is 0.83 kWh. The total energy consumption of the AGV-drone joint in-plant supply with pick-up service by drones for the shift including 25 milkruns and xix production lines is 20.62 kWh, which means a 12.5% savings in energy consumption cost. Table 3 shows the virtual CO_2, SO_2, CO, HC, NO_X, and PM emission.

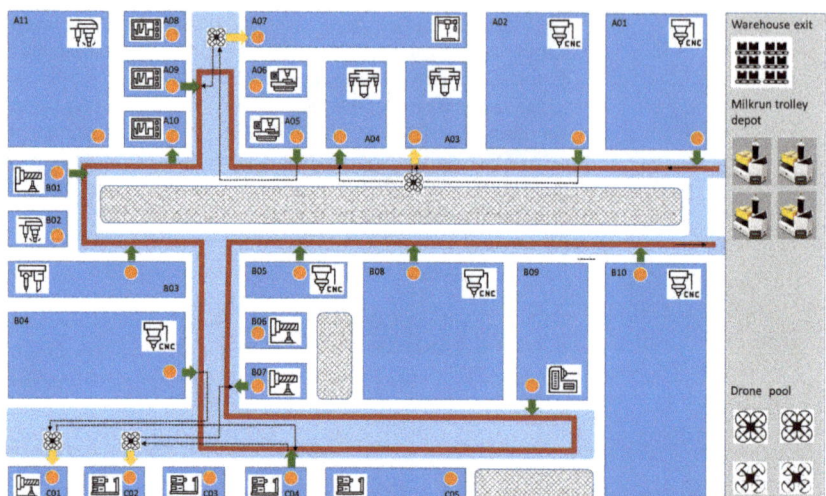

Figure 4. The scheduled and performed in-plant supply operations, where suitable delivery operations are performed by drone (green arrow is for pick-up and delivery operations of AGV-based milkrun trolley, yellow arrow is for pick-up and delivery operations of drone, black arrow shows the direction of the milkrun route, orange dots are for pick-up and delivery locations of machines).

Table 3. The virtual CO_2, SO_2, CO, HC, NO_X, and PM emission of AGV-based milkrun trolley and drone in the case of Scenario 3.

Electricity Generation Source	Emission (g)					
	CO_2	SO_2	CO	HC	NO_X	PM
Lignite	21,738.626	0.660	18.150	9.900	98.174	0.825
Coal	18,314.895	0.577	15.118	8.250	81.675	0.619
Oil	15,118.039	0.454	12.684	6.909	68.557	0.577
Natural gas	10,291.816	0.330	8.621	4.702	45.911	0.392
Photovoltaic	1753.115	0.041	1.506	0.825	8.167	0.062
Biomass	928.120	0.021	0.784	0.433	4.228	0.041
Nuclear	598.122	<0.001	0.495	0.268	2.722	0.021
Water	536.247	<0.001	0.454	0.247	2.454	0.021
Wind	536.247	<0.001	0.454	0.247	2.454	0.021
Mix 1: 40% Oil–60%Biomass	6604.087	0.193	5.543	3.023	29.959	0.255
Mix 2: 60% Coal–40%Water	11,203.435	0.346	9.252	5.048	49.986	0.379

4.4. Scenario 4: AGV-Drone Joint In-Plant Supply: Shuttle Supply Services by the Drone

The length of the AGV-based milkrun trolley per milkrun route is 253.12 m, the required transportation time is 600.8 s, and the required loading and unloading time is 502.9 s (see Table A9 and Figure 5). The length of the drone per milkrun route is 650.7 m, the required transportation time is 309.8 s, and the required loading and unloading time is 96 s (see Table A10 and Figure 5).

The virtual CO_2, SO_2, CO, HC, NO_X, and PM emission of the AGV-based in-plant supply can be calculated based on the total length of the milkrun routes and the loading of trolleys. The energy consumption of the AGV-based milkrun trolley is 93.2 Wh in the case of the analyzed route; the total energy consumption for the shift including 25 milkruns and six production lines is 13.98 kWh. The energy consumption of the drone is 17.14 Wh in the case of the analyzed route; the total energy consumption for the shift including 25 milkruns and six production lines is 2.57 kWh. The total energy consumption of the AGV-drone joint in-plant supply with pick-up service by drones for the shift including 25 milkruns and six production lines is 16.55 kWh, which means a 30% savings in energy consumption cost. Table 4 shows the virtual CO_2, SO_2, CO, HC, NO_X, and PM emission.

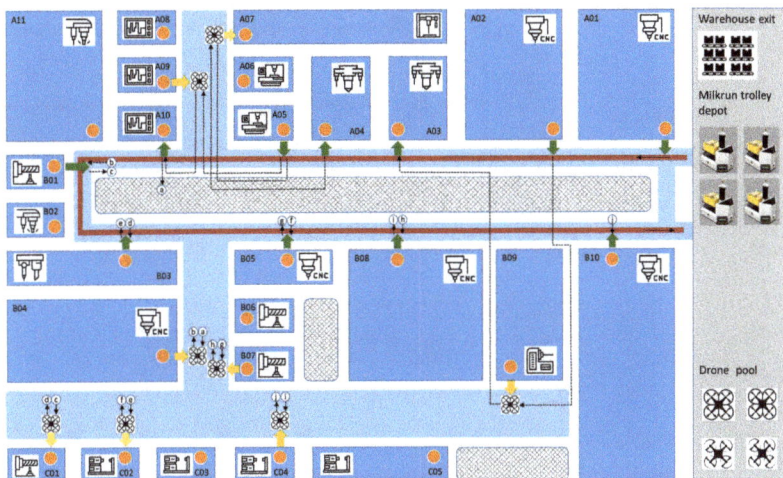

Figure 5. The scheduled and performed in-plant supply operations, where all suitable services are performed by drone as shuttle services (green arrow is for pick-up and delivery operations of AGV-based milkrun trolley, yellow arrow is for pick-up and delivery operations of drone, black arrow shows the direction of the milkrun route, orange dots are for pick-up and delivery locations of machines, letters from a-j represent the routes performed by the drone).

Table 4. The virtual CO_2, SO_2, CO, HC, NO_X, and PM emission of AGV-based milkrun trolleys in g in the case of Scenario 5.

Electricity Generation Source	Emission (g)					
	CO_2	SO_2	CO	HC	NO_X	PM
Lignite	17,459.254	0.530	14.577	7.951	78.848	0.663
Coal	14,709.504	0.464	12.142	6.626	65.596	0.497
Oil	12,141.967	0.364	10.187	5.549	55.061	0.464
Natural gas	8265.814	0.265	6.924	3.777	36.873	0.315
Photovoltaic	1408.004	0.033	1.209	0.663	6.560	0.050
Biomass	745.414	0.017	0.629	0.348	3.396	0.033
Nuclear	480.378	<0.001	0.398	0.215	2.187	0.017
Water	430.684	<0.001	0.364	0.199	1.971	0.017
Wind	430.684	<0.001	0.364	0.199	1.971	0.017
Mix 1: 40% Oil–60%Biomass	5304.035	0.155	4.452	2.428	24.061	0.205
Mix 2: 60% Coal–40%Water	8997.975	0.278	7.430	4.055	40.146	0.304

4.5. Scenario 5: AGV-Drone Joint In-Plant Supply: Milkrun Supply Services by the Drone

The length of the AGV-based milkrun trolley per milkrun route is 253.12 m, the required transportation time is 600.8 s, and the required loading and unloading time is 494.8 s (see Table A11 and Figure 6). The length of the drone per milkrun route is 254.24 m, the required transportation time is 121 s, and the required loading and unloading time is 96 s (see Table A12 and Figure 6).

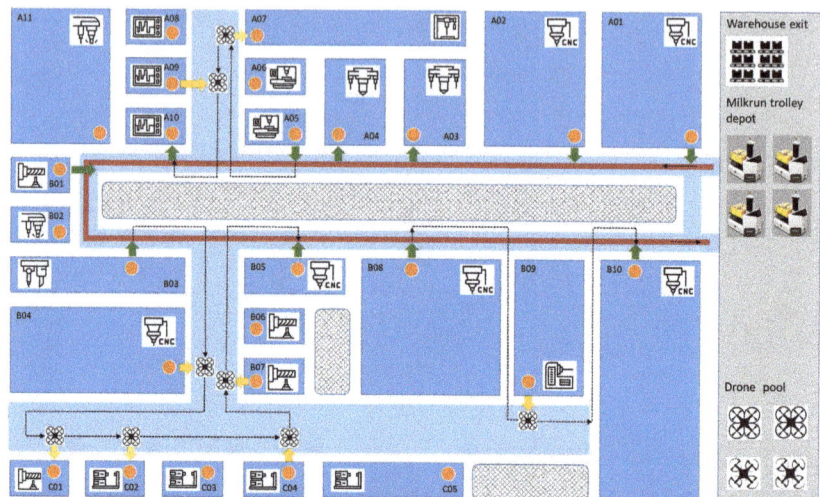

Figure 6. The scheduled and performed in-plant supply operations, where all suitable services are performed by drone as milkruns (green arrow is for pick-up and delivery operations of AGV-based milkrun trolley, yellow arrow is for pick-up and delivery operations of drone, black arrow shows the direction of the milkrun route, orange dots are for pick-up and delivery locations of machines).

The virtual CO_2, SO_2, CO, HC, NO_X, and PM emission of the AGV-based in-plant supply can be calculated based on the total length of the milkrun routes and the loading of trolleys. The energy consumption of the AGV-based milkrun trolley is 91.85 Wh in the case of the analyzed route; the total energy consumption for the shift including 25 milkruns and six production lines is 13.77 kWh. The energy consumption of the drone is 7.04 Wh in the case of the analyzed route; the total energy consumption for the shift including 25 milkruns and six production lines is 1.05 kWh. The total energy consumption of the AGV-drone joint in-plant supply with pick-up service by drones for the shift including 25 milkruns and six production lines is 14.82 kWh, which means a 37% savings in energy consumption cost. The virtual CO_2, SO_2, CO, HC, NO_X, and PM emission depending on the generation source of the electricity is shown in Table 5.

Table 5. The virtual CO_2, SO_2, CO, HC, NO_X, and PM emission of AGV-based milkrun trolley and drone depending on the electricity generation source in CO_2 emission in g in the case of Scenario 5.

Electricity Generation Source	Emission					
	CO_2	SO_2	CO	HC	NO_X	PM
Lignite	15,635.705	0.475	13.054	7.121	70.613	0.593
Coal	13,173.156	0.415	10.874	5.934	58.745	0.445
Oil	10,873.787	0.326	9.123	4.970	49.310	0.415
Natural gas	7402.483	0.237	6.201	3.382	33.022	0.282
Photovoltaic	1260.944	0.030	1.083	0.593	5.875	0.045
Biomass	667.559	0.015	0.564	0.312	3.041	0.030
Nuclear	430.204	0.000	0.356	0.193	1.958	0.015
Water	385.701	0.000	0.326	0.178	1.765	0.015
Wind	385.701	0.000	0.326	0.178	1.765	0.015
Mix 1: 40% Oil–60%Biomass	4750.050	0.139	3.988	2.175	21.549	0.184
Mix 2: 60% Coal–40%Water	8058.174	0.249	6.655	3.632	35.953	0.273

5. Discussion

The conventional models of milkrun-based in-plant supply solutions consider only the AGV-based solutions [17,23]. Drones open new perspectives to support the flexible material handling solutions in the case of small-sized components. Within the frame of this article, the new approach focuses on the potentials of AGV-drone joint supply, and this is the superiority of the proposed model.

Within the frame of this research work, the energy efficiency of AGV-drone joint in-plant supply of production lines is discussed. An AGV-drone joint in-plant supply solution is described focusing on the potentials of pick-up and delivery operation of small components suitable for drone-based supply. The mathematical description of AGV-drone joint in-plant supply solutions focuses on the evaluation of different models from the perspective of energy efficiency and GHG emission. The numerical analysis of the different scenarios shows the significant impact of different AGV-drone joint in-plant supply solutions on energy efficiency. As the different scenarios show, the AGV-drone joint in-plant supply solutions can lead to a 5 to 30% energy consumption reduction, depending on the cooperation level of AGV-based milkrun trolley and drone.

The potential implications of the study can be summarized as follows:

- The conventional AGV-based in-plant supply solutions can be improved by the application of drones. The proposed approach makes it possible to model and analyze the joint AGV-drone service of manufacturing and assembly resources.
- In the case of electrical materials handling resources, it is possible to calculate the virtual GHG emission, which depends on the electricity generation source. By taking into account the virtual GHG emission, it is possible to compute the GHG emission caused by electricity generation required for the materials handling operation in the discussed in-plant system, which gives a more realistic picture of the ecological footprint of the logistics process.
- The AGV-drone joint in-plant supply solution can lead to a decreased energy consumption and GHG emission. This energy consumption and GHG emission reduction depends on the type of the service processes (e.g., shuttle or milkrun service by drones, electricity generation source, master schedule of the production plant). In the analyzed scenarios, the energy consumption and emission reduction was 5% in the case of drone-based pick-up operations, while it was about 30% in the case of integrated pick-up delivery services using milkrun drone routes instead of shuttle services of drones.

More generally, this paper focuses on the mathematical description of supply processes of production lines including capacity and energy-related constraints. Why is so much effort being put into this research? The importance of in-plant supply and intralogistics has increased in the last few years from the conventional in-advance design processes to the dynamic, real-time processes, where the parameters of in-plant supply solutions (capacities, resources, scheduling) can be dynamically changed depending on the demands of production resources.

The added value of the paper is the description of the AGV-drone joint in-plant supply solution, which makes it possible to analyze the impact of different solutions on the energy efficiency of the in-plant material handling operations. The scientific contribution of this paper for researchers in this field is the mathematical modelling of the AGV-drone joint in-plant supply solution. The results can be generalized because the model can be applied for different production environments and warehouses [37], and it can also be applied in the case of value chains and supply chains [32]. The described method makes it possible to support managerial decisions, because by depending on the results of analysis of different potential solutions, it is possible to influence the supply strategies and the investment decisions.

However, there are also limitations of the study. This study took capacities and energy consumption-related parameters and deterministic parameters into consideration. Fuzzy models can be used to analyze the impact of stochastic parameters on energy efficiency.

Other future research direction is the integration of milkrun design and the material handling equipment selection for production workplaces [40].

Funding: This research received no external funding.

Data Availability Statement: Not applicable.

Conflicts of Interest: The author declares no conflict of interest.

Appendix A

Table A1. Layout parameters of the production lines: location of production resources.

ID	Name	Coordinate [m]		ID	Name	Coordinate [m]	
		X	Y			X	Y
A01	CNC Milling 012	96	50	B03	Milling 035	19	32
A02	CNC Milling 014	80	50	B04	CNC Milling 048	23	19
A03	CNC Drilling 032	58	50	B05	CNC Milling 048	42	32
A04	CNC Drilling 034	47	50	B06	Turning X26	36	24
A05	CNC Honing 051	42	50	B07	Turning X28	36	17
A06	CNC Honing 052	36	57	B08	CNC Milling 049	58	32
A07	CNC Honing 054	36	65	B09	Turning 217	73	17
A08	Inspection 082	24	65	B10	CNC Milling 126	88	32
A09	Inspection 083	24	57	C01	Finishing C01	8	4
A10	Inspection 084	24	50	C02	Finishing C02	18	4
A11	Shaping 095	14	50	C03	Finishing C03	29	4
B01	Turning X22	8	46	C04	Finishing C04	41	4
B02	Milling 024	8	27	C05	Finishing C05	62	4

Table A2. Weight of components to be picked up or delivered.

ID	Weight [kg]	Type of Service	ID	Weight (kg)	Type of Service
A01	12	PUbAGV [1]	B04	0.5	PUbDRONE [4]
A02	24	PUbAGV [1]	B05	9	DbAGV [3]
A03	2	DbDRONE [2]	B07	2.4	PUbDRONE [4]
A04	8	DbAGV [3]	B08	1	PUbDRONE [4]
A05	12	PUbAGV [1]	B09	5	PUbAGV [1]
A07	1.5	DbDRONE [2]	B10	8	PUbAGV [1]
A09	2.1	PUbDRONE [4]	C01	0.8	DbDRONE [2]
A10	9	DbAGV [3]	C02	2	DbDRONE [2]
B01	1.2	PUbDRONE [4]	C04	1.6	PUbDRONE [4]
B03	25	PUbAGV [1]	-	-	-

[1] PUbAGV = Pick-up service task suitable for AGV. [2] DbDRONE = Delivery service task suitable for drone. [3] DbAGV = Delivery service task suitable for AGV. [4] PUbDRONE = Pick-up service task suitable for drone.

Table A3. Specific CO_2, SO_2, CO, HC, NO_X, and PM emission in CO_2 emission in g/kWh [39].

Electricity Generation Source	Emission					
	CO_2	SO_2	CO	HC	NO_X	PM
Lignite	1054	0.032	0.880	0.480	4.760	0.040
Coal	888	0.028	0.733	0.400	3.960	0.030
Oil	733	0.022	0.615	0.335	3.324	0.028
Natural gas	499	0.016	0.418	0.228	2.226	0.019
Photovoltaic	85	0.002	0.073	0.040	0.396	0.003
Biomass	45	0.001	0.038	0.021	0.205	0.002
Nuclear	29	$<10^{-3}$	0.024	0.013	0.132	0.001
Water	26	$<10^{-3}$	0.022	0.012	0.119	0.001
Wind	26	$<10^{-3}$	0.022	0.012	0.119	0.001

Appendix B

Table A4. The scheduled and performed in-plant supply operations by AGV-based milkrun trolley in the case of Scenario 1.

ID and Name of Production Resource	Type of Service	TL * (kg)	ID and Name of Production Resource	Type of Service	TL * (kg)
Warehouse	-	226	B04 CNC Milling 048	PUbDRONE [4]	282.3
A01 CNC Milling 012	PUbAGV [1]	238	C01 Finishing C01	DbDRONE [2]	281.5
A02 CNC Milling 014	PUbAGV [1]	262	C02 Finishing C02	DbDRONE [2]	279.5
A03 CNC Drilling 032	DbDRONE [2]	260	C04 Finishing C04	PUbDRONE [4]	281.1
A04 CNC Drilling 034	DbAGV [3]	252	B09 Turning 217	PUbDRONE [4]	286.1
A05 CNC Honing 051	PUbAGV [1]	264	B07 Turning X28	PUbDRONE [4]	288.5
A07 CNC Honing 054	DbDRONE [2]	262.5	B05 CNC Milling 048	DbAGV [3]	279.5
A09 Inspection 083	PUbDRONE [4]	264.6	B08 CNC Milling 049	PUbDRONE [4]	280.5
A10 Inspection 084	DbAGV [3]	255.6	B10 CNC Milling 126	PUbAGV [1]	288.5
B01 Turning X22	PUbDRONE [4]	256.8	Warehouse	-	288.5
B03 Milling 035	PUbAGV [1]	281.8	-	-	-

[1] PUbAGV = Pick-up service task suitable for AGV. [2] DbDRONE = Delivery service task suitable for drone. [3] DbAGV = Delivery service task suitable for AGV. [4] PUbDRONE = Pick-up service task suitable for drone. * TL = total load.

Table A5. The scheduled and performed in-plant supply operations by AGV-based milkrun trolley in the case of Scenario 2.

ID and Name of Production Resource	Type of Service	TL * (kg)	ID and Name of Production Resource	Type of Service	TL * (kg)
Warehouse	-	226	A10 Inspection 084	DbAGV [3]	261.5
A01 CNC Milling 012	PUbAGV [1]	238	B03 Milling 035	PUbAGV [1]	287.8
A02 CNC Milling 014	PUbAGV [1]	262	C01 Finishing C01	DbDRONE [2]	287.5
A03 CNC Drilling 032	DbDRONE [2]	265	C02 Finishing C02	DbDRONE [2]	287.9
A04 CNC Drilling 034	DbAGV [3]	258	B05 CNC Milling 048	DbAGV [3]	280.5
A05 CNC Honing 051	PUbAGV [1]	270	B10 CNC Milling 126	PUbAGV [1]	288.5
A07 CNC Honing 054	DbDRONE [2]	268.5	Warehouse	-	288.5

[1] PUbAGV = Pick-up service task suitable for AGV. [2] DbDRONE = Delivery service task suitable for drone. [3] DbAGV = Delivery service task suitable for AGV. * TL = total load.

Table A6. The scheduled and performed in-plant pick-up services by the drone in the case of Scenario 2.

ID and Name of Production Resource			Type of Service	Weight (kg)
From	Pick-Up Location	To		
A02 CNC Milling 014	B09 Turning 217	A03 CNC Drilling 032	PUbDRONE [1]	5
A03 CNC Drilling 032	B08 CNC Milling 049	A04 CNC Drilling 034	PUbDRONE [1]	1
A07 CNC Honing 054	A09 Inspection 083	A10 Inspection 084	PUbDRONE [1]	2.1
A10 Inspection 084	B01 Turning X22	B03 Milling 035	PUbDRONE [1]	1.2
B03 Milling 035	B04 CNC Milling 048	C01 Finishing C01	PUbDRONE [1]	0.5
C01 Finishing C01	B07 Turning X28	C02 Finishing C02	PUbDRONE [1]	2.4
C02 Finishing C02	C04 Finishing C04	B05 CNC Milling 048	PUbDRONE [1]	1.6

[1] PUbDRONE = Pick-up service task suitable for drone.

Table A7. The scheduled and performed in-plant supply operations by AGV-based milkrun trolley in the case of Scenario 3.

ID and Name of Production Resource	Type of Service	TL * (kg)	ID and Name of Production Resource	Type of Service	TL * (kg)
Warehouse	-	226	B04 CNC Milling 048	PUbDRONE [3]	281.5
A01 CNC Milling 012	PUbAGV [1]	238	C04 Finishing C04	PUbDRONE [3]	281.1
A02 CNC Milling 014	PUbAGV [1]	260	B09 Turning 217	PUbAGV [1]	286.1
A04 CNC Drilling 034	DbAGV [2]	252	B07 Turning X28	PUbDRONE [3]	288.5
A05 CNC Honing 051	PUbAGV [1]	262.5	B05 CNC Milling 048	DbAGV [2]	279.5
A09 Inspection 083	PUbDRONE [3]	264.6	B08 CNC Milling 049	PUbDRONE [3]	280.5
A10 Inspection 084	DbAGV [2]	255.6	B10 CNC Milling 126	PUbAGV [1]	288.5
B01 Turning X22	PUbDRONE [3]	256.8	Warehouse	-	-
B03 Milling 035	PUbAGV [1]	281.8	-	-	-

[1] PUbAGV = Pick-up service task suitable for AGV. [2] DbAGV = Delivery service task suitable for AGV. [3] PUbDRONE = Pick-up service task suitable for drone. * TL = total load.

Table A8. The scheduled and performed in-plant delivery services by the drone in the case of Scenario 3.

ID and Name of Production Resource			Type of Service	Weight (kg)
From	Pick-Up Location	To		
A02 CNC Milling 014	A03 CNC Drilling 032	A04 CNC Drilling 034	DbDRONE [1]	2
A05 CNC Honing 051	A07 CNC Honing 054	A09 Inspection 083	DbDRONE [1]	1.5
B04 CNC Milling 048	C01 Finishing C01	C04 Finishing C04	DbDRONE [1]	0.8
C04 Finishing C04	C02 Finishing C02	B07 Turning X28	DbDRONE [1]	2

[1] DbDRONE = Delivery service task suitable for drone.

Table A9. The scheduled and performed in-plant supply operations by AGV-based milkrun trolley in the case of Scenario 4.

ID and Name of Production Resource	Type of Service	TL * (kg)	ID and Name of Production Resource	Type of Service	TL * (kg)
Warehouse	-	226	B01 Turning X22	PUbDRONE [4]	261.5
A01 CNC Milling 012	PUbAGV [1]	238	B03 Milling 035	PUbAGV [1]	284.5
A02 CNC Milling 014	PUbAGV [1]	262	B05 CNC Milling 048	DbAGV [3]	275.5
A03 CNC Drilling 032	DbDRONE [2]	265	B08 CNC Milling 049	PUbDRONE [4]	278.9
A04 CNC Drilling 034	DbAGV [3]	255.5	B10 CNC Milling 126	PUbAGV [1]	288.5
A05 CNC Honing 051	PUbAGV [1]	267.5	Warehouse	-	-
A10 Inspection 084	DbAGV [3]	260.6	-	-	-

[1] PUbAGV = Pick-up service task suitable for AGV. [2] DbDRONE = Delivery service task suitable for drone. [3] DbAGV = Delivery service task suitable for AGV. [4] PUbDRONE = Pick-up service task suitable for drone. * TL = total load.

Table A10. The scheduled and performed in-plant supply services by the drone in the case of Scenario 4.

ID and Name of Production Resource			Type of Service	Weight (kg)
From	Pick-Up Location	To		
A02 CNC Milling 014	B09 Turning 217	A03 CNC Drilling 032	PUbDRONE [1]	5
A04 CNC Drilling 034	A07 CNC Honing 054	A05 CNC Honing 051	DbDRONE [2]	1.5
A05 CNC Honing 051	A09 Inspection 083	A10 Inspection 084	PUbDRONE [1]	2.1
A10 Inspection 084	B04 CNC Milling 048	B01 Turning X22	PUbDRONE [1]	0.5
B01 Turning X22	C01 Finishing C01	B03 Milling 035	DbDRONE [2]	0.8
B03 Milling 035	C02 Finishing C02	B05 CNC Milling 048	DbDRONE [2]	2
B05 CNC Milling 048	B07 Turning X28	B08 CNC Milling 049	PUbDRONE [1]	2.4
B08 CNC Milling 049	C04 Finishing C04	B10 CNC Milling 126	PUbDRONE [1]	1.6

[1] PUbDRONE = Pick-up service task suitable for drone. [2] DbDRONE = Delivery service task suitable for drone.

Table A11. The scheduled and performed in-plant supply operations by AGV-based milkrun trolley in the case of Scenario 5.

ID and Name of Production Resource	Type of Service	TL * (kg)	ID and Name of Production Resource	Type of Service	TL * (kg)
Warehouse	-	226	B01 Turning X22	PUbDRONE [4]	256.8
A01 CNC Milling 012	PUbAGV [1]	238	B03 Milling 035	PUbAGV [1]	279
A02 CNC Milling 014	PUbAGV [1]	262	B05 CNC Milling 048	DbAGV [3]	274.5
A03 CNC Drilling 032	DbDRONE [2]	260	B08 CNC Milling 049	PUbDRONE [4]	275.5
A04 CNC Drilling 034	DbAGV [3]	252	B10 CNC Milling 126	PUbAGV [1]	288.5
A05 CNC Honing 051	PUbAGV [1]	262.5	Warehouse	-	-
A10 Inspection 084	DbAGV [3]	255.6	-	-	-

[1] PUbAGV = Pick-up service task suitable for AGV. [2] DbDRONE = Delivery service task suitable for drone.
[3] DbAGV = Delivery service task suitable for AGV. [4] PUbDRONE = Pick-up service task suitable for drone.
* TL = total load.

Table A12. The scheduled and performed in-plant supply operations by the drone in the case of Scenario 5.

ID and Name of Production Resource	Type of Service	Load (kg)	TL * (kg)
Route 1			
A05 CNC Honing 051	DLAGV [1]	1.5	1.5
A07 CNC Honing 054	DbDRONE [2]	−1.5	0
A09 Inspection 083	PUbDRONE [3]	2.1	2.1
A10 Inspection 084	DAAGV [4]	−2.1	0
Route 2			
B03 Milling 035	DLAGV [1]	2.8	2.8
B04 CNC Milling 048	PUbDRONE [3]	0.5	3.3
C01 Finishing C01	DbDRONE [2]	−0.8	2.5
C02 Finishing C02	DbDRONE [2]	−2	0.5
C04 Finishing C04	PUbDRONE [3]	1.6	2.1
B07 Turning X28	PUbDRONE [3]	2.4	4.5
B05 CNC Milling 048	DAAGV [4]	−4.5	0
Route 3			
B08 CNC Milling 049	DLAGV [1]	0	0
B09 Turning 217	PUbDRONE [3]	5	5
B10 CNC Milling 126	DAAGV [4]	−5	0

[1] DLAGV = Drone left AGV. [2] DbDRONE = Delivery service task suitable for drone. [3] PUbDRONE = Pick-up service task suitable for drone. [4] DAAGV = Drone arrived to AGV. * TL = total load.

References

1. Teschemacher, U.; Reinhart, G. Ant Colony Optimization Algorithms to Enable Dynamic Milkrun Logistics. *Procedia CIRP* **2017**, *63*, 762–767. [CrossRef]
2. Kilic, H.S.; Durmusoglu, M.B.; Baskak, M. Classification and modeling for in-plant milk-run distribution systems. *Int. J. Adv. Manuf. Technol.* **2012**, *62*, 1135–1146. [CrossRef]
3. Unmanned Systems: Drone Service Market. Available online: https://www.fortunebusinessinsights.com/drone-services-market-102682 (accessed on 15 April 2023).
4. Wang, X.; Jiang, R.; Qi, M. A robust optimization problem for drone-based equitable pandemic vaccine distribution with uncertain supply. *Omega* **2023**, *119*, 102872. [CrossRef] [PubMed]
5. Juned, M.; Sangle, P.; Gudheniya, N.; Haldankar, P.V.; Tiwari, M.K. Designing the drone based end-to-end local supply chain distribution network. *IFAC-PapersOnLine* **2022**, *55*, 743–748. [CrossRef]
6. Farajzadeh, F.; Moadab, A.; Valilai, O.F.; Houshmand, M. A Novel Mathematical Model for a Cloud-Based Drone Enabled Vehicle Routing Problem considering Multi-Echelon Supply Chain. *IFAC-PapersOnLine* **2020**, *53*, 15035–15040. [CrossRef]
7. Damoah, I.S.; Ayakwah, A.; Tingbani, I. Artificial intelligence (AI)-enhanced medical drones in the healthcare supply chain (HSC) for sustainability development: A case study. *J. Clean. Prod.* **2021**, *328*, 129598. [CrossRef]
8. Yin, Y.; Li, D.; Wang, D.; Ignatius, J.; Cheng, T.C.E.; Wang, S. A branch-and-price-and-cut algorithm for the truck-based drone delivery routing problem with time windows. *Eur. J. Oper. Res.* **2023**, *309*, 1125–1144. [CrossRef]

9. Satoglu, S.I.; Sahin, I.E. Design of a just-in-time periodic material supply system for the assembly lines and an application in electronics industry. *Int. J. Adv. Manuf. Technol.* **2013**, *65*, 319–332. [CrossRef]
10. Korytkowski, P.; Karkoszka, R. Simulation-based efficiency analysis of an in-plant milk-run operator under disturbances. *Int. J. Adv. Manuf. Technol.* **2016**, *82*, 827–837. [CrossRef]
11. Mindlina, J.; Tempelmeier, H. Performance analysis and optimisation of stochastic flow lines with limited material supply. *Int. J. Prod. Res.* **2022**, *60*, 5293–5306. [CrossRef]
12. Droste, M.; Hasselmann, V.-R.; Deuse, J. Optimization of in-plant milkrun systems: Development of a parameter-based model to optimize the provision of materials. *Product. Manag.* **2012**, *17*, 25–28.
13. Saysaman, A.; Chutima, P. Transportation of part supply improvement in agricultural machinery assembly plant. *IOP Conf. Ser. Mater. Sci. Eng.* **2018**, *311*, 012010. [CrossRef]
14. Serkowsky, J.; Kotzab, H.; Fischer, J. Integrating Regional Food Manufacturers into Grocery Retail Supply Chains in Germany. In *Dynamics in Logistics*; Lecture Notes in Logistics; Springer: Cham, Switzerland, 2022; pp. 15–25.
15. Kholil, M.; Hendri, D.M.R.; Bagus, Y.R. Improving the Efficiency of the Milkrun Truck Suppliers in Cikarang Area by Merging the Payload Cycles and Optimizing the Milkrun Route Using the Saving Matrix Methods. *J. Phys. Conf. Ser.* **2019**, *1175*, 012201. [CrossRef]
16. Bányai, T.; Telek, P.; Landschützer, C. Milkrun based in-plant supply—An automotive approach. *Lect. Notes Mech. Eng.* **2018**, *1*, 170–185.
17. Strachotová, D.; Pavlištík, J. The Assessment of Efficiency of In-Plant Milk-Run Distribution System in Cable Manufacturing for Automotive Industry. In Proceedings of the 29th International Business Information Management Association Conference—Education Excellence and Innovation Management through Vision 2020: From Regional Development Sustainability to Global Economic Growth, Vienna, Austria, 3–4 May 2017; pp. 105–109.
18. Ucar, C.; Bayrak, T. Improving in-plant logistics: A case study of a washing machine manufacturing facility. *Int. J. Ind. Eng. Theory Appl. Pract.* **2015**, *22*, 195–212.
19. Pawlewski, P. Interactive layout in the redesign of intralogistics systems. *Adv. Intell. Syst. Comput.* **2019**, *835*, 462–473.
20. Martini, A.; Rohe, A.; Stache, U.; Trenker, F. Factors influencing internal milkrun systems—A method for calculating the intensity of influence of dimensioning parameters. *WT Werkstattstech.* **2015**, *105*, 65–71. [CrossRef]
21. Satoh, I. A formal approach for milk-run transport logistics. *IEICE Trans. Fundam. Electron. Commun. Comput. Sci.* **2008**, *E91-A*, 3261–3268. [CrossRef]
22. Rempel, W.; Stache, U.; Zoller, C.S. Integral and automated planning of tugger train systems. *ZWF Z. Fuer Wirtsch. Fabr.* **2021**, *116*, 496–500.
23. Miqueo, A.; Gracia-Cadarso, M.; Torralba, M.; Gil-Vilda, F.; Yagüe-Fabra, J.A. Multi-Model In-Plant Logistics Using Milkruns for Flexible Assembly Systems under Disturbances: An Industry Study Case. *Machines* **2023**, *11*, 66. [CrossRef]
24. Abele, E.; Ander, R.; Brungs, F.; Mosch, C. Modeling of capacity of milkrun processes in lean production by material flow simulation. *Product. Manag.* **2011**, *16*, 48–51.
25. Chuah, K.H.; Yingling, J.C. Analyzing inventory/transportation cost tradeoffs for milkrun parts delivery systems to large JIT assembly plants. *SAE Tech. Pap.* **2001**, 2600.
26. Vieira, A.; Dias, L.S.; Pereira, G.B.; Oliveira, J.A.; Carvalho, M.S.; Martins, P. 3D Micro Simulation of Milkruns and Pickers in Warehouses Using SIMIO. In Proceedings of the Modelling and Simulation 2014—European Simulation and Modelling Conference, ESM 2014, Porto, Portugal, 22–24 October 2014; pp. 261–269.
27. Banyai, T.; Francuz, A. A new design approach for milkrun-based in-plant supply in manufacturing systems. *J. Mach. Eng.* **2022**, *22*, 91–115. [CrossRef]
28. Veres, P.; Bányai, T.; Illés, B. Optimization of in-plant production supply with black hole algorithm. *Solid State Phenom.* **2017**, *261*, 503–508. [CrossRef]
29. Hormes, F.; Lieb, C.; Fottner, J.; Günthner, W.A. In-plant Milkrun control-morphological classification of static und dynamic approaches. *ZWF Z. Fuer Wirtsch. Fabr.* **2017**, *112*, 778–782.
30. Domingo, R.; Alvarez, R.; Peña, M.M.; Calvo, R. Materials flow improvement in a lean assembly line: A case study. *Assem. Autom.* **2007**, *27*, 141–147. [CrossRef]
31. Bocewicz, G.; Bozejko, W.; Wójcik, R.; Banaszak, Z. Milk-run routing and scheduling subject to a trade-off between vehicle fleet size and storage capacity. *Manag. Prod. Eng. Rev.* **2019**, *10*, 41–53.
32. Kovács, A. Optimizing the storage assignment in a warehouse served by milkrun logistics. *Int. J. Prod. Econ.* **2011**, *133*, 312–318. [CrossRef]
33. Kitamura, T.; Okamoto, K. Automated Route Planning for Milk-Run Transport Logistics Using Model Checking. In Proceedings of the 2012 3rd International Conference on Networking and Computing, ICNC 2012, Okinawa, Japan, 5–7 December 2012; pp. 240–246.
34. Fedorko, G.; Vasil, M.; Bartosova, M. Use of simulation model for measurement of MilkRun system performance. *Open Eng.* **2019**, *9*, 600–605. [CrossRef]
35. Van der Gaast, J.P.; de Koster, R.B.M.; Adan, I.J.B.F. Optimizing product allocation in a polling-based milkrun picking system. *IISE Trans.* **2019**, *51*, 486–500. [CrossRef]

36. Alnahhal, M.; Ridwan, A.; Noche, B. In-plant milk run decision problems. In Proceedings of the 2nd IEEE International Conference on Logistics Operations Management 2014, Rabat, Morocco, 5–7 June 2014; pp. 85–92.
37. Bányai, T. Impact of the Integration of First-Mile and Last-Mile Drone-Based Operations from Trucks on Energy Efficiency and the Environment. *Drones* **2022**, *6*, 249. [CrossRef]
38. Bányai, Á. Energy Consumption-Based Maintenance Policy Optimization. *Energies* **2021**, *14*, 5674. [CrossRef]
39. Gubán, M.; Udvaros, J. A Path Planning Model with a Genetic Algorithm for Stock Inventory Using a Swarm of Drones. *Drones* **2022**, *6*, 364. [CrossRef]
40. Telek, P.; Kostal, P. Material handling equipment selection algorithm for production workplaces. *Adv. Logist. Syst. Theory Pract.* **2022**, *16*, 37–46. [CrossRef]

Disclaimer/Publisher's Note: The statements, opinions and data contained in all publications are solely those of the individual author(s) and contributor(s) and not of MDPI and/or the editor(s). MDPI and/or the editor(s) disclaim responsibility for any injury to people or property resulting from any ideas, methods, instructions or products referred to in the content.

Review

Perspective on the Development of Energy Storage Technology Using Phase Change Materials in the Construction Industry: A Review

Sandra Cunha [1,*], Antonella Sarcinella [2], José Aguiar [1] and Mariaenrica Frigione [2]

[1] Department of Civil Engineering, Environment and Construction (CTAC), Centre for Territory, Campus de Azurém, University of Minho, 4800-058 Guimaraes, Portugal; aguiar@civil.uminho.pt
[2] Innovation Engineering Department, University of Salento, 73100 Lecce, Italy; antonella.sarcinella@unisalento.it (A.S.); mariaenrica.frigione@unisalento.it (M.F.)
* Correspondence: sandracunha@civil.uminho.pt

Abstract: The construction industry is responsible for high energetic consumption, especially associated with buildings' heating and cooling needs. This issue has attracted the attention of the scientific community, governments and authorities from all over the world, especially in the European Union, motivated by recent international conflicts which forced the countries to rethink their energy policies. Over the years, energy consumption has been based on non-renewable energy sources such as natural gas, oil and coal. Nowadays, it is urgent to implement solutions that aim to minimize these high energetic consumptions and act based on clean and renewable energy sources. In recent years, phase change materials (PCM) have become an area of high interest and development, since they allow to minimize the energy consumption in buildings, based in solar energy, due to their thermal storage capacity. The main objective of this work consists of a perspective of the evolution of the development and application of thermal storage technology through the incorporation of PCM in the construction sector, focusing on the last 10 years of research, showing the most recent developments of its application in construction materials, such as mortars, concrete, incorporation in porous aggregates, naturally based materials, carbon-based materials, boards, blocks and solar thermal systems.

Keywords: energy efficiency; thermal storage technology; phase change materials

Citation: Cunha, S.; Sarcinella, A.; Aguiar, J.; Frigione, M. Perspective on the Development of Energy Storage Technology Using Phase Change Materials in the Construction Industry: A Review. *Energies* **2023**, *16*, 4806. https://doi.org/10.3390/en16124806

Academic Editors: Chiara Martini and Claudia Toro

Received: 18 April 2023
Revised: 9 June 2023
Accepted: 15 June 2023
Published: 19 June 2023

Copyright: © 2023 by the authors. Licensee MDPI, Basel, Switzerland. This article is an open access article distributed under the terms and conditions of the Creative Commons Attribution (CC BY) license (https://creativecommons.org/licenses/by/4.0/).

1. Introduction

Currently, humanity faces challenges never experienced before, related to the little or no sustainable development we have witnessed, which is further aggravated by conflicts and wars that increasingly expose and leave the most fragile populations vulnerable.

The United Nations Organization has been working and establishing policies and objectives that allow combating the main problems of today, with the 2030 Agenda being a clear example of this effort [1,2]. The 2030 Agenda is a broad and ambitious program that addresses various dimensions of sustainable development, establishing 17 main goals. The seventh goal, "Ensure access to affordable, reliable, sustainable and modern energy for all" is directly related to energy consumption worldwide, indicating that it is necessary to establish universal, reliable, modern and affordable access to energy, substantially increasing the participation of renewable energies. On the other hand, it will be necessary to improve energy efficiency, increase research into energy technologies and modernize the technology for providing energy services [1]. Thus, taking into account the high energy consumption verified in the construction industry, the development of energy storage technology using phase change materials (PCM), based on solar energy in the construction industry and especially applied to construction materials, can constitute an important line of research and development to achieve these objectives and their implementation goals. It is important to note that solar energy is a clean energy source without carbon dioxide emissions and waste generation, is widely available and free [3], and also allows to contribute to carbon neutrality [4]. It should also be noted that lower energy consumption based on low-cost energy sources with no impact on the environment will lead to a decrease in the population living with difficulties in heating and cooling buildings (situation of energy poverty). On the other hand, it will also be possible to contribute to sustainable cities and communities, in order to make cities and communities more inclusive, safe, resilient and sustainable. Finally, it will be possible to move towards a more innovative, resilient, inclusive and sustainable construction industry, which aims for the development of quality and sustainable buildings, supporting economic development and human well-being. Thus, this manuscript addresses the follow United Nations Sustainable Development Goals:

- ODS 1: End energy poverty;
- ODS 7: Affordable and clean energy;
- ODS 9: Resilient, sustainable and innovative infrastructures;
- ODS 11: Sustainable cities and communities.

The aim of this work is to provide a perspective on the development of energy storage technology using phase change materials in the construction industry, addressing energy consumption in the construction sector and the development of thermal storage technologies using phase change materials, as well as a broad and current demonstration of the application of this technology to different construction materials.

Scope of the Review Paper

The methodology followed in the creation of this review was based on the premise of giving priority to articles published in the last 10 years (since 2012), using numerical and experimental approaches, with the objective of keeping the review up to date.

The review is divided into seven sections. The Introduction presents the policies and objectives defined by the European Union, namely, the objectives for sustainable development, in which the theme of this review is inserted. Section 2 presents the energy consumption in the construction industry, i.e., the final energy consumption in households by type of fuel, the purpose of the energy uses in the residential sector and also the European countries most vulnerable to energy poverty. Section 3 describes the evolution in the last years of the thermal energy storage technology, showing the number of papers published and the countries more active in researching this topic. Section 4 is dedicated to phase change materials, being subdivided into several subsections, addressing the main criteria for the correct application of PCM in buildings, its classification and existing incorporation techniques, which makes the article comprehensive in this domain. These concepts are directly related to the information presented in the next Section. Section 5, in fact, provides an in-depth approach to the use of PCM in different construction materials and building applications, presenting different subsections covering mortar, concrete, functionalized aggregates, carbon-based materials, naturally based materials, boards, bricks and solar thermal systems. Thus, the immense potential of this thermal storage technology is demonstrated, using several PCM's and several incorporation techniques. Section 6 is devoted to the cost analysis of these types of constructive solutions. Finally, Section 7 highlights detailed recommendations and conclusions, and the future prospects and challenges in thermal storage technology.

2. Energy Consumption in the Construction Industry

Nowadays, the world energy consumption is increasing, due to world population growth [5], technological development [6] and industrial intensity [7]. In the last 30 years, the total energy supply from all sources (oil, biofuels and waste, hydro, wind and solar, nuclear, natural gas and coal) and the carbon dioxide (CO_2) emissions associated with energy production increased about 60% [8]. At the present time the utilization of renewable energy sources such as hydro, wind and solar energy represents only 5.21% of the total energy supply. However, it is important to bear in mind that since 1990 there has been an increase in the use of energy from renewable sources and, according to the most recent data (2020), the verified increase has been around 300% [8].

The construction industry is a huge consumer of energy. This consumption starts from the extraction of raw materials, production of construction materials, building constructions, maintenance, demolition and finally disposal of the generated waste. Thus, the construction industry possesses several ways to contribute to the minimization of energy consumption, from reducing the extraction and use of natural raw materials [9,10], to reducing the amount of construction and demolition waste in landfills [11,12], and also reducing the energy demand of buildings during their operating cycle through the adoption of more efficient constructive solutions [13–15].

Figure 1 shows the final energy consumption by sector in 2021 in Europe: it indicates that 30% of energy is associated with the transport sector, 29% with the residential sector (households), 27% with the industry sector and 14% with commercial and public services [16]. It is important to note that Europe is dependent on non-European countries in terms of energy, with this dependence increasing slightly, since data from 2012 point to a dependence of around 55% and data from 2021 to a dependence in terms of imported energy of around 56%. Italy and Portugal are countries that have high energy import dependency rates, above the European average, being around 74% and 69% in 2021, respectively [17].

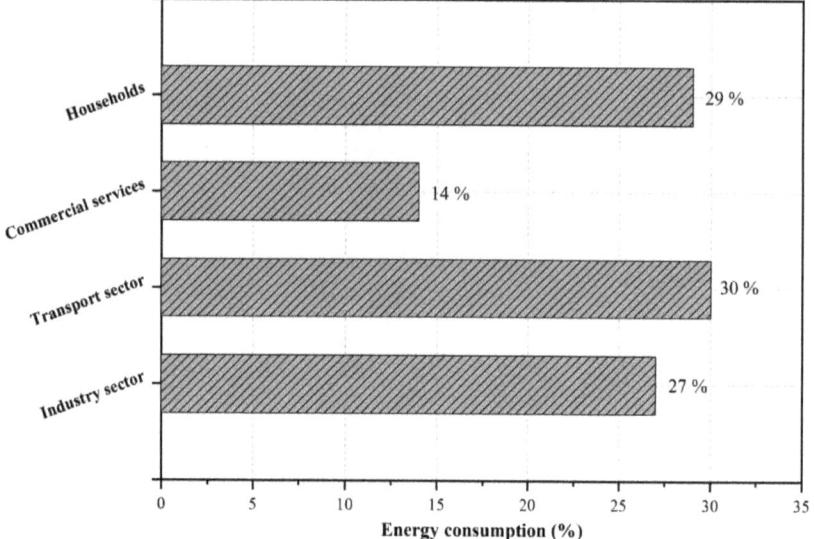

Figure 1. Final energy consumption by sector in Europe.

The energy consumption in households is based on different types of fuels (Figure 2). Natural gas, electricity and solid biofuels represent the most important fuel sources, contributing around 33%, 25% and 17% to energy consumption in the residential sector [18]. It is also important to note that in the last 10 years there has been a slight increase in the use of these energy sources and between 2012 and 2021 the consumption of natural gas, electricity and solid biofuels increased by around 3%, 4% and 2% [18]. Bearing in mind that these are non-renewable energy sources, it is important that the construction industry begins to think about and invest in constructive solutions that can be implemented in buildings which contribute to the minimization of energy consumption but are based on renewable energy sources.

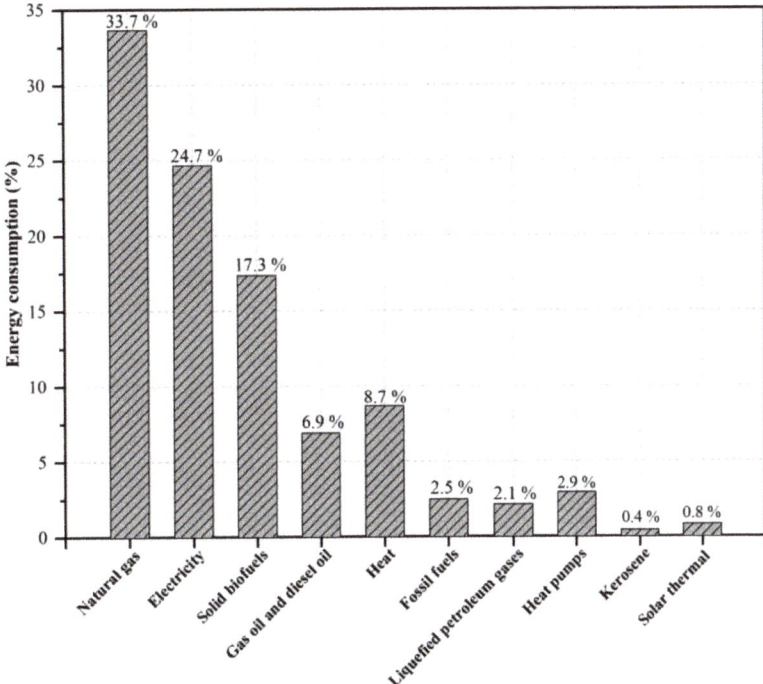

Figure 2. Final energy consumption in households by type of fuel.

The energy consumption in the residential sector is associated with different purposes such as the space heating, space cooling, water heating, cooking, lighting and other end-uses. According to Figure 3, in 2020 in the European Union, 63.2% of the consumed energy in the residential sector was associated with space climatization needs [19].

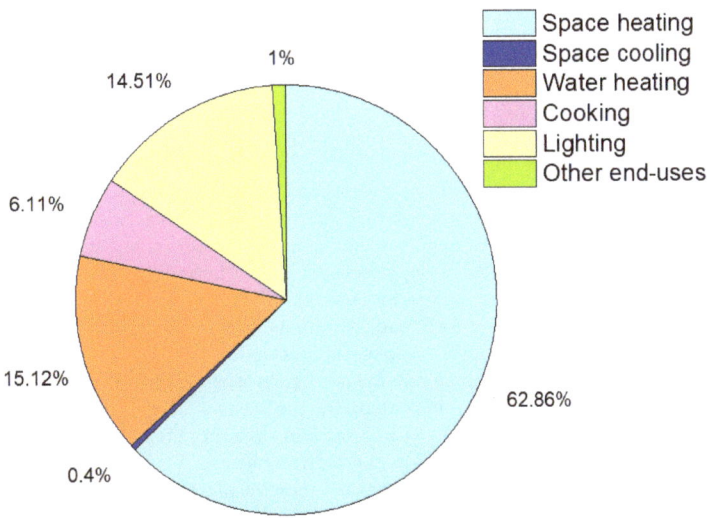

Figure 3. Final energy consumption in the residential sector by use.

Taking into account the enormous heating and cooling needs of buildings, the heating, ventilation and air conditioning (HVAC) systems have huge utilization rates, which consequently implies large energy consumption to maintain comfortable conditions inside buildings; thus, in recent years the concept of energy poverty has emerged, affecting the European Union's population, essentially the most vulnerable families. In Figure 4, it is possible to see a geographical distribution of the countries most vulnerable to energy poverty [20]. Due to the cost increase of natural gas, electricity and oil, an increase in the number of households living in energy poverty across Europe is expected. Thus, the high energy consumption in the construction industry, but especially during the buildings' utilization due to the enormous need for heating and cooling, constitutes one of the highest challenges for the sector's development, as well for the sustainable development of the planet. For this reason, it is essential that sector stakeholders start to develop and invest in constructive technologies and functional construction materials that contribute to minimizing energy poverty and increasing the energy efficiency of buildings. This can be achieved by investing in thermal storage technologies and in the incorporation of phase change materials (PCM) in construction materials.

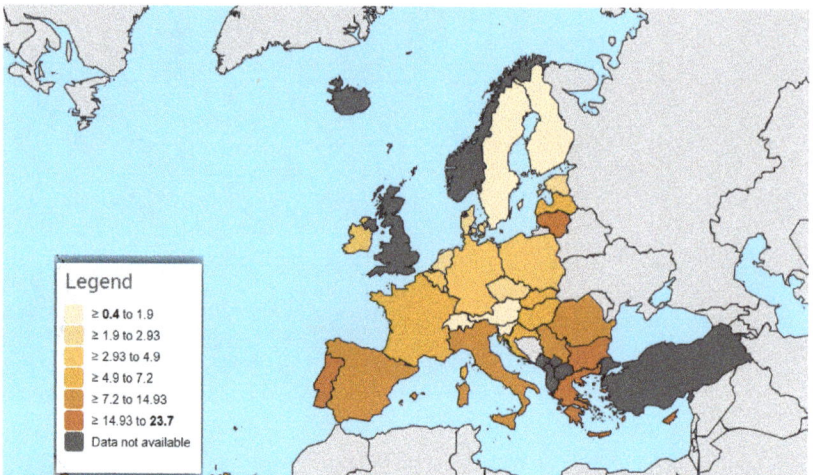

Figure 4. Geographical distribution of the population unable to keep home adequately warm by poverty status [20].

3. Evolution of Thermal Energy Storage Technology

The technology known as Thermal Energy Storage (TES) is a method that can store thermal energy, making it available when needed. This process is made possible by a medium that takes over the excess thermal energy. In general, there are three methods based on this technology: Sensible Heat Thermal Energy Storage (SH-TES), Thermo-Chemical Thermal Energy Storage (TC-TES) and Latent Heat Thermal Energy Storage (LH-TES). In the SH-TES mode, thermal energy is retained in a medium by varying (traditionally increasing) its temperature. A classic example is hot water storage (i.e., hydro-accumulation) but other materials are used, such as: rock, iron, aluminum, air, steam and hydrogen [21]. In general, it is important that the material used in this type of application has a high specific heat capacity per unit mass to store more thermal energy, high conductivity, and thermal stability as well as chemical stability and low thermal expansion. The TC-TES method is characterized by a reversible chemical reaction during which two or more reacting chemical compounds absorb and release thermal energy. Basically, during the reaction, endothermic dissociation, storage of the reaction products and, finally, an exothermic reaction of the dissociated products take place. Then, in the final stage of the reaction, the initial materials are recreated so that the process can be repeated [22]. The most used materials for this kind of application are $MgSO_4 \cdot 7H_2O$ that reacts with $MgSO_4$ using H_2O as working fluid, $Ca(OH)_2$ that reacts with CaO using H_2O as working fluid, $CaSO_4 \cdot 2H_2O$ that reacts with $CaSO_4$ using H_2O as working fluid, and $FeCO_3$ that reacts with FeO using CO_2 as working fluid [23]. In the LH-TES, a phase change material is employed as a medium to store and release thermal energy according to its physical state. In most cases, this is a solid-to-liquid transformation and vice versa (although there can be other transformations, such as solid-to-gas, solid-to-solid and liquid-to-gas). Phase change materials generally used are paraffin waxes, polymers, fatty acids, esters, eutectic salts and hydrated salts [24]. Among these three modes, recently LH-TES has attracted a considerable attention. The main reason is related to the isothermal nature of the phase exchange process, but also to its low weight per unit storage capacity and compactness. In addition, compared with the storage materials used in SH-TES and TC-TES technologies, phase change materials have better thermal properties, such as stable phase exchange temperature and high latent heat [25,26]. These reasons have strongly contributed to its affirmation and diffusion. The interest in this technology is also confirmed by the increasing number of scientific publications on this subject. Looking only at the last 10 years (2012–2022), it is possible to observe how research on "phase change materials" has progressively grown (Figure 5a) and in which country the scientific interest has been predominant (Figure 5b).

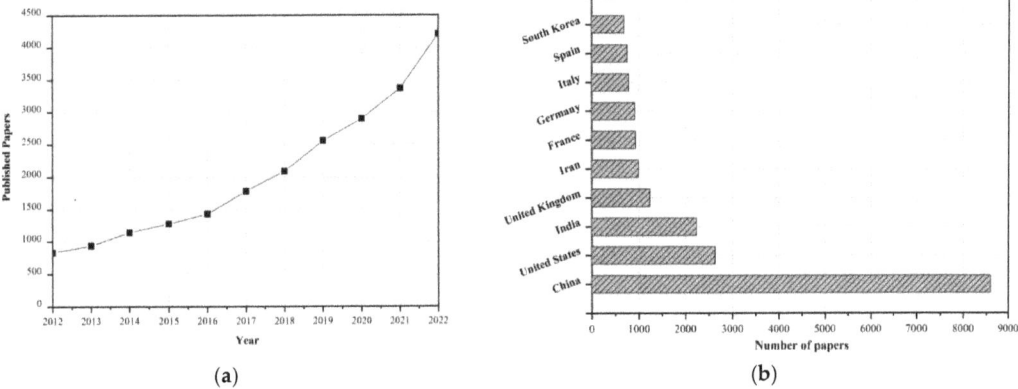

Figure 5. (a) Number of papers published within 2012–2022 on "Phase Change Material"; (b) the 10 countries that have published the most on this topic. Source: Scopus Database. Searched on the day: 3 February 2023.

To date, PCMs are considered one of the most viable strategies for saving energy, and the sector in which they are most widely used is the building industry [27]. Even in this case, the growing number of publications related to the use of PCMs applied in construction has demonstrated the enormous interest in them. This can be seen with a simple bibliometric analysis considering the number of papers published in the last 10 years (2012–2022) on "Phase Change Material used in buildings" (Figure 6a) and in which state this interest was greatest (Figure 6b).

The trend of using PCMs to improve the energy efficiency in buildings is also demonstrated by the keywords that appear most often in scientific articles: thermal comfort, energy efficiency, building, thermal performance and building envelope are the top five keywords appearing most often. Figure 7 shows the network visualization among the keywords found most often in those articles that the Scopus database generated by searching the "phase change material" topic.

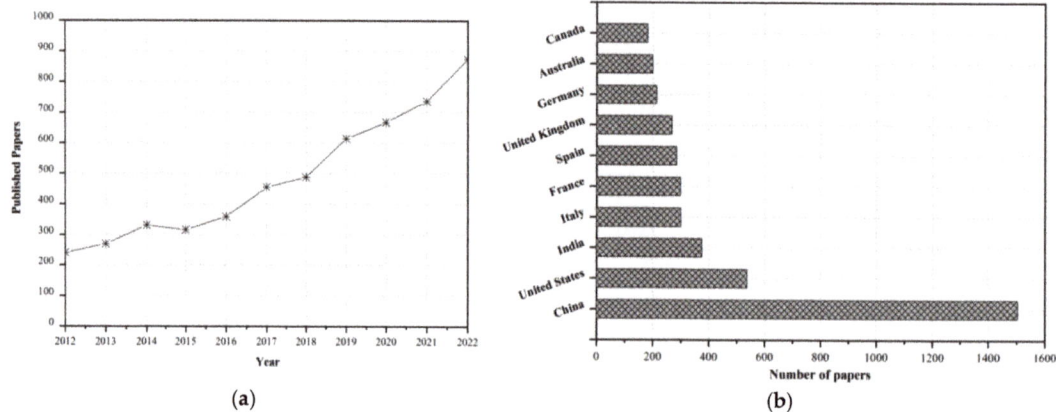

Figure 6. (a) Number of papers published within 2012–2022 on "Phase Change Material used in buildings"; (b) the 10 countries that have published the most on this topic. Source: Scopus Database. Searched on the day: 3 February 2023.

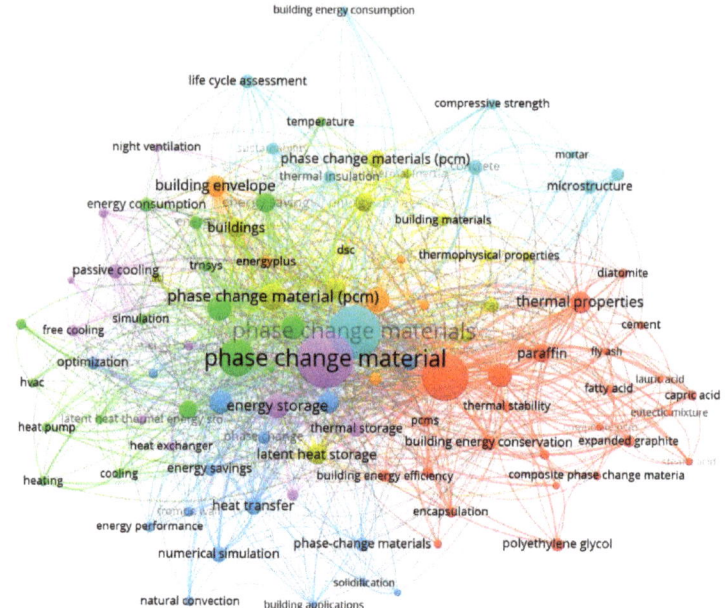

Figure 7. Network visualization of the keywords.

This knowledge will benefit future researchers in choosing keywords that will make it easier to spot previously published material on a certain topic. These keywords also highlight the increasing interest of researchers in this area because it is demonstrated that the application of a PCM brings with it several advantages from a thermal point of view and, consequently, this has important economic benefits that cannot be underestimated.

4. Phase Change Material

A Phase Change Material (PCM) is a substance that can absorb, store, and release a large amount of energy in the form of heat during a phase transition. These materials can exist in different phases, such as solid, liquid and gas, and can change their phase when heated or cooled. During this phase change, the material can either absorb heat (melting) or release heat (freezing), which allows it to be used for various applications such as thermal energy storage, temperature regulation and heat transfer [28,29]. Figure 8 shows schematically the physical operation of a PCM and how, as temperature increases or decreases, its energy content changes.

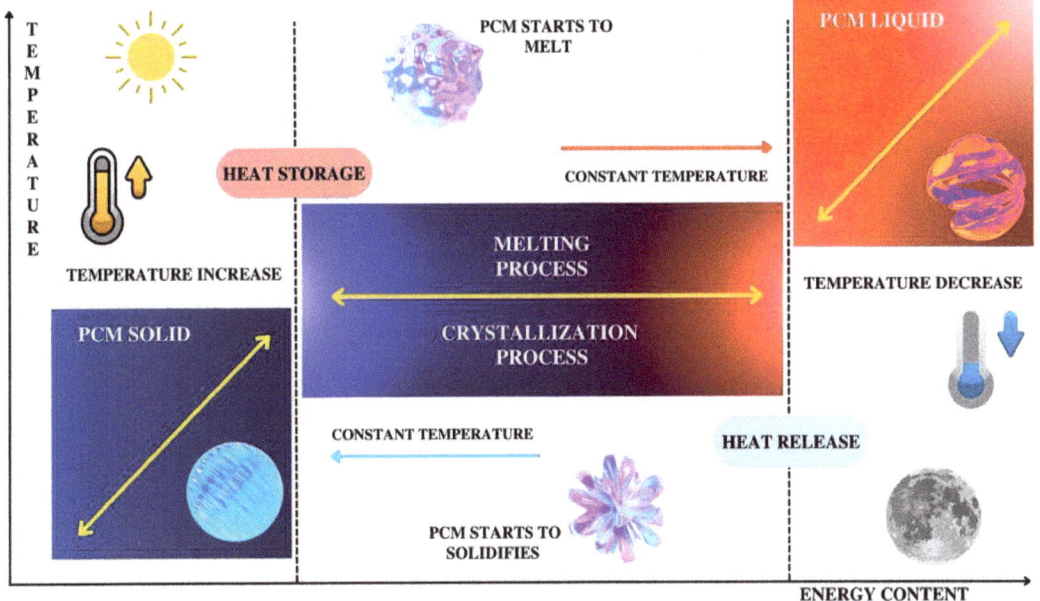

Figure 8. Phase Change Material process.

In general, PCMs have high energy storage density, meaning they can store more energy per unit mass than conventional materials such as water or concrete. This makes them useful for a range of applications, including building insulation [24], refrigeration [30] and renewable energy systems [31]. It is commonly agreed that the ability of PCMs to store and release thermal energy can help improve energy efficiency, reduce the use of fossil fuels and contribute to a more sustainable future [32]. There are several substances which are used as PCMs; each of them has a different melting and crystallization temperature, making them suitable for different temperature ranges and applications.

4.1. PCMs: A Novel Classification

PCMs can be classified based mainly on their melting temperature, which determines the range of temperatures at which they can store or release heat. Commonly, PCMs classification include organic, inorganic and eutectic PCM. However, very recently a new and interesting class of PCMs has emerged which is worth reporting, namely, bio-based PCMs [33–35]. They are briefly illustrated in Figure 9.

The difference between these PCMs, including advantages and disadvantages, can be described as follows:

- Organic PCMs are typically made from paraffin wax or non-paraffin substances, such as fatty acids, esters, alcohols and glycols, and have a melting temperature range of 0 °C to 150 °C [36]. They are commonly used in building materials and implemented into different construction materials, as will be seen later. Organic PCMs are characterized by a high energy storage capacity and a high latent heat of fusion. A very important aspect is that of supercooling. Organic PCMs do not undergo this phenomenon during solidification and have a good nucleation efficiency. They are available in a wide range of melting temperatures; this makes them suitable for use in a variety of applications. Typically, organic PCMs are considered non-corrosive, non-toxic, and environmentally friendly, making them safe for use in many applications, including food storage and transportation [34]. Nevertheless, these kinds of PCMs have some disadvantages that should be considered, such as low thermal conductivity, meaning that they may not transfer heat efficiently, and may require more material to achieve the same thermal storage capacity (even though progress has been made to remediate this issue [37]). Some of the organic PCMs are potentially flammable or combustible, which could pose a safety risk in some applications and may require additional safety measures [38]. Their limited durability is another problem: organic PCMs can degrade over time due to chemical reactions, which can reduce their effectiveness and lifespan. This may require more frequent replacement or maintenance of the PCM system. Finally, referring to their costs, organic PCMs can be more expensive than some other types of PCMs, which could be an important factor in the choice of a PCM for a given application [39].

- Inorganic PCMs are generally composed of salt hydrates or metals and have a higher melting temperature range, i.e., from 100 °C to 1000 °C. They are commonly used in industrial applications, such as metal casting or high-temperature energy storage and in solar applications [35,40]. Their temperature range makes them more durable than organic PCMs as they do not degrade quickly over time. This means that they may require less frequent replacement or maintenance, which can be cost-saving over the lifetime of a PCM system. Referring to their costs, inorganic PCMs are less costly compared to some other types of PCMs, also considering that they can be produced from readily available materials, which can help keep costs low. Furthermore, inorganic PCMs have a high thermal conductivity and they are non-flammable [41,42]. However, they have a corrosive nature and therefore can rarely (depending on applications) stand alone. In many cases, they must be contained in other materials/elements to avoid the damaging of materials with which they come in contact. In addition, but only in the case of salt hydrates, they may encounter phase segregation and the phenomenon of supercooling, which consequently decreases the energy storage capacity [43].

- Eutectic PCMs are mixtures of two or more substances that have a lower melting temperature than each of the individual components. They are commonly used in refrigeration and air conditioning systems since they are characterized by a low melting point. The latter, moreover, can be customized to meet specific requirements of a given application by adjusting the ratio of the component materials. Eutectic PCMs have a high energy-storage capability and they are chemically stable. They have a long lifespan and result in non-toxic final materials. However, their use may be limited by the availability and cost of the component materials, and they may not be suitable for all applications [44].

- Bio-based PCMs are very close to organic PCMs but they are derived from natural materials originating from animals or plants, such as oils, fats and starches, i.e., not from petroleum refining. They can have a wide range of melting temperatures and they represent an eco-friendly alternative to traditional PCMs [45,46]. They are able, in fact, to ensure biodegradability, sustainability, lack of flammability and non-toxicity. Fatty acids can be easily and cheaply produced from animal fat or oily plants; in addition, non-edible or waste materials can be employed to produce them, helping to avoid wasting food. Referring to their main characteristics as PCM materials, they have small volume changes during the phase change process, small corrosion activity, and thermal and chemical stability [47]. The melting temperature of the most common fatty acid-based PCMs is approximately in the same range as paraffins and, likewise, they have poor thermal conduction capacity. However, several research works have shown that the addition of metal nanoparticles or other conductive materials (i.e., graphite nanoplatelets, carbon nanotubes, etc.) can overcome this problem [48].

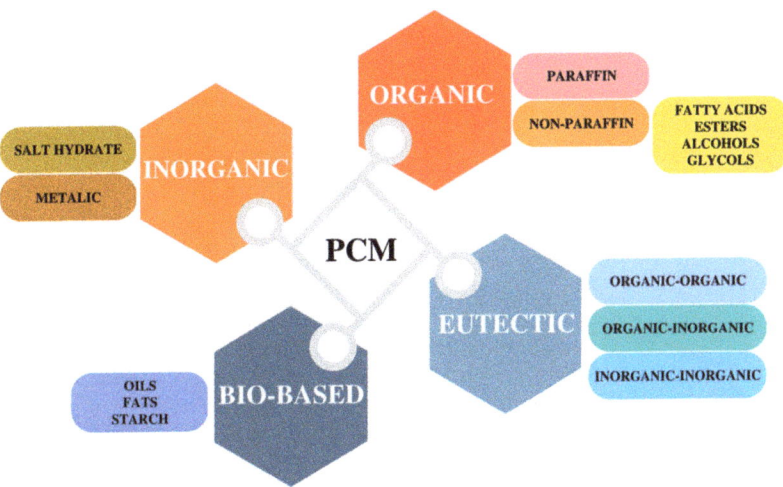

Figure 9. Classification of PCMs.

4.2. Application of PCMs in Building Materials and Associated Challenges

It is possible to incorporate PCMs into building materials in several ways [27]. The presence of many techniques makes it possible to insert PCMs into a wide variety of construction materials, as will be seen in Section 5. So far, the techniques used are direct embedding, immersion, (micro, macro and nano) encapsulation, form-stable and shape-stabilization.

Direct embedding and immersion are the simplest and least expensive methods. The former involves embedding PCM directly into the mix design of a mortar or concrete [49]; the latter, on the other hand, involves immersing a (relatively porous) building material in a PCM, which is then absorbed into its pores by capillary action. Absorption of a PCM can take place at atmospheric pressure or under vacuum [50]. These techniques showed, however, some issues such as leakage of the PCM causing alterations in the final properties of the material in which it was embedded, also leading to possible corrosion problems, incompatibility between the combined materials and increased flammability of the final material [51].

The encapsulation method is divided into three sub-methods: micro-encapsulation, macro-encapsulation and nano-encapsulation. These methods differ in the size of the capsules involved: the size of nano-capsules is less than 1 μm, of micro-capsules between 1 μm and 1 mm in diameter, and macro-capsules from 1 mm upwards. Nano-capsules are only recently becoming widespread, and for this reason their production (which is complex) has not yet been optimized on a large scale [50,52]. However, their conception and diffusion are linked to overcoming certain difficulties related to the use of micro- and macro-capsules. In fact, problems such as low thermal conductivity, supercooling and phase segregation phenomena seem to affect micro- and macro-encapsulated PCMs [53]. To create the micro-capsules, a small amount of PCM is enclosed in another material (which can be a polymer, a lipid or an inorganic compound) that acts as a barrier between the PCM and its surroundings. Then, these micro-capsules can be mixed with other materials, such as mortar or concrete, to form composite materials with enhanced thermal properties [54,55]. At the same time, macro-capsules are made by a PCM or a combination of different PCMs enclosed in a shell made of a suitable material (metals, polymers or composites) that can be integrated into plaster, concrete or gypsum board. These have been the most common techniques so far, although they are very expensive methods [56]. For this reason, and for all the issues mentioned so far, two other techniques have become widespread in the past decade and have shown considerable promise: the form-stable and shape-stabilization methods. These are based on the same idea: to create a composite material (consisting of a matrix and a PCM) that retains its shape even during phase transition. In fact, the matrix must provide structural stability and prevent dispersion of the PCM during its melting or heating [50,57]. When these two methods became popular, they differed in two respects: in the type of matrix used and in the type of methodology employed to produce the final material. The shape-stabilization method involved the use of polymer or metal matrices. Both the matrix and PCM were melted and then mixed. Once they were given the desired shape, they were brought to a solid state and then used as a composite PCM [52,58]. On the other hand, the form-stable method is characterized by porous matrices, such as silica, vermiculite, perlite, diatomite, calcium carbonate, etc. [52,57]. In this case, the final composite material can be obtained by immersion or vacuum impregnation of the active PCM phase [50,59]. As research went on, this distinction gradually narrowed, making, in fact, these two methods coincide; today, they can be considered synonymous [27]. Furthermore, in the present day, this overlap can be translated in only one way: if the term "shape-stabilized" is encountered in an experimental work, it means that it is a porous matrix impregnated (with or without a vacuum) with PCM. Over the past decade, the number of experimental research works involving the use of this technique has steadily increased, and this is due to several advantages of the method: it is very simple and low-cost, requires modest equipment, creates a stable composite PCM reducing its leakage over the melting temperature, is thermally reliable and has a high heat transfer rate [60].

Each of the techniques described so far have arisen from the need to overcome certain problems. Some of these have already been mentioned but in Table 1 they are briefly described and analyzed.

Table 1. Issues related to the incorporation of PCMs in building materials.

Problems	Description	Possible Solutions
Leakage	Leakage of PCM into building envelope causes a number of problems: the PCM loses its storage capacity, and its dispersions can cause the corrosion of the surrounding materials (especially if it is in contact with reinforcing steel). Therefore, mechanical properties start to decrease. Last but not least, leakage is also associated with possible aesthetic defects.	PCM capsules were created to contain PCM and prevent its dispersion. However, these (especially the macro-capsules) need extra attention (e.g., against nails being hung on the wall or other housework involving holes) [61]. Composite materials generated by the shape-stabilized or form-stable method seem not to incur this problem. It was observed that the matrix can contain the PCM even when it undergoes phase change, avoiding its dispersion [62].
Low thermal conductivity	The conductivity of organic PCMs is very low (around 0.2–0.3 W/m K). This has consequences for the PCM's ability to retain and release heat.	PCM capsules were designed to improve the thermal conductivity of PCM, and after them, more success was achieved by composite materials (shape-stabilized or form-stable) because additives can be used, such as: expanded graphite, graphene, metal particles, carbon-fiber, or multi-wall nanotubes [37,63].
Supercooling	This phenomenon occurs when the solidification temperature is lower than the melting temperature. For this reason, the melting temperature should coincide with the crystallization temperature; otherwise, there is a risk that the stored latent heat will not be released. However, this phenomenon occurs only in inorganic PCMs.	The following solutions have been developed to avoid this problem: the addition of nucleating agents or metal additives, or the use of nanofluid PCMs [64].

Table 1. *Cont.*

Problems	Description	Possible Solutions
Phase segregation	This phenomenon usually occurs when there are more than one PCM that have different densities and that, due to gravity, separate causing the fusion process to take place at different times. PCMs that are affected by this problem are hydrated salts or eutectic compounds.	This problem can be solved by adding additives such as a thickening agent or gelling material [27].
Flammability	This problem is mainly associated with organic PCMs that are highly flammable.	The only possible solution so far is the addition of an additive known as a flame retardant. Studies have shown how the addition of this additive reduces fire hazards at the cost of a slight change in the thermal properties of the PCM [65].
Thermal stability	The thermal stability of a PCM indicates its ability to keep its thermal characteristics intact after numerous melting/crystallization cycles. This ability is critical for not losing the thermal properties of storing/releasing latent thermal energy within a building.	Studies conducted to verify the thermal stability of a PCM have observed that composite PCMs produced through shape-stabilized or form-stable methods are able to maintain the thermal properties of PCMs embedded in the matrix intact [66,67].
Suitable PCM	In general, the PCM that is applied in a certain environment must be suitable for that specific climate. The problem arises when the application of a PCM is required in a climate characterized by temperatures (high or low) that are too extreme. In fact, the summer and winter seasons are the most challenging ones.	In the recent period, the use of combinations of PCMs (which have different characteristic temperatures) to increase the range within which the phase change takes place is spreading [68,69].

4.3. Criteria for PCM Selection to Be Applied to Buildings

After listing the problems associated with the use of PCMs in building materials, it becomes easy to understand that there is no such thing as the perfect one. However, there are certain criteria that can be used to select the most suitable PCM for the type of application for which it is intended. Each individual PCM is characterized by thermal, chemical, physical and kinetic properties. However, nowadays, properties such as environmental sustainability, availability and cost are considered extremely important as well. Figure 10 shows the main criteria that should be followed when applying PCMs in buildings, according to [50,52,70,71].

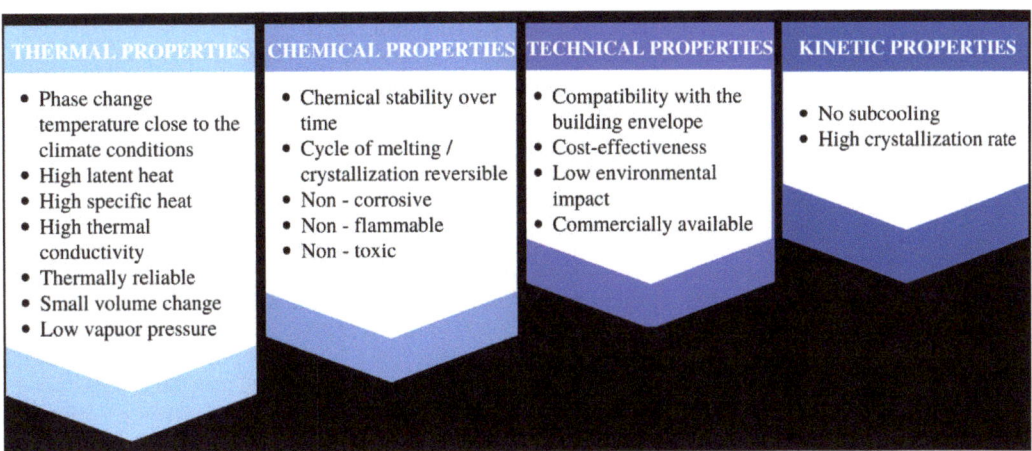

Figure 10. Criteria for selecting a suitable PCM for building materials.

In addition to the criteria given in Figure 10, some experimental calculations can also be relied upon to select a suitable PCM for the type of application. In [72], for example, a simple calculation is proposed that can determine the melting/crystallization temperature a PCM should have in a given context and also the thickness of the wall containing the PCM. A more modern tool for identifying the most appropriate PCM is algorithms. In recent years, several algorithms have been developed (e.g., Analytical Hierarchy Process (AHP), COmplex PRoportional Evaluation (COPRAS), VlseKriterijumska Optimizacija I Kompromisno Resenje (VIKOR), Technique for Order Performance by Similarity to an Ideal Solution (TOPSIS), ELimination and Choice Translating REality (ELECTRE), etc.) [73,74]. These algorithms can select a PCM from those available based on multi-criteria decision making. Therefore, this tool considers not only the target environmental conditions but also other factors concerning the properties of the PCM according to the priorities that are set at the beginning of the search.

5. Applications of Thermal Storage Technology in Buildings Using Phase Change Materials

Construction materials that employ thermal storage technology through the use of phase change materials can be applied in a wide variety of constructive solutions, whose working principle is shown in Figure 8. Thus, PCMs can be included in walls [75–79], ceilings [80–82], floors [83–85] and even in solar thermal systems [86–88]. It is also possible to choose different construction materials, such as: mortars [75,89–91], concrete [92–95], functionalized aggregates [96–100], naturally based materials [101–103], carbon-based materials [104–107], boards [108–111], bricks [77,112–114] and others, according to the specific needs of the building and the available budget for the intervention.

5.1. Mortars

The development of mortars incorporating PCM has attracted the attention of several researchers worldwide [54,75,76,89–91,115–119]. Several incorporation techniques and different types of PCM have been used. The incorporation of PCMs in mortars has also been developed, becoming one of the most used inclusion techniques [75,76,115–118]; it has, however very high raw material acquisition costs [59]. Cunha et al. [75,76,115,116] developed mortars based on gypsum, aerial lime, hydraulic lime and cement doped with different microcapsulated PCM contents: for each type of binder, a reference mortar without any PCM and mortars with 20%, 40% and 60% by weight of PCM as mortar aggregate were developed and characterized. The results of mechanical properties revealed a decrease in the mortars' flexural and compressive strengths, due to a high water/binder ratio which led to a greater porosity related to the incorporation of a higher PCM content. Regarding the thermal behavior, the same research team indicated that all PCM mortars reveal higher thermal regulation, due to the decrease in maximum temperature, increasing the minimum temperatures and lowering energy needs. Illampas et al. [91] used PCM microcapsules in cementitious repair mortars, producing a mortar without PCM microcapsules and three mortars with different PCM contents (5%, 10 and 20% by weight of the powder components). The study revealed that a higher PCM content in mortars originated an increase in the open porosity and a decrease in the flexural strength, compressive strength and elastic modulus, due to the lower strength and stiffness of the PCM microcapsules when compared to the aggregate particles. On the other hand, the thermal performance of the PCM mortars was improved due to the decrease in thermal conductivity and thermal diffusivity and increase in the specific heat capacity, which contributed to the attenuation and time shifting of temperature peaks.

Other studies present an alternative to the utilization of microencapsulated PCM, in order to decrease the PCM cost, due to the treatment that material suffers during the encapsulation. The utilization of a pure PCM, based on direct incorporation technique or immersion can be an alternative since the use of a non-encapsulated PCM causes a decrease in the production cost of the mortars [59].

Cunha et al. [54,90] developed cement mortars with pure PCM incorporation through the direct incorporation technique, demonstrating a decrease in the compressive strength with the presence of a higher PCM content, due to the delay in the cement hydration process. These studies also proved that the PCM did not move from the mortar matrix, being contained in the pores and coating the structural matrix of the mortars.

The immersion technique was approached by Lopez-Arias et al. [89], who developed three different mortars with different water/binder ratios (0.45, 0.55, 0.65) and consequently, different porosity. In this study, the PCM was added to the hardened structure of the mortars through the immersion technique and with resource to a vacuum application, during different periods of immersion (15 min, 1 h and 4 h). The test results showed that the mortars with higher porosity had a higher absorbed PCM content; on the other hand, it was observed that the PCM content increases with the higher immersion time of the samples in PCM. It was also possible to verify that a higher PCM content in the mortars resulted in a higher thermal conductivity and compressive strength. Thus, the samples with higher water/binder ratio (0.65) and 4 h of PCM vacuum immersion presented the most interesting behavior and higher PCM content.

5.2. Porous Aggregates

The necessity to prevent PCM from being dispersed in the main building material has made inclusion in porous aggregates increasingly common. This is done, as was seen in Section 4.2, through shape-stabilized or form-stable methods by direct or vacuum impregnation. Initially, expanded perlite (EP) [96,120], vermiculite [121,122], diatomite [123] and montmorillonite [124,125] were the most used porous aggregates.

Expanded perlite and vermiculite are widely used materials in building applications and are famous for their thermal insulation. Zhang et al. [126], developed a composite material based on EP and paraffin through the vacuum impregnation method. To improve the thermal conductivity of the final material, they added carbon nanotubes (CNTs), producing four different samples containing different CNTs mass fraction (0, 1.83, 3.62, 5.27 wt%). A complete characterization of these materials showed good chemical and thermal stability and a thermal conductivity significantly improved by CNTs that, consequently, enhanced the storage and release properties. In another work, Zhang et al. [97], proposed a new composite PCM made through the vacuum impregnation method as a potential PCM for building energy reserve. They used a mixture of lauric-palmic-stearic acid as PCM impregnating vermiculite as supporting matrix. The composite material demonstrated excellent chemical and thermal stability but low thermal conductivity. Adding 2 wt% of EG increased the thermal conductivity by 68%, going on to also improve the thermal and enthalpy properties of the final material. In a more recent work, Rathore et al. [66], prepared a composite PCM, with the aim to produce a PCM capable of regulating the indoor temperature of a building, using expanded vermiculite and EG and combining both with a commercial and low-cost PCM (i.e., OM37), using the vacuum impregnation. Good results of thermal and chemical stability are achieved in this work, as well as significant results achieved from the point of view of thermal performance: the realized PCM can regulate the internal temperature by lowering the temperature peaks. Costa et al. [98] prepared a form-stable based on diatomite-vermiculite and paraffin for cement mortars. Different amounts of PCM (5 to 25 wt%) were used. Again, characterization of the composite materials demonstrated improved thermal performance and the ability of the porous aggregates used to retain PCM.

In this paper [57], several porous materials used as a matrix to contain PCMs are presented. Nine porous materials are considered: kaolin, diatomite, sepiolite, montmorillonite, perlite, SiO_2, attapulgite, vermiculite and fly ash. They are all presented as suitable media to hold PCMs and then be incorporated into various applications that require temperature control. However, all these materials, before being used, require preparation to improve their absorbability, such as acid/alkali treatment, ionic exchange method, calcination or hydrothermal method, according to their physical and crystal structure. These treatments affect not only production time but also costs, and that is why some researchers have proposed new materials that are naturally porous, such as some types of stones. Frigione et al. [99,100] proposed a novel porous aggregate, called Lecce Stone (LS), a biocalcarenite composed mainly of $CaCO_3$ with a total open porosity of 30.33 ± 0.99%. Due to these porosimetric characteristics, LS was successfully impregnated with Poly-Ethylene Glycol (PEG) as PCM. The study of thermal properties showed that PEG, once incorporated into LS through vacuum impregnation, had characteristic temperatures and enthalpy suitable for incorporation into mortars for thermal control of buildings. This composite material (i.e., LS/PEG) also demonstrated excellent chemical stability and good thermal performance especially in cementitious mortars in a temperate climate, such as the Mediterranean [127]. Another virtuous aspect of this research is represented by the fact that the LS came from processing waste. In recent years, as a matter of fact, this trend of reusing waste materials as PCM carriers is growing. Uthaichotirat at al. [128] studied the thermal and sound properties of concrete containing highly porous aggregates from manufacturing waste impregnated with paraffin. They used autoclaved aerated concrete (AAC) coming from waste which are extremely interesting because they have a large content of voids (almost 80% of their volume). Therefore, they do not require any further processing to make them suitable as matrix. The presence of these porous materials impregnated with the PCM and used as aggregates in concrete shown improved compressive and flexural strength and improved thermal properties. However, as the amount of the PCM increased, the sound insulation decreased because of the porosity reduction with the presence of the PCM filling the pores.

5.3. Concrete

The utilization of PCM in concrete was also developed by several authors evaluating the properties in fresh and hardened state, with recourse to a different incorporation technique [92–94,129–132]. Different studies indicate different behaviors in mechanical terms. However, the positive influence of the PCM in the temperature control is attested by all, which influences the concrete heat of hydration or the environment in which the concrete was applied.

Most of the studies carried out focused on the incorporation of encapsulated PCM. The study of a PCM macroencapsulation solution in concrete was evaluated by several authors [92,129]. Cui et al. [92] adopted a PCM macroencapsulation solution in metallic capsules, in order to replace the aggregate in concrete reinforced with steel fibers. The effect of different contents of steel fibers and different thickness of the metallic macrocapsules (0.3 mm and 1 mm) were evaluated. The microcapsules used presented a diameter of 19 mm. The test results showed that the thermal conductivity and compressive strength of the concrete activated with the PCM macrocapsules increased with an increase in the steel fiber content and metallic macrocapsules thickness. Dong et al. [129] also developed a study in concrete with incorporation of PCM macrocapsules with 22 mm of diameter, observing that the PCM incorporation leads to a lower peak indoor air temperature and fluctuation of indoor temperature. However, a decrease in the compressive strength of the concrete doped with PCM was observed when the PCM macrocapsules content increased in the concrete mix.

The utilization of microencapsulated PCM solutions also attracted several researchers [93,130–132]. Cellat et al. [93] developed a PCM microcapsule coated with a capsule in polystyrene and a core with a eutectic mixture (capric acid and myristic acid). The PCM microcapsules were added to a different concrete mixture, allowing to observe that the PCM did not affect the hydration reaction; however, the peak temperature of fresh concrete in the first 40 h was lowered due to the absorption of heat by the PCM. A decrease in the concrete compressive strength with the PCM incorporation was also verified. Additionally, metal surfaces and concrete samples in contact with rebar were examined to determine the corrosion products, observing that the PCM incorporation did not affect the corrosive effect on metal surfaces in concrete. Jayalath et al. [130] also studied the properties of concretes with PCM microcapsule incorporation, as replacement of fine aggregate. The test results allowed to observe a decrease in the thermal conductivity and an increase in the heat capacity due to the PCM incorporation. However, a loss in the compressive strength was also observed, due to the PCM low density when compared with the fine aggregates.

D'Alessandro et al. [94] developed a concrete solution with incorporation of two types of PCM (microcapsules and macrocapsules). Several concrete compositions with different PCM microcapsules and macrocapsules contents were developed. It was possible to observe that the PCM incorporation produced a concrete with higher thermal performance. As expected, the PCM incorporation resulted in a decrease in the compressive strength, as mentioned for other studies. However, this decrease in the mechanical performance does not compromise their use as structural material.

However, the costs of acquiring and developing PCM nano, micro or macrocapsules continue to be quite high, especially when applied in concrete, taking into account the enormous consumption of raw materials. Thus, a new study has emerged in which pure PCM (non-encapsulated PCM) is incorporated into construction and demolition waste, the waste being used as a substitute for natural aggregate for the concrete production [95].

5.4. Carbon Based Materials

Thermal conductivity is a determining factor in evaluating the thermal performance of PCMs. The magnitude of thermal conductivity determines the rate of energy storage/consumption and therefore affects energy utilization efficiency. Normally, carbon materials have high thermal conductivities, so their use as a support material for PCMs is an effective way to improve the thermal conductivity of PCMs [101,133]. Carbon-based materials such as graphene, expanded graphite, carbon nanotubes, carbon nanofibers and carbon nanosheets with high thermal conductivity can effectively increase the thermal transfer rate of PCMs [133].

Porous carbon-based materials are varied and can be obtained at different costs; for example, carbon fiber and expanded graphite are cheaper compared to other porous materials such as metal foam or porous ceramics [133]. Thus, the studies developed based on carbon materials are based on different materials. However, the high cost of these materials still limits their potential application.

Some authors adopted graphite as support material to improve the thermal properties of PCM [104,134–137]. Graphite is one form of carbon, is black and soft, and is constituted by a planar and multi-layered structure. The carbon molecules in each layer are systematically arranged in the form of a honeycomb [133]. Karthik et al. [104] developed a composite consisting of paraffin wax and graphite using a low-cost and small-scale process, observing an increase in the thermal conductivity when compared to the paraffin wax. The obtained results also demonstrated a composite compressive strength higher than the graphite foam. The developed composite can also be considered as a potential material for various thermal energy storage applications in buildings, vehicles and solar thermal harvesting.

Graphene is a crystal with a unique thermal conductivity, considered the mother of all carbon materials due to its superior thermal and mechanical properties [133]. Thus, it has become an interesting candidate for phase change material applications [105,138–141]. Mehrali et al. [105] studied a composite material using a palmitic acid as a PCM impregnated by vacuum in graphene oxide as supporting material. The thermal test showed that the PCM composite material exhibits good thermal reliability and chemical stability. The PCM composite thermal conductivity increases about three times. Thus, the composite material was considered adequate for thermal energy storage applications. However, even though it has numerous advantages related to its thermal properties, graphene is not widely used due to its high cost, which can make its practical applications unsustainable.

Some works have also been developed using carbon nanotubes [106,142–145]. These materials are extremely thin and long cylindrical structures, consisting of carbon atoms. There are two main types of carbon nanotubes: single-walled and multi-walled. Single-walled nanotubes consist of just one layer of graphene (a flat sheet of carbon atoms arranged in a hexagonal lattice), while multi-walled nanotubes contain multiple layers of graphene stacked on top of each other. These materials are synthesized at high temperatures, usually between 600 °C to 1000 °C, whose properties can be affected by impurities [133]. Yang et al. [106] created a composite material composed by stearic acid as PCM and carbon nanotubes as porous material support, evaluating their thermal performance. The results allow to conclude that the phase change temperatures varied slightly while the latent heat decreased with the increased carbon nanotubes content. On the other hand, the thermal conductivity of the composites is higher than the pure PCM, which indicates the material's potential for thermal management.

Carbon nanofibers are materials composed of thin carbon fibers, with diameters ranging from 1 to 100 nanometers. These materials present high mechanical strength, excellent electrical and thermal conductivity, low density, large surface area and high chemical stability. Liu et al. [107] prepared a new composite material using steric acid as a PCM and carbon fiber as a nanoparticle. The results of this study allow to observe high latent heat and good thermal and chemical stabilities of the composite material even submitted to 200 melt/freeze cycles, which indicates a potential application for solar energy storage application such as solar heat storage tank.

5.5. Naturally Based Materials

The search for inexpensive, easily available and environmentally friendly composite materials with high enthalpy is still a great challenge; thus, the incorporation of PCM in natural materials has been attracting the interest of the academic community. Considering the high cost of carbon-based materials, some researchers carbonized natural materials in order to obtain biological porous carbon to combine with PCM [146–149]. Several solutions based on various natural materials have been developed such as the application of PCM in flower stems [101,150], potatoes and radishes [149], corn stalks [146], rice straw [103,147], watermelon rinds [148], wood [151–154] and earth construction [102].

Wen et al. [146] used stearic acid as phase change material; this is a saturated fatty acid abundant in nature, obtained from different vegetable and animal oils and fats with low acquisition cost. In this study, carbonized maize straw was used a low-cost matrix, impregnated by vacuum. Corn straw is a porous waste material of an economical nature, since corn is one of the most planted crops in the world, existing in large quantities. The test results revealed a good chemical compatibility between stearic acid and the maize straw, leading to a higher thermal conductivity. Zhang et al. [103] developed a study based on the PCM incorporation into a rice carbonized matrix, once again a very implemented culture with large quantities available. The selected PCM was a eutectic mixture based on palmitic and lauric acid, two natural compounds very abundant in nature. The developed composite material presents a potential to be used in construction industry, namely, in building envelopes, due to their high thermal energy storage and good thermal and form stability. Wang et al. [101] prepared form-stable composite PCMs for building temperature regulation based on daisy stem as support material. The selection of this raw material was justified by the high absorption capacity of daisy stems. The presented results allowed to observe that the PCM composites presented a good thermal reliability without enthalpy alteration after a high number of heating and cooling cycles. Regarding thermal efficiency, a decrease in temperature fluctuation was observed and on the recorded differential between the interior and ambient temperature of the cardboard test cells.

Even adopting low-cost and widely available natural materials, biomass-based materials require additional treatments, which normally require high energy consumption, compromising the energy gains obtained with their application. Thus, other studies have emerged based on the incorporation of PCM into natural materials without the need for prior treatment. Cunha et al. [102] developed different compressed earth blocks with direct incorporation of different content of pure and non-encapsulated PCM. The soil used was collected in the north region of Portugal and was stabilized with a lower content of Portland cement. The test results showed that the incorporation of PCM in compressed earth blocks can be carried out successfully, without danger of PCM leakage. The PCM incorporation leads to a decrease in water absorption, due to the partial or total occupation of earth blocks by the PCM; however, a decrease in the mechanical behavior (compressive strength) was also observed related to the lower water content added during the mixture procedure, which can negatively affect the cement hydration.

5.6. Boards

One of the earliest applications of PCMs in construction was involving boards (plaster board or gypsum board). The main reasons of this application were related to the wide use of these boards in buildings (due to their low cost) and for their location (in the interior side of the wall) in the building system [155]. In this work, A. Oliver [109], studied the thermal behavior of a gypsum board containing 45 wt% of a commercial micro-capsulated PCM (Micronal DS 5001X, produced by BASF, Ludwigshafen, Germany) with a phase change temperature around 26 °C and an enthalpy of about 110 J/g. It was concluded that the panel with the presence of PCM was able to store five times more energy than a thermal brick wall, nine times more energy than a standard brick wall and three times more energy than a normal gypsum board (without the PCM). This was one of many papers that highlighted the advantages of the use of PCMs in boards. Kuznik et al. [156] performed a full-scale experiment to evaluate the presence of a PCM in plasterboards. These were applied to the walls and ceiling of an office, and that room was monitored for one year. Similarly, another identical room but without the PCM was monitored for the same period. A commercial paraffin-based microencapsulated PCM was employed (ENERGAIN by the Dupont de Nemours Society). The results demonstrated an improvement in thermal comfort.

Lai et al. [110] evaluated the incorporation of micro-encapsulated PCM into gypsum board and then they studied the physical properties, heat transfer and thermal storage behavior. A commercial paraffin-based PCM (MPCM 28-D by Microtek Laboratories, Inc., Moraine, OH, USA) with a melting point around 28 °C and a melting heat of about 180–195 J/g was employed. In this work, different experimental test cells were prepared using different PCM content (23%, 30% and 40% of mass fraction), and different temperatures (hot and cold) were used to investigate the thermal performance of the gypsum boards. Numerical and experimental simulations were performed. The results showed that increasing the content of PCM does not necessarily increase its energy storage capacity. In fact, in this case, the best percentage of PCM was at 30%. In addition, it was noted that at low temperatures the phenomenon of supercooling was observed while at temperatures close to the melting temperature of PCM, an improvement in thermal properties was recorded. More recently, Mourid et al. [111], conducted a full-scale experimental work integrating a commercial micro-capsulated paraffin-based PCM (Energain man, produced by BASFufactured by DuPont™) into wallboards. This experiment was performed in Morocco, which is characterized by a Mediterranean climate, in the winter and spring seasons. The wallboards had been placed in different ways within the different rooms (walls or ceilings). The results obtained showed an interesting aspect: the boards placed on the ceiling performed better than the others on the walls. This should make it clear that the placement of a PCM in each environment must be customized, taking into account the specific thermal properties of a building (air passage, exposure to the sun, number of inhabitants, amount and height of buildings around, etc.).

5.7. Bricks

Numerical and experimental research have found that inserting PCMs into bricks increases the thermal mass and improves the thermal properties of the walls, ensuring the indoor comfort while saving energy. Gao et al. [112] found that, usually, PCMs are inserted into bricks following two techniques: macro-encapsulation and form-stable. The amount of PCMs that can be inserted using these two techniques is around 16.44% and 20.55%, respectively.

In the literature, there are several experimental works concerning the study of PCMs in bricks, and the experiments almost always involve the construction of scaled test cells placed in climatic chambers in which certain climatic conditions are simulated. Zhu et al. [113], studied the inner surface index of a brick wall with and without the PCM. This study demonstrated an attenuation of the internal surface temperature amplitude for that wall added with the PCM brick. This attenuation was 3.8–4.4 times less that of the reference brick wall; the lag time ratio of the internal surface was 8–12 times that of the reference brick wall. Vicente et al. [114] prepared three different specimens and tested them in a climatic chamber. The first specimen was the reference (M1), the second was with the PCM (M2) and the third was with the PCM and insulation material (M3). The results showed that the maximum peak of the temperature was decreased by about 50% (using the M2) and 80% (using M3) compared with the reference brick wall; the time lag of the brick walls with PCM was increased by 2 h compared with the wall without the PCM. Saxena et al. [77] incorporated two different commercial PCM (Eicosane and OM35) into hollow brick using the typical climatic conditions of Delhi during summer. They recorded a reduction of about 5–6 °C compared to the conventional brick wall and an energy saving of 8% using the Eicosane PCM and of 12% using the OM35. Abbas et al. [157] performed a test using natural conditions outdoors consisting of two identical rooms using PCM bricks. The results illustrated that the inner surface temperature of the brick wall with the PCM was reduced of about 4.7 °C, the time lag was increased by 2 h, the temperature fluctuation was reduced by 23.84 % and the damping factor was 70% compared with the reference wall. In general, the literature that has been developed on this topic reports that the application of PCMs in brick walls should be developed because it is believed to be an excellent method of incorporating PCMs into a building. Moreover, studies should be conducted based on actual climatic conditions, and more study of these full-scale applications is desired.

5.8. Solar Thermal Systems

Nowadays, photovoltaic panels for electricity generation have become a common practice all over the world due to their contribution to decreasing electrical energy consumption. Many countries in the European Union, namely, Portugal, have created funding programs in order to economically support families to be able to install this type of systems in their buildings [158]. Thus, worldwide there was an increase in the installation of photovoltaic panels of about 12% between 2018 and 2019; however, an increase of more than 10 times the currently installed photovoltaic capacity is expected by 2050. However, the lifetime of these systems can be reduced due to the high incident solar energy, which causes an increase in the panels' operating temperature. The temperature surface of the photovoltaic panel can reach temperatures of 40 °C above the environmental temperature [159]. Thus, it has been necessary to find ways to refrigerate and regulate the photovoltaic panels' temperature. PCM applications can be seen as a suitable solution [3].

The application of PCMs in photovoltaic panels requires specific characteristics quite different from those required for applications in traditional building materials. Thus, several researchers, especially in recent years, have been dedicated to studying the transition temperature, which is a dominant aspect for photovoltaic panels' thermal management to achieve high electrical and thermal power [160–162]. For an effective management, the transition temperature of the PCM must be lower than the photovoltaic panel temperature, being also limited by the summer ambient temperature during the night and by the panel temperature during the winter [163,164]. Hasan et al. [86] used two different PCMs, a eutectic mixture of capric–palmitic acid and a salt hydrate (calcium chloride hexahydrate) applied to a metallic PCM container attached to the back of the photovoltaic panel. Three different panels (reference, capric-palmitic panel and slat hydrate panel) were placed outdoors facing south in Dublin, Ireland and Vehari, Pakistan, enabling the analysis of cold and hot climatic conditions. The test results and simulations allow to observe a better performance of the salt hydrates when compared to the eutectic mixture. Salt hydrate PCM coupled to the photovoltaic panel successfully reduced panel temperature by 10 °C in Ireland and 21.5 °C in Pakistan. The two PCMs used conducted a panel's higher performance in hot and stable climatic conditions. However, applications other than the cooling panel systems have also been studied. Sharma et al. [87] evaluated the influence of the PCM's presence on the electrical efficiency of photovoltaic panels, observing that the PCM use increases the electrical efficiency of the system by 7.7%.

The low thermal conductivity of PCMs is currently still a challenge regarding exploring these materials as a solution for thermal regulation of photovoltaic panels. Thus, Abdulmunem et al. [88] studied the incorporation effects of carbon nanotube nanoparticles as additives to the PCM and copper foam matrix on the PV panel performance. The test results allow to observe that the carbon nanotubes' incorporation within the PCM and copper foam greatly improved the effective thermal properties of the material and consequently the electrical performance of the panels, when compared to the panels without passive cooling materials.

Other applications in solar thermal systems can arise, combining for example the use of PCM with solar thermal water heating systems, in order to increase the thermal storage capacity of solar energy for long periods of time [165–167], or even solar dryers for application in other industries, such as the food industry, in the drying of fruits and vegetables [168]. Huang et al. [167] performed a numerical analysis and experimental test of a solar water heating system, developed based on the use of capillary pipes placed above and below a prefabricated concrete skeleton with vacancies occupied by PCM macrocapsules. A reference model and a PCM model were simulated, indicating that the floor's energy storage capacity with PCM is greatly enhanced with the benefit of saving water tank space.

Table 2 summarizes the main information about the various works presented earlier.

Table 2. Summary of PCM incorporation in construction material studies.

Study	PCM Type	PCM Properties	PCM Incorporation Technique	Construction Material	Main Results
Lopez-Arias et al. [89]	Organic—Paraffin	Temperature transition of 70 °C; Enthalpy of 120 J/g; Density of 0.968 g/cm^3.	Immersion	Mortars	Higher PCM content leads to a higher thermal conductivity and higher compressive strength.
Cunha et al. [75,76]	Organic—Paraffin	Temperature transition of 24 °C; Enthalpy of 150 kJ/kg; Density of 880 kg/m^3.	Microencapsulation	Mortars	Higher PCM content leads to a higher water/binder ratio, lower flexural and compressive strength. However, a decrease in extreme temperatures and a decrease in heating and cooling needs was observed.
Cunha et al. [90]	Organic—Paraffin	Temperature transition of 22 °C; Enthalpy of 200 kJ/kg; Density of 760 kg/m^3.	Direct incorporation	Mortars	Higher PCM content leads to a lower water/binder ratio, a higher liquid/binder ratio, lower compressive strength and better thermal performance, due to the decrease in heating and cooling needs.
Illampas et al. [91]	Organic—Paraffin	Temperature transition of 37 °C; Enthalpy of 190 J/g.	Microencapsulation	Mortars	Higher PCM content leads to an increase in the open porosity and a decrease in the flexural strength, compressive strength and elastic modulus. The thermal performance of the PCM mortars was improved, decreasing the thermal conductivity and thermal diffusivity, increasing the specific heat capacity and attenuation of the temperature peaks.

Table 2. Cont.

Study	PCM Type	PCM Properties	PCM Incorporation Technique	Construction Material	Main Results
Kheradmand et al. [117]	Organic—Paraffin	Temperature transition of 26 °C; Enthalpy of 110,000 J/kg; Apparent density of 350 kg/m^3.	Microencapsulation	Mortars	Higher PCM content leads to an increase in water absorption and a decrease in the compressive strength.
Shadnia et al. [118]	Organic—Paraffin	Temperature transition of 28 °C; Enthalpy of 180–195 kJ/kg; Density of 900 kg/m^3.	Microencapsulation	Mortars	Higher PCM content leads to a decrease in the compressive strength. The PCM incorporation can effectively reduce the transport of heat through geopolymer mortar.
Aguayo et al. [119]	Organic—Paraffin	PCM-M temperature transition of 24.3 °C and enthalpy of 100 J/g. PCM-E temperature transition of 23.4 °C and enthalpy of 159 J/g.	Microencapsulation	Mortars	Higher PCM-E content leads to an increase in the compressive and flexural strengths until a certain replacement level of PCM. Higher PCM-E content leads to an decrease in the compressive and flexural strengths until a certain replacement level of PCM-M.
Rathore et al. [66]	Organic—Paraffin	Temperature transition of 38.23 °C and enthalpy of 206.32 J/g for the fusion process; Temperature transition of 29.40 °C and enthalpy of 229 J/g for the freezing process	Form-stable through vacuum impregnation	Porous aggregates	They used EV as main supporting matrix and EG to act both as a support matrix and to improve thermal conductivity. The latter was enhanced, indeed; the leakage phenomenon was avoided and the thermal performance was significantly appreciable.
Frigione et al. [99,100]	Organic—Polymer	Temperature transition of 42.8 °C and enthalpy of 129.3 J/g for the fusion process; Temperature transition of 23.6 °C and enthalpy of 129.8 J/g for the freezing process	Form-stable through vacuum impregnation	Porous aggregates	The porous matrix was a natural stone from waste production. The composite form-stable PCM was used as aggregate for mortar based on different binders. A decrease in the flexural and compressive strength was detected.
Zhang et al. [126]	Organic—Paraffin	Temperature transition of 44.32 °C and enthalpy of 177.54 J/g for the fusion process; Temperature transition of 48.30 °C and enthalpy of 181.31 J/g for the freezing process	Form-stable through vacuum impregnation	Porous aggregates	They used EP as supporting matrix and added carbon nanotubes to improve thermal conductivity. The final material showed good chemical and thermal stability.
Cui et al. [92]	Organic—Paraffin	Temperature transition of 23 °C; Enthalpy of 188 kJ/kg; Density at solid state of 833.8 kg/m^3; Density at liquid state of 786.7 kg/m^3.	Macroencapsulation	Concrete	Higher PCM content leads to an increase in water absorption, thermal conductivity and compressive strength.
Cellat et al. [93]	Eutetic Mixture (capric acid and myristic acid)	Temperature transition of 26 °C; Entalpy of 155.44 J/g.	Microencapsulation	Concrete	The PCM microcapsules' incorporation did not affect the hydration reaction in concrete; however, the peak temperature of fresh concrete decreases due to the absorption of heat by the PCM. The PCM addition leads to a decrease in compressive strength and does not affect the corrosive effect on metal surfaces in concrete.

Table 2. Cont.

Study	PCM Type	PCM Properties	PCM Incorporation Technique	Construction Material	Main Results
D'Alessandro et al. [94]	Organic—Paraffin	Temperature transition of 18 °C; Microcapsules particle size between 14 and 24 µm; Macrocapsules particle size between 3 and 5 mm.	Microencapsulation and macroencapsulation.	Concrete	Higher PCM content leads to an increase in the thermal performance and a decrease in the compressive strength.
Jia et al. [95]	Organic—Paraffin	Temperature transition of 22 °C; Enthalpy of 200 kJ/kg; Density of 760 kg/m^3.	Direct incorporation	Concrete	Higher PCM content leads to a decrease in the compressive strength.
Dong et al. [129]	Organic—Paraffin	Temperature transition of 29.2 °C and enthalpy of 246.4 J/g for the fusion process; Temperature transition of 22.7 °C and enthalpy of 249.7 J/g for the freezing process.	Macroencapsulation	Concrete	Higher PCM content leads to an increase in the thermal performance and a decrease in the compressive strength.
Jayalath et al. [130]	Organic—Paraffin	Temperature transition of 23 °C; Enthalpy of 100 kJ/kg; Density 250–350 kg/m^3	Microencapsulation	Concrete	Higher PCM content leads to an increase in the thermal performance and a decrease in the compressive strength.
Karthik et al. [134]	Organic—Paraffin	Enthalpy of 206 J/g; Density of 0.91 g/cm^3; Thermal conductivity at 25 °C of 0.24 W/mK.	Form-stabilization	Carbon materials—Graphite foam	PCM incorporation leads to an increase in the thermal conductivity and compressive strength.
Mehrali et al. [105]	Organic—Non-paraffin	Temperature transition (melting process) of 61.14 °C; Enthalpy (melting process) of 202 kJ/kg; Temperature transition (freezing process) of 59.84 °C; Enthalpy (freezing process) of 208.87 kJ/kg.	Form-stabilization	Carbon materials—Graphene oxide	The PCM composite material exhibits good thermal reliability, good chemical stability, and higher thermal conductivity.
Yang et al. [106]	Organic—Paraffin	Temperature transition (melting process) of 69.45 °C; Enthalpy (melting process) of 210.9 kJ/kg; Temperature transition (freezing process) of 66.96 °C; Enthalpy (freezing process) of 211.9 kJ/kg.	Form-stabilization	Carbon materials—Nanotubes	Higher carbon nanotube content leads a slight change in temperature transition and a decrease in enthalpy. The thermal conductivity of PCM composite is higher than the pure PCM.
Liu et al. [107]	Organic—Non-paraffin	Temperature transition of 68 °C; Enthalpy of 229.4 J/g.	Form-stabilization	Carbon materials—Nanofibers	High latent heat and good thermal and chemical stabilities of the composite material even submitted to 200 melt/freeze cycles.
Wang et al. [101]	Organic—Paraffin	Temperature transition of 40.1 °C; Enthalpy of 213.6 J/g.	Form-stabilization	Natural materials—Daisy stems	Good PCM thermal reliability and higher thermal efficiency with the PCM composites' application, due to the decrease in temperature fluctuation inside the test cells.
Cunha et al. [102]	Organic—Paraffin	Temperature transition of 22 °C; Enthalpy of 200 kJ/kg; Density of 760 kg/m^3.	Direct incorporation	Natural materials—Compressed earth bricks	Higher PCM content leads to a decrease in the water absorption, compressive strength and modulus of elasticity.

Table 2. Cont.

Study	PCM Type	PCM Properties	PCM Incorporation Technique	Construction Material	Main Results
Zhang et al. [103]	Organic—Non-paraffin	Temperature transition (melting process) of 35.7 °C; Entalpy (melting process) of 171.8 J/g; Temperature transition (freezing process) of 28.2 °C; Entalpy (freezing process) of 160.5 J/g.	Form-stabilization	Natural materials—Carbonized rice	High thermal energy storage and good thermal and form stability.
Wen et al. [146]	Organic—Non-paraffin	Temperature transition (melting process) of 69.23 °C; Entalpy (melting process) of 208.16 J/g; Temperature transition (freezing process) of 65.78 °C; Entalpy ((melting process) of 207.44 J/g.	Form-stabilization	Natural materials—Carbonized maize straw	Good chemical compatibility between PCM and carbonized maize straw matrix and higher thermal conductivity.
Liu et al. [148]	Inorganic—Salt hydrates	Temperature transition of 63.2 °C; Enthalpy of 255.9 J/g.	Form-stabilization	Natural materials—Watermelon rind	High thermal conductance, good shape stability and excellent thermal cycle stability.
Liang et al. [151]	Organic—Non-paraffin	Temperature transition (melting process) between 45.8–63.7 °C; Entalpy (melting process) between 96.6–144.7 J/g; Temperature transition (freezing process) between 22.2–39.6 °C; Entalpy (freezing process) between 77.3–167.3 J/g.	Immersion	Natural materials—Wood flour	Good thermal reliability and chemical stability.
Oliver [109]	Organic—Paraffin	Temperature transition of 26 °C; Enthalpy of 110 J/g.	Microencapsulation	Gypsum Boards	This board with the presence of PCM was able to store five times more energy than a thermal brick wall, nine times more energy than a standard brick wall and three times more energy than a normal gypsum board (without the PCM).
Kuznik et al. [156]	Organic—Paraffin	Temperature transition of melting process 13.6 °C and enthalpy of 107.5 J/g. Temperature transition of freezing process 23.5 °C and enthalpy of 104.5 J/g.	Microencapsulation	Gypsum Boards	The gypsum board with PCM was applied into the walls and ceiling of an office. Monitoring was performed for one year and compared to another identical room without the PCM. The results demonstrated an improvement in thermal comfort.
Vicente et al. [114]	Organic—Paraffin	Temperature transition of 18 °C; Enthalpy of 134 J/g	Microencapsulation	Bricks	The bricks containing the PCM macrocapsules were able to decrease by about 50% the maximum peak of temperature and to reach a time lag of 2 h compared with the wall without the PCM.

Table 2. Cont.

Study	PCM Type	PCM Properties	PCM Incorporation Technique	Construction Material	Main Results
Abbas et al. [157]	-	Temperature range of 38–43 °C for the fusion process and temperature range of 43–37 °C for the freezing process; enthalpy of 174 J/g	Microencapsulation	Bricks	The thermal performance was studied in a natural outdoor environment. The results illustrated that the brick wall with the PCM reduced by about 4.7 °C the maximum peak of temperature, increased the lag time by 2 h and reduced the temperature fluctuation by 23.84% compared with the reference wall.
Hasan et al. [86]	Eutetic mixture (capric acid and palmitic acid) and a salt hydrated	Eutectic mixture: - Temperature transition of 22.5 °C; - Enthalpy of 173 kJ/kg. Salt hydrated: - Temperature transition of 29.8 °C; - Enthalpy of 191 kJ/kg.	Macroencapsulation	Solar thermal systems	The PCM decreased the photovoltaic panel temperature.
Sharma et al. [87]	Organic—Paraffin	Temperature transition of 42 °C; Enthalpy of 165 kJ/kg.	Macroencapsulation	Solar thermal systems	The PCM increased the electrical efficiency of the photovoltaic panel.
Abdulmunem et al. [88]	Organic—Paraffin	Temperature transition of 59.01 °C; Enthalpy of 154.42 J/g	Macroencapsulation	Solar thermal systems	The incorporation of carbon nanotubes nanoparticles as additives to the PCM increased the average electrical efficiency of the photovoltaic panels.
Huang et al. [167]	Organic—Non-paraffin	Temperature transition of 29.3 °C; Enthalpy of 162 kJ/kg.	Macroencapsulation	Solar thermal systems	The use of the PCM as a composite energy storage layer in a solar water floor heating system greatly enhanced heat storage capacity of the floor, saving water tank space.

6. Technology Cost Analysis

The cost analysis of this type of thermal storage technology is still an area under development, taking into account the diversity of external temperature laws and the diversity of constructive solutions with PCM integration, being able to vary the type of PCM used, its content, incorporation technique and location in the constructive solution. Thus, the number of existing studies that take an approach in terms of cost analysis is much lower than the number of studies developed in which the various additive solutions with PCM are tested in physical, mechanical and thermal terms.

Gholamibozanjani and Farid [169] constructed two identical test cells, with external dimensions of 2.7 × 2.7 × 2.7 m^3. In one test cell, an air-PCM heat storage unit was installed (active thermal storage system) and in the other, the PCM was integrated into the wallboards (passive thermal storage system). It was observed that the test cell with air-PCM heat storage unit consumed less energy, saved more costs and maintained comfortable conditions inside. However, the reduction in cost was not proportional to the decrease in energy consumption. On one of the analyzed days, a reduction of 20% in energy consumption corresponded to a reduction cost of 32%. It was also possible to verify that the active thermal storage system solution was more efficient than the passive thermal storage system solution. Cunha et al. [76] estimated the energy consumption per month and the cost related to the energy consumption of small test cell coating with different mortars based on different binders and activated with microencapsulated PCM. It was observed that the PCM use suppressed the heating and cooling needs in typical spring and autumn months in the north part of Portugal. The results allow to conclude that mortars with PCM incorporation lead to a cost decrease higher than 52% for all binders analyzed. M'hamdi et al. [170] studied the influence of using different PCM types in different types of buildings located in three different climatic areas in the north of Africa. An environmental and economic analysis was performed showing that the PCM utilization can reduce by 10% the energy cost and by 707 kg/year the carbon dioxide emissions. Panayiotou et al. [171] studied the PCM influence in the Mediterranean region. The energy savings achieved by PCM incorporation were between 21.7 and 28.6%. However, for the PCM combined with a common thermal insulation topology, the energy savings per year were 66.2%. A Life Cycle Cost analysis showed that the PCM solution presented a payback period of 14 years, while the payback period of the combined solution was reduced to 7 years.

There are only a very limited number of studies in which economic analyses are carried out [76,172–175]. Mi et al. [172] developed one of the only studies which used a dynamic payback period concept, considering concepts such as the money time value. The authors compared the obtained values for the dynamic payback period with the static payback period for different discount rate levels.

The cost-effectiveness of constructive solutions with PCM is highly sensitive to inflation and discount rates, as these solutions are more economical as the inflation rate increases and the discount rate decreases [175]. Thus, due to the wide variation in inflation rates currently being experienced in Europe, it is quite difficult to establish specific conclusions about the cost analysis of these types of solutions.

7. Conclusions

In recent years, especially in the last 10 years, research in the area of phase change materials has evolved a lot, largely due to the enormous environmental and economic challenges that the world has been going through. The political measures implemented worldwide, but especially in Europe, have also been a driving force to increasingly seek more sustainable and efficient construction solutions based on thermal storage technologies.

Currently, there are several possibilities for incorporating PCMs in construction materials, based on different incorporation techniques. However, it is important to note that a detailed study on the type of PCM to be incorporated must be carried out in advance based on the problems related to the incorporation of PCMs in building materials and the criteria for PCM selection to be applied to buildings.

So far, the form-stable method turns out to be the most versatile one because it is only necessary to have a porous material as a substrate and to impregnate it with any PCM (according to the type of application intended for). However, it would be necessary to reduce additives used to improve the properties of the PCM because the consequence is the reduction of the PCM amount.

This article discussed the possibilities of incorporating PCMs in mortars, concrete, porous aggregates, carbon-based materials, naturally based materials, boards, brick and solar thermal systems, as well as an approach to the economic analysis of the application of this type of technology. Despite the various advances in the development of thermal storage technology using phase change materials, some issues still require further investigation and others need to be investigated, namely:

- Applications of constructive solutions for the exterior of buildings;
- Development of economic analysis;
- Development of life cycle analysis of constructive solutions.

Studying the thermal behavior of a PCM in the real world contributes greatly to the understanding of these materials. Therefore, it is recommended as part of future development that more full-scale experiments be performed.

Author Contributions: Conceptualization, S.C., A.S., J.A. and M.F.; methodology, S.C., A.S., J.A. and M.F.; investigation, S.C., A.S., J.A. and M.F.; resources, S.C., A.S., J.A. and M.F.; data curation, S.C., A.S., J.A. and M.F.; writing—original draft preparation, S.C. and A.S.; writing—review and editing, S.C., A.S., J.A. and M.F.; visualization, S.C., A.S., J.A. and M.F. All authors have read and agreed to the published version of the manuscript.

Funding: This research was funded by FUNDAÇÃO PARA A CIÊNCIA E TECNOLOGIA (FCT), grant number UIDB/04047/2020.

Data Availability Statement: Not applicable.

Acknowledgments: This work was supported by FCT/MCTES through national funds (PIDDAC) under the R&D Unit Centre for Territory, Environment and Construction (CTAC) under reference UIDB/04047/2020.

Conflicts of Interest: The authors declare no conflict of interest.

References

1. Regional Information Center for Western Europe. Available online: https://unric.org/pt/Objetivos-de-Desenvolvimento-Sustentavel/ (accessed on 22 February 2023).
2. United Nations, General Assembly. *Transforming Our World: The 2030 Agenda for Sustainable Development, Seventieth Session;* United Nations: New York, NY, USA, 2015.
3. Preet, S. A review on the outlook of thermal management of photovoltaic panel using phase change material. *Energy Clim. Chang.* **2021**, *2*, 100033. [CrossRef]
4. Dai, B.; Liu, C.; Liu, S.; Wang, D.; Wang, Q.; Zou, T.; Zhou, X. Life cycle techno-enviro-economic assessment of dual-temperature evaporation transcritical CO_2 high-temperature heat pump systems for industrial waste heat recovery. *Appl. Therm. Eng.* **2023**, *219*, 119570. [CrossRef]
5. Lupi, V.; Marsiglio, S. Population growth and climate change: A dynamic integrated climate-economy-demography model. *Ecol. Econ.* **2021**, *184*, 107011. [CrossRef]
6. Pylaeva, I.S.; Podshivalova, M.V.; Alola, A.A.; Podshivalov, D.V.; Demin, A.A. A new approach to identifying high-tech manufacturing SMEs with sustainable technological development: Empirical evidence. *J. Clean. Prod.* **2022**, *363*, 132322. [CrossRef]
7. Dai, S.; Qian, Y.; He, W.; Wang, C.; Shi, T. The spatial spillover effect of China's carbon emissions trading policy on industrial carbon intensity: Evidence from a spatial difference-in-difference metho. *Struct. Chang. Econ. Dyn.* **2022**, *63*, 139–149. [CrossRef]
8. International Energy Agency. Available online: https://www.iea.org/data-and-statistics/ (accessed on 24 February 2023).
9. Yan, K.; Lan, H.; Li, Q.; Ge, D.; Li, Y. Optimum utilization of recycled aggregate and rice husk ash stabilized base material. *Constr. Build. Mater.* **2022**, *348*, 128627. [CrossRef]
10. Ismail, A.; Younis, K.; Maruf, S. Recycled aggregate concrete made with silica fume: Experimental investigation. *Civ. Eng. Archit.* **2020**, *8*, 1136–1143. [CrossRef]
11. Tang, Y.; Xiao, J.; Zhang, H.; Duan, Z.; Xia, B. Mechanical properties and uniaxial compressive stress-strain behavior of fully recycled aggregate concrete. *Constr. Build. Mater.* **2022**, *323*, 126546. [CrossRef]
12. Zhang, H.; Xiao, J.; Tang, Y.; Duan, Z.; Poon, C. Long-term shrinkage and mechanical properties of fully recycled aggregate concrete: Testing and modelling. *Cem. Concr. Compos.* **2022**, *130*, 104527. [CrossRef]
13. Abu-Hamdeh, N.H.; Khoshaim, A.; Alzahrani, M.A.; Hatamleh, R.I. Study of the flat plate solar collector's efficiency for sustainable and renewable energy management in a building by a phase change material: Containing paraffin-wax/Graphene and Paraffin-wax/graphene oxide carbon-based fluids. *J. Build. Eng.* **2022**, *57*, 104804. [CrossRef]
14. Yang, H.; Xu, Z.; Cui, H.; Bao, X.; Tang, W.; Sang, G.; Chen, X. Cementitious composites integrated phase change materials for passive buildings: An overview. *Constr. Build. Mater.* **2022**, *361*, 129635. [CrossRef]
15. Liu, Q.; Zhang, J.; Liu, J.; Sun, W.; Xu, H.; Liu, C. Self-healed inorganic phase change materials for thermal energy harvesting and management. *Appl. Therm. Eng.* **2023**, *219 Pt A*, 119423. [CrossRef]
16. Eurostat. Final Energy Consumption by Sector. Available online: https://ec.europa.eu/eurostat/databrowser/view/TEN00124/default/table?lang=en&category=nrg.nrg_quant.nrg_quanta.nrg_bal (accessed on 24 February 2023).
17. Eurostat. Energy Imports Dependency. Available online: https://ec.europa.eu/eurostat/databrowser/view/NRG_IND_ID/default/table?lang=en&category=nrg.nrg_quant.nrg_quanta.nrg_ind.nrg_ind_ (accessed on 24 February 2023).
18. Eurostat. Final Energy Consumption in Households by Type of Fuel. Available online: https://ec.europa.eu/eurostat/databrowser/view/TEN00125/default/table?lang=en&category=nrg.nrg_quant.nrg_quanta.nrg_bal (accessed on 24 February 2023).
19. Eurostat Statistics Explained. Available online: https://ec.europa.eu/eurostat/statistics-explained/index.php?title=Energy_consumption_in_households#Energy_products_used_in_the_residential_sector (accessed on 15 November 2022).
20. Eurostat. Population Unable to Keep Home Adequately Warm by Poverty Status. Available online: https://ec.europa.eu/eurostat/databrowser/view/sdg_07_60/default/table?lang=en (accessed on 24 February 2023).

21. Fleuchaus, P.; Godschalk, B.; Stober, I.; Blum, P. Worldwide Application of Aquifer Thermal Energy Storage–A Review. *Renew. Sustain. Energy Rev.* **2018**, *94*, 861–876. [CrossRef]
22. Chinnasamy, V.; Palaniappan, S.K.; Raj, M.K.A.; Rajendran, M.; Cho, H. Thermal Energy Storage and Its Applications. *Mater. Sol. Energy Convers. Mater. Methods Appl.* **2021**, 353–377. [CrossRef]
23. Koohi-Fayegh, S.; Rosen, M.A. A Review of Energy Storage Types, Applications and Recent Developments. *J. Energy Storage* **2020**, *27*, 101047. [CrossRef]
24. Sharshir, S.W.; Joseph, A.; Elsharkawy, M.; Hamad, M.A.; Kandeal, A.; Elkadeem, M.; Thakur, A.K.; Ma, Y.; Moustapha, M.E.; Rashad, M. Thermal Energy Storage Using Phase Change Materials in Building Applications: A Review of the Recent Development. *Energy Build.* **2023**, *285*, 112908. [CrossRef]
25. Jouhara, H.; Żabnieńska-Góra, A.; Khordehgah, N.; Ahmad, D.; Lipinski, T. Latent Thermal Energy Storage Technologies and Applications: A Review. *Int. J. Thermofluids* **2020**, *5*, 100039. [CrossRef]
26. Lu, S.; Lin, Q.; Liu, Y.; Yue, L.; Wang, R. Study on Thermal Performance Improvement Technology of Latent Heat Thermal Energy Storage for Building Heating. *Appl. Energy* **2022**, *323*, 119594. [CrossRef]
27. Liu, L.; Hammami, N.; Trovalet, L.; Bigot, D.; Habas, J.-P.; Malet-Damour, B. Description of Phase Change Materials (PCMs) Used in Buildings under Various Climates: A Review. *J. Energy Storage* **2022**, *56*, 105760. [CrossRef]
28. Xu, J.; Zhang, X.; Zou, L. A Review: Progress and Perspectives of Research on the Functionalities of Phase Change Materials. *J. Energy Storage* **2022**, *54*, 105341. [CrossRef]
29. Mofijur, M.; Mahlia, T.M.; Silitonga, A.S.; Ong, H.C.; Silakhori, M.; Hasan, M.H.; Putra, N.; Rahman, S.M.A. Phase Change Materials (PCM) for Solar Energy Usages and Storage: An Overview. *Energies* **2019**, *12*, 3167. [CrossRef]
30. Nguyen, V.N.; Lanh Le, T.; Quang Duong, X.; Vang Le, V.; Tuyen Nguyen, D.; Quy Phong Nguyen, P.; Rajamohan, S.; Vu Vo, A.; Son Le, H. Application of Phase Change Materials in Improving the Performance of Refrigeration Systems. *Sustain. Energy Technol. Assess.* **2023**, *56*, 103097. [CrossRef]
31. Singh, P.; Khanna, S.; Newar, S.; Sharma, V.; Reddy, K.S.; Mallick, T.K.; Becerra, V.; Radulovic, J.; Hutchinson, D.; Khusainov, R. Solar Photovoltaic Panels with Finned Phase Change Material Heat Sinks. *Energies* **2020**, *13*, 2558. [CrossRef]
32. Aridi, R.; Yehya, A. Review on the Sustainability of Phase-Change Materials Used in Buildings. *Energy Convers. Manag. X* **2022**, *15*, 100237. [CrossRef]
33. Mehrizi, A.A.; Karimi-Maleh, H.; Naddafi, M.; Karimi, F. Application of Bio-Based Phase Change Materials for Effective Heat Management. *J. Energy Storage* **2023**, *61*, 106859. [CrossRef]
34. Tao, J.; Luan, J.; Liu, Y.; Qu, D.; Yan, Z.; Ke, X. Technology Development and Application Prospects of Organic-Based Phase Change Materials: An Overview. *Renew. Sustain. Energy Rev.* **2022**, *159*, 112175. [CrossRef]
35. Junaid, M.F.; ur Rehman, Z.; Čekon, M.; Čurpek, J.; Farooq, R.; Cui, H.; Khan, I. Inorganic Phase Change Materials in Thermal Energy Storage: A Review on Perspectives and Technological Advances in Building Applications. *Energy Build.* **2021**, *252*, 111443. [CrossRef]
36. Cabeza, L.F.; Castell, A.; de Barreneche, C.; De Gracia, A.; Fernández, A. Materials Used as PCM in Thermal Energy Storage in Buildings: A Review. *Renew. Sustain. Energy Rev.* **2011**, *15*, 1675–1695. [CrossRef]
37. Tan, S.; Zhang, X. Progress of Research on Phase Change Energy Storage Materials in Their Thermal Conductivity. *J. Energy Storage* **2023**, *61*, 106772. [CrossRef]
38. Su, Y.; Fan, Y.; Ma, Y.; Wang, Y.; Liu, G. Flame-Retardant Phase Change Material (PCM) for Thermal Protective Application in Firefighting Protective Clothing. *Int. J. Therm. Sci.* **2023**, *185*, 108075. [CrossRef]
39. Ikutegbe, C.A.; Farid, M.M. Application of Phase Change Material Foam Composites in the Built Environment: A Critical Review. *Renew. Sustain. Energy Rev.* **2020**, *131*, 110008. [CrossRef]
40. Karthick, A.; Manokar Athikesavan, M.; Pasupathi, M.K.; Manoj Kumar, N.; Chopra, S.S.; Ghosh, A. Investigation of Inorganic Phase Change Material for a Semi-Transparent Photovoltaic (STPV) Module. *Energies* **2020**, *13*, 3582. [CrossRef]
41. Mohamed, S.A.; Al-Sulaiman, F.A.; Ibrahim, N.I.; Zahir, M.H.; Al-Ahmed, A.; Saidur, R.; Yılbaş, B.S.; Sahin, A.Z. A Review on Current Status and Challenges of Inorganic Phase Change Materials for Thermal Energy Storage Systems. *Renew. Sustain. Energy Rev.* **2017**, *70*, 1072–1089. [CrossRef]
42. Xie, N.; Huang, Z.; Luo, Z.; Gao, X.; Fang, Y.; Zhang, Z. Inorganic Salt Hydrate for Thermal Energy Storage. *Appl. Sci.* **2017**, *7*, 1317. [CrossRef]
43. Kenisarin, M.; Mahkamov, K. Salt Hydrates as Latent Heat Storage Materials: Thermophysical Properties and Costs. *Sol. Energy Mater. Sol. Cells* **2016**, *145*, 255–286. [CrossRef]
44. Singh, P.; Sharma, R.; Ansu, A.; Goyal, R.; Sarı, A.; Tyagi, V. A Comprehensive Review on Development of Eutectic Organic Phase Change Materials and Their Composites for Low and Medium Range Thermal Energy Storage Applications. *Sol. Energy Mater Sol. Cells* **2021**, *223*, 110955. [CrossRef]
45. Okogeri, O.; Stathopoulos, V.N. What about Greener Phase Change Materials? A Review on Biobased Phase Change Materials for Thermal Energy Storage Applications. *Int. J. Thermofluids* **2021**, *10*, 100081. [CrossRef]
46. Nazari, M.; Jebrane, M.; Terziev, N. Bio-Based Phase Change Materials Incorporated in Lignocellulose Matrix for Energy Storage in Buildings—A Review. *Energies* **2020**, *13*, 3065. [CrossRef]
47. Ravotti, R.; Fellmann, O.; Lardon, N.; Fischer, L.J.; Stamatiou, A.; Worlitschek, J. Analysis of Bio-Based Fatty Esters PCM's Thermal Properties and Investigation of Trends in Relation to Chemical Structures. *Appl. Sci.* **2019**, *9*, 225. [CrossRef]

48. Cellat, K.; Beyhan, B.; Güngör, C.; Konuklu, Y.; Karahan, O.; Dündar, C.; Paksoy, H. Thermal Enhancement of Concrete by Adding Bio-Based Fatty Acids as Phase Change Materials. *Energy Build.* **2015**, *106*, 156–163. [CrossRef]
49. Faraj, K.; Khaled, M.; Faraj, J.; Hachem, F.; Castelain, C. A Summary Review on Experimental Studies for PCM Building Applications: Towards Advanced Modular Prototype. *Energies* **2022**, *15*, 1459. [CrossRef]
50. Frigione, M.; Lettieri, M.; Sarcinella, A. Phase Change Materials for Energy Efficiency in Buildings and Their Use in Mortars. *Materials* **2019**, *12*, 1260. [CrossRef]
51. Rathore, P.K.S.; Shukla, S.K. Enhanced Thermophysical Properties of Organic PCM through Shape Stabilization for Thermal Energy Storage in Buildings: A State of the Art Review. *Energy Build.* **2021**, *236*, 110799. [CrossRef]
52. Memon, S.A. Phase Change Materials Integrated in Building Walls: A State of the Art Review. *Renew. Sustain. Energy Rev.* **2014**, *31*, 870–906. [CrossRef]
53. Rathore, P.K.S.; Shukla, S.K.; Gupta, N.K. Potential of Microencapsulated PCM for Energy Savings in Buildings: A Critical Review. *Sustain. Cities Soc.* **2020**, *53*, 101884. [CrossRef]
54. Cunha, S.; Lima, M.; Aguiar, J.B. Influence of adding phase change materials on the properties of cement mortars. *Constr. Build. Mater.* **2016**, *127*, 100081. [CrossRef]
55. Cunha, S.; Aguiar, J.B.; Ferreira, V.; Tadeu, A. Influence of Adding Encapsulated Phase Change Materials in Aerial Lime Based Mortars. *Adv. Mater. Res.* **2013**, *687*, 255–261. [CrossRef]
56. Cárdenas-Ramírez, C.; Jaramillo, F.; Gómez, M. Systematic Review of Encapsulation and Shape-Stabilization of Phase Change Materials. *J. Energy Storage* **2020**, *30*, 101495. [CrossRef]
57. Lv, P.; Liu, C.; Rao, Z. Review on Clay Mineral-Based Form-Stable Phase Change Materials: Preparation, Characterization and Applications. *Renew. Sustain. Energy Rev.* **2017**, *68*, 707–726. [CrossRef]
58. Zhang, Y.P.; Lin, K.P.; Yang, R.; Di, H.F.; Jiang, Y. Preparation, Thermal Performance and Application of Shape-Stabilized PCM in Energy Efficient Buildings. *Energy Build.* **2006**, *38*, 1262–1269. [CrossRef]
59. Cunha, S.; Aguiar, J. Phase change materials and energy efficiency of buildings: A review of knowledge. *J. Energy Storage* **2020**, *27*, 101083. [CrossRef]
60. Zhang, Y.; Jia, Z.; Moqeet Hai, A.; Zhang, S.; Tang, B. Shape-Stabilization Micromechanisms of Form-Stable Phase Change Materials-A Review. *Compos. Part A Appl. Sci. Manuf.* **2022**, *160*, 107047. [CrossRef]
61. Hassan, A.; Shakeel Laghari, M.; Rashid, Y. Micro-Encapsulated Phase Change Materials: A Review of Encapsulation, Safety and Thermal Characteristics. *Sustainability* **2016**, *8*, 1046. [CrossRef]
62. Wu, M.Q.; Wu, S.; Cai, Y.F.; Wang, R.Z.; Li, T.X. Form-Stable Phase Change Composites: Preparation, Performance, and Applications for Thermal Energy Conversion, Storage and Management. *Energy Storage Mater.* **2021**, *42*, 380–417. [CrossRef]
63. Xu, C.; Zhang, H.; Fang, G. Review on Thermal Conductivity Improvement of Phase Change Materials with Enhanced Additives for Thermal Energy Storage. *J. Energy Storage* **2022**, *51*, 104568. [CrossRef]
64. Shamseddine, I.; Pennec, F.; Biwole, P.; Fardoun, F. Supercooling of Phase Change Materials: A Review. *Renew. Sustain. Energy Rev.* **2022**, *158*, 112172. [CrossRef]
65. Png, Z.M.; Soo, X.Y.D.; Chua, M.H.; Ong, P.J.; Suwardi, A.; Tan, C.K.I.; Xu, J.; Zhu, Q. Strategies to Reduce the Flammability of Organic Phase Change Materials: A Review. *Sol. Energy* **2022**, *231*, 115–128. [CrossRef]
66. Rathore, P.K.S.; Shukla, S. Kumar Improvement in Thermal Properties of PCM/Expanded Vermiculite/Expanded Graphite Shape Stabilized Composite PCM for Building Energy Applications. *Renew. Energy* **2021**, *176*, 295–304. [CrossRef]
67. Gencel, O.; Sarı, A.; Ustaoglu, A.; Hekimoglu, G.; Erdogmus, E.; Yaras, A.; Sutcu, M.; Cay, V.V. Eco-Friendly Building Materials Containing Micronized Expanded Vermiculite and Phase Change Material for Solar Based Thermo-Regulation Applications. *Constr. Build. Mater.* **2021**, *308*, 125062. [CrossRef]
68. Sarcinella, A.; de Aguiar, J.L.B.; Frigione, M. Physical Properties of an Eco-Sustainable, Form-Stable Phase Change Material Included in Aerial-Lime-Based Mortar Intended for Different Climates. *Materials* **2022**, *15*, 1192. [CrossRef]
69. Sarcinella, A.; de Aguiar, J.L.B.; Frigione, M. Physical Properties of Eco-Sustainable Form-Stable Phase Change Materials Included in Mortars Suitable for Buildings Located in Different Continental Regions. *Materials* **2022**, *15*, 2497. [CrossRef] [PubMed]
70. Nazir, H.; Batool, M.; Bolivar Osorio, F.J.; Isaza-Ruiz, M.; Xu, X.; Vignarooban, K.; Phelan, P.; Inamuddin; Kannan, A.M. Recent Developments in Phase Change Materials for Energy Storage Applications: A Review. *Int. J. Heat Mass Transf.* **2019**, *129*, 491–523. [CrossRef]
71. Chandel, S.S.; Agarwal, T. Review of Current State of Research on Energy Storage, Toxicity, Health Hazards and Commercialization of Phase Changing Materials. *Renew. Sustain. Energy Rev.* **2017**, *67*, 581–596. [CrossRef]
72. Peippo, K.; Kauranen, P.; Lund, P.D. A Multicomponent PCM Wall Optimized for Passive Solar Heating. *Energy Build.* **1991**, *17*, 259–270. [CrossRef]
73. Yang, K.; Zhu, N.; Chang, C.; Wang, D.; Yang, S.; Ma, S. A Methodological Concept for Phase Change Material Selection Based on Multi-Criteria Decision Making (MCDM): A Case Study. *Energy* **2018**, *165*, 1085–1096. [CrossRef]
74. Mukhamet, T.; Kobeyev, S.; Nadeem, A.; Memon, S.A. Ranking PCMs for Building Façade Applications Using Multi-Criteria Decision-Making Tools Combined with Energy Simulations. *Energy* **2021**, *215*, 119102. [CrossRef]
75. Cunha, S.; Aguiar, J.B.; Ferreira, V.M.; Tadeu, A. Mortars Based in different binders with incorporation of phase change materials: Physical and mechanical properties. *Eur. J. Environ. Civ. Eng.* **2015**, *19*, 1216–1233. [CrossRef]

76. Cunha, S.; Aguiar, J.B.; Tadeu, A. Thermal performance and cost analysis of PCM mortars based in different binders. *Constr. Build. Mater.* **2016**, *122*, 637–648. [CrossRef]
77. Saxena, R.; Rakshit, D.; Kaushik, S. Phase change material (PCM) incorporated bricks for energy conservation in composite climate: A sustainable building solution. *Sol. Energy* **2019**, *183*, 276–284. [CrossRef]
78. Ahmad, M.; Bontemps, A.; Sallée, H.; Quenard, D. Experimental investigation and computer simulation of thermal behaviour of wallboards containing a phase change material. *Energy Build.* **2016**, *38*, 357–366. [CrossRef]
79. Bahrar, M.; Djamai, Z.; Mankibi, M.; Larbi, A.; Salvia, M. Numerical and experimental study on the use of microencapsulated phase change materials (PCMs) in textile reinforced concrete panels for energy storage. *Sustain. Cities Soc.* **2018**, *41*, 455–468. [CrossRef]
80. Chou, H.; Chen, C.; Nguyen, V. A new design of metal-sheet cool roof using PCM. *Energy Build.* **2013**, *57*, 42–50. [CrossRef]
81. Mustafa, J.; Alqaed, S.; Sharifpur, M. PCM embedded radiant chilled ceiling as a solution to shift the cooling peak load-focusing on solidification process acceleration. *J. Build. Eng.* **2022**, *57*, 104894. [CrossRef]
82. Bogatu, D.; Kazanci, O.B.; Olesen, B.W. An experimental study of the active cooling performance of a novel radiant ceiling panel containing phase change material (PCM). *Energy Build.* **2021**, *243*, 110981. [CrossRef]
83. Kitagawa, H.; Asawa, T.; Kubota, T.; Trihamdani, A.R. Numerical simulation of radiant floor cooling systems using PCM for naturally ventilated buildings in a hot and humid climate. *Build. Environ.* **2022**, *226*, 109762. [CrossRef]
84. Entrop, A.; Brouwers, H.; Reinders, A. Experimental research on the use of micro-encapsulated phase change materials to store solar energy in concrete floors and to save energy in Dutch houses. *Sol. Energy* **2011**, *85*, 1007–1020. [CrossRef]
85. Larwa, B.; Cesari, S.; Bottarelli, M. Study on thermal performance of a PCM enhanced hydronic radiant floor heating system. *Energy* **2021**, *225*, 120245. [CrossRef]
86. Hasan, A.; McCormack, J.S.; Sarwar, J.; Norton, B. Increased photovoltaic performance through temperature regulation by phase change materials: Material comparison in different climates. *Sol. Energy* **2015**, *115*, 264–276. [CrossRef]
87. Sharma, S.; Tahir, A.; Reddy, K.S.; Malick, T.K. Performance enhancement of a building-integrated concentrating photovoltaic system using phase change material. *Sol. Energy Mater. Sol. Cells* **2016**, *149*, 29–39. [CrossRef]
88. Abdulmunem, A.R.; Samin, P.M.; Rahman, H.A.; Hussien, H.A.; Ghazali, H. A novel thermal regulation method for photovoltaic panels using porous metals filled with phase change material and nanoparticle additives. *J. Energy Storage* **2021**, *39*, 102621. [CrossRef]
89. Lopez-Arias, M.; Francioso, V.; Velay-Lizancos, M. High thermal inertia mortars: New method to incorporate phase change materials (PCMs) while enhancing strength and thermal design models. *Constr. Build. Mater.* **2023**, *370*, 130621. [CrossRef]
90. Cunha, S.; Leite, P.; Aguiar, J.B. Characterization of innovative mortars with direct incorporation of phase change materials. *J. Energy Storage* **2020**, *30*, 101439. [CrossRef]
91. Illampas, R.; Rigopoulos, I.; Ioannou, I. Influence of microencapsulated Phase Change Materials (PCMs) on the properties of polymer modified cementitious repair mortar. *J. Build. Eng.* **2021**, *40*, 102328. [CrossRef]
92. Cui, H.; Zou, J.; Gong, Z.; Zheng, D.; Bao, X.; Chen, X. Study on the thermal and mechanical properties of steel fibre reinforced PCM-HSB concrete for high performance in energy piles. *Constr. Build. Mater.* **2022**, *350*, 128822. [CrossRef]
93. Cellat, K.; Tezcan, F.; Beyhan, B.; Kardaş, G.; Paksoy, H. A comparative study on corrosion behavior of rebar in concrete with fatty acid additive as phase change material. *Constr. Build. Mater.* **2017**, *143*, 490–500. [CrossRef]
94. D'Alessandro, A.; Pisello, A.; Fabiani, C.; Ubertini, F.; Cabeza, L.F.; Cotana, F. Multifunctional smart concretes with novel phase change materials: Mechanical and thermo-energy investigation. *Appl. Energy* **2018**, *212*, 1448–1461. [CrossRef]
95. Zhiyou, J.; Cunha, S.; Aguiar, J.B. Influence of the incorporation of construction and demolition waste with the incorporation of phase change material on the mechanical properties of concrete. In Proceedings of the Congress Construção 2022, Guimarães, Portugal, 5–7 December 2022; Volume I, pp. 104–114. (In Portuguese).
96. He, Y.; Zhang, X.; Zhang, Y. Preparation Technology of Phase Change Perlite and Performance Research of Phase Change and Temperature Control Mortar. *Energy Build.* **2014**, *85*, 506–514. [CrossRef]
97. Zhang, N.; Yuan, Y.; Li, T.; Cao, X.; Yang, X. Study on Thermal Property of Lauric–Palmitic–Stearic Acid/Vermiculite Composite as Form-Stable Phase Change Material for Energy Storage. *Adv. Mech. Eng.* **2015**, *7*, 1687814015605023. [CrossRef]
98. Costa, J.A.C.; Martinelli, A.E.; do Nascimento, R.M.; Mendes, A.M. Microstructural Design and Thermal Characterization of Composite Diatomite-Vermiculite Paraffin-Based Form-Stable PCM for Cementitious Mortars. *Constr. Build. Mater.* **2020**, *232*, 117167. [CrossRef]
99. Frigione, M.; Lettieri, M.; Sarcinella, A.; Barroso de Aguiar, J. Sustainable Polymer-Based Phase Change Materials for Energy Efficiency in Buildings and Their Application in Aerial Lime Mortars. *Constr. Build. Mater.* **2020**, *231*, 117149. [CrossRef]
100. Frigione, M.; Lettieri, M.; Sarcinella, A.; Barroso de Aguiar, J.L. Applications of Sustainable Polymer-Based Phase Change Materials in Mortars Composed by Different Binders. *Materials* **2019**, *12*, 3502. [CrossRef]
101. Wang, C.; Cheng, C.; Jin, T.; Dong, H. Water evaporation inspired biomass-based PCM from daisy stem and paraffin for building temperature regulation. *Renew. Energ.* **2022**, *194*, 211–219. [CrossRef]
102. Cunha, S.; Campos, A.; Aguiar, J.; Martins, F. A study of phase change material (PCM) on the physical and mechanical properties of compressed earth bricks (CEBs). *Malays. Constr. Res. J.* **2022**, *38*, 1–19.

103. Zhang, X.; Huang, Z.; Yin, Z.; Zhang, W.; Huang, Y.; Liu, Y.; Fang, M.; Wu, X.; Min, X. Form stable composite phase change materials from palmitic-lauric acid eutectic mixture and carbonized abandoned rice: Preparation, characterization, and thermal conductivity enhancement. *Energy Build.* **2017**, *154*, 46–54. [CrossRef]
104. Karthik, M.; Faik, A.; D'Aguanno, B. Graphite foam as interpenetrating matrices for phase change paraffin wax: A candidate composite for low temperature thermal energy storage. *Sol. Energy Mater. Sol. Cell.* **2017**, *172*, 324–334. [CrossRef]
105. Mehrali, M.; Latibari, S.T.; Mehrali, M.; Mahlia, T.M.I.; Metselaar, H.S.C. Preparation and properties of highly conductive palmitic acid/graphene oxide composites as thermal energy storage materials. *Energy* **2013**, *58*, 628–634. [CrossRef]
106. Yang, L.; Zhang, N.; Yuan, Y.; Cao, X.; Xiang, B. Thermal performance of stearic acid/carbon nanotube composite phase change materials for energy storage prepared by ball milling. *Int. J. Energy Res.* **2019**, *43*, 6327–6336. [CrossRef]
107. Liu, Y.; Wang, N.; Ding, Y. Preparation and properties of composite phase change material based on solar heat storage system. *J. Energy Storage* **2021**, *40*, 102805. [CrossRef]
108. Cunha, S.; Aguiar, I.; Aguiar, J. Phase change materials composite boards and mortars: Mixture design, physical, mechanical and thermal behavior. *J. Energy Storage* **2022**, *53*, 105135. [CrossRef]
109. Oliver, A. Thermal characterization of gypsum boards with PCM included: Thermal energy storage in buildings through latent heat. *Energy Build.* **2012**, *48*, 1–7. [CrossRef]
110. Lai, C.; Chen, R.H.; Lin, C.-Y. Heat Transfer and Thermal Storage Behaviour of Gypsum Boards Incorporating Micro-Encapsulated PCM. *Energy Build.* **2010**, *42*, 1259–1266. [CrossRef]
111. Mourid, A.; El Alami, M.; Kuznik, F. Experimental Investigation on Thermal Behavior and Reduction of Energy Consumption in a Real Scale Building by Using Phase Change Materials on Its Envelope. *Sustain. Cities Soc.* **2018**, *41*, 35–43. [CrossRef]
112. Gao, Y.; Meng, X. A Comprehensive Review of Integrating Phase Change Materials in Building Bricks: Methods, Performance and Applications. *J. Energy Storage* **2023**, *62*, 106913. [CrossRef]
113. Zhu, Q.; Zhang, W.W.; Zhang, Y.; He, J.P. Experiments of Thermal Response of Phase Change Energy Storage Material on Block Walls. *J. Nanjing Tech. Univ. (Nat. Sci. Ed.)* **2016**, *38*, 119–124.
114. Vicente, R.; Silva, T. Brick Masonry Walls with PCM Macrocapsules: An Experimental Approach. *Appl. Therm. Eng.* **2014**, *67*, 24–34. [CrossRef]
115. Cunha, S.; Aguiar, J.B.; Ferreira, V.M. Durability of mortars with incorporation of phase change materials microcapsules. *Rom. J. Mater.* **2017**, *47*, 166–175.
116. Cunha, S.; Aguiar, J.B.; Pacheco-Torgal, F. Effect of temperature on mortars with incorporation of phase change materials. *Constr. Build. Mater.* **2015**, *98*, 89–101. [CrossRef]
117. Kheradmand, M.; Abdollahnejad, Z.; Pacheco-Torgal, F. Alkali-activated cement-based binder mortars containing phase change materials (PCMs): Mechanical properties and cost analysis. *Eur. J. Environ. Civil Eng.* **2018**, *24*, 1068–1090. [CrossRef]
118. Shadnia, R.; Zhang, L.; Li, P. Experimental study of geopolymer mortar with incorporated PCM. *Constr. Build. Mater.* **2015**, *84*, 95–102. [CrossRef]
119. Aguayo, M.; Das, S.; Maroli, A.; Kabay, N.; Mertens, J.C.E.; Rajan, S.D.; Sant, G.; Chawla, N.; Neithalat, N. The influence of microencapsulated phase change material (PCM) characteristics on the microstructure and strength of cementitious composites: Experiments and finite element simulations. *Cem. Concr. Compos.* **2016**, *73*, 29–41. [CrossRef]
120. Yao, C.; Kong, X.; Li, Y.; Du, Y.; Qi, C. Numerical and Experimental Research of Cold Storage for a Novel Expanded Perlite-Based Shape-Stabilized Phase Change Material Wallboard Used in Building. *Energy Convers. Manag.* **2018**, *155*, 20–31. [CrossRef]
121. Wen, R.; Zhang, X.; Huang, Y.; Yin, Z.; Huang, Z.; Fang, M.; Liu, Y.; Wu, X. Preparation and Properties of Fatty Acid Eutectics/Expanded Perlite and Expanded Vermiculite Shape-Stabilized Materials for Thermal Energy Storage in Buildings. *Energy Build.* **2017**, *139*, 197–204. [CrossRef]
122. Xie, N.; Luo, J.; Li, Z.; Huang, Z.; Gao, X.; Fang, Y.; Zhang, Z. Salt Hydrate/Expanded Vermiculite Composite as a Form-Stable Phase Change Material for Building Energy Storage. *Sol. Energy Mater. Sol. Cells* **2019**, *189*, 33–42. [CrossRef]
123. Yang, Y.; Shen, Z.; Wu, W.; Zhang, H.; Ren, Y.; Yang, Q. Preparation of a Novel Diatomite-Based PCM Gypsum Board for Temperature-Humidity Control of Buildings. *Build. Environ.* **2022**, *226*, 109732. [CrossRef]
124. Fang, X.; Zhang, Z. A Novel Montmorillonite-Based Composite Phase Change Material and Its Applications in Thermal Storage Building Materials. *Energy Build.* **2006**, *38*, 377–380. [CrossRef]
125. Fang, X.; Zhang, Z.; Chen, Z. Study on Preparation of Montmorillonite-Based Composite Phase Change Materials and Their Applications in Thermal Storage Building Materials. *Energy Convers. Manag.* **2008**, *49*, 718–723. [CrossRef]
126. Zhang, X.; Wen, R.; Huang, Z.; Tang, C.; Huang, Y.; Liu, Y.; Fang, M.; Wu, X.; Min, X.; Xu, Y. Enhancement of Thermal Conductivity by the Introduction of Carbon Nanotubes as a Filler in Paraffin/Expanded Perlite Form-Stable Phase-Change Materials. *Energy Build.* **2017**, *149*, 463–470. [CrossRef]
127. Sarcinella, A.; De Aguiar, J.L.B.; Lettieri, M.; Cunha, S.; Frigione, M. Thermal Performance of Mortars Based on Different Binders and Containing a Novel Sustainable Phase Change Material (PCM). *Materials* **2020**, *13*, 2055. [CrossRef]
128. Uthaichotirat, P.; Sukontasukkul, P.; Jitsangiam, P.; Suksiripattanapong, C.; Sata, V.; Chindaprasirt, P. Thermal and Sound Properties of Concrete Mixed with High Porous Aggregates from Manufacturing Waste Impregnated with Phase Change Material. *J. Build. Eng.* **2020**, *29*, 101111. [CrossRef]
129. Dong, Z.; Cui, H.; Tang, W.; Chen, D.; Wen, H. Development of Hollow Steel Ball Macro-Encapsulated PCM for Thermal Energy Storage Concrete. *Materials* **2016**, *9*, 59. [CrossRef]

130. Jayalath, A.; Nicolas, R.S.; Sofi, M.; Shanks, R.; Ngo, T.; Aye, L.; Mendis, P. Properties of cementitious mortar and concrete containing micro-encapsulated phase change materials. *Constr. Build. Mater.* **2016**, *120*, 408–417. [CrossRef]
131. Lecompte, T.; Le Bideau, P.; Glouannec, P.; Nortershauser, D.; Le Masson, S. Mechanical and thermo-physical behaviour of concretes and mortars containing phase change material. *Energy Build.* **2015**, *94*, 52–60. [CrossRef]
132. Thiele, A.M.; Sant, G.; Pilon, L. Diurnal thermal analysis of microencapsulated PCM-concrete composite walls. *Energy Convers. Manag.* **2015**, *93*, 215–227. [CrossRef]
133. Singh, P.; Sharma, R.; Khalid, M.; Goyal, R.; Sarı, A.; Tyagi, V. Evaluation of carbon based-supporting materials for developing form-stable organic phase change materials for thermal energy storage: A review. *Sol. Energy Mater Sol. Cells* **2022**, *246*, 111896. [CrossRef]
134. Zeng, J.; Chen, Y.; Shu, L.; Yu, L.; Zhu, L.; Song, L.; Cao, Z.; Sun, L. Preparation and thermal properties of exfoliated graphite/erythritol/mannitol eutectic composite as form-stable phase change material for thermal energy storage. *Sol. Energy Mater. Sol. Cell.* **2018**, *178*, 84–90. [CrossRef]
135. Zhang, N.; Yuan, Y.; Wang, X.; Cao, X.; Yang, X.; Hu, S. Preparation and characterization of lauric–myristic–palmitic acid ternary eutectic mixtures/expanded graphite composite phase change material for thermal energy storage. *Chem. Eng. J.* **2013**, *231*, 214–219. [CrossRef]
136. He, Y.; Zhang, X.; Zhang, Y.; Song, Q.; Liao, X. Utilization of lauric acid-myristic acid/expanded graphite phase change materials to improve thermal properties of cement mortar. *Energy Build.* **2016**, *133*, 547–558. [CrossRef]
137. Yu, Q.; Zhang, C.; Lu, Y.; Kong, Q.; Wei, H.; Yang, Y.; Gao, Q.; Wu, Y.; Sciacovelli, A. Comprehensive performance of composite phase change materials based on eutectic chloride with SiO2 nanoparticles and expanded graphite for thermal energy storage system. *Renew. Energy* **2021**, *172*, 1120–1132. [CrossRef]
138. Peng, L.; Sun, Y.; Gu, X.; Liu, P.; Bian, L.; Wei, B. Thermal conductivity enhancement utilizing the synergistic effect of carbon nanocoating and graphene addition in palmitic acid/halloysite FSPCM. *Appl. Clay Sci.* **2021**, *206*, 106068. [CrossRef]
139. Mehrali, M.; Latibari, S.T.; Mehrali, M.; Metselaar, H.S.C.; Silakhori, M. Shape-stabilized phase change materials with high thermal conductivity based on paraffin/graphene oxide composite. *Energy Convers. Manag.* **2013**, *67*, 275–282. [CrossRef]
140. Li, B.; Liu, T.; Hu, L.; Wang, Y.; Nie, S. Facile preparation and adjustable thermal property of stearic acid–graphene oxide composite as shape-stabilized phase change material. *Chem. Eng. J.* **2013**, *215*, 819–826. [CrossRef]
141. Yang, J.; Qi, G.; Liu, Y.; Bao, R.; Liu, Z.; Yang, W.; Xie, B.; Yang, M. Hybrid graphene aerogels/phase change material composites: Thermal conductivity, shape-stabilization and light-to-thermal energy storage. *Carbon* **2016**, *100*, 693–702. [CrossRef]
142. Le, V.T.; Ngo, C.L.; Le, Q.T.; Ngo, T.T.; Nguyen, D.N.; Vu, M.T. Surface modification and functionalization of carbon nanotube with some organic compounds. *Adv. Nat. Sci. Nanosci. Nanotechnol.* **2013**, *4*, 035017. [CrossRef]
143. Fikri, M.A.; Pandey, A.K.; Samykano, M.; Kadirgama, K.; George, M.; Saidur, R.; Selvaraj, J.; Rahim, N.A.; Sharma, K.; Tyagi, V.V. Thermal conductivity, reliability, and stability assessment of phase change material (PCM) doped with functionalized multi-wall carbon nanotubes (FMWCNTs). *J. Energy Storage* **2022**, *50*, 104676. [CrossRef]
144. Meng, X.; Zhang, H.; Sum, L.; Xu, F.; Jiao, Q.; Zhao, Z.; Zhang, Z.; Zhou, H.; Sawada, Y.; Liu, Y. Preparation and thermal properties of fatty acids/CNTs composite as shape-stabilized phase change materials. *J. Therm. Anal. Calorim.* **2013**, *111*, 377–384. [CrossRef]
145. Feng, Y.; Wei, R.; Huang, Z.; Zhang, X.; Wang, G. Thermal properties of lauric acid filled in carbon nanotubes as shape-stabilized phase change materials. *Phys. Chem. Chem. Phys.* **2018**, *20*, 7772–7780. [CrossRef]
146. Wen, R.; Liu, Y.; Yang, C.; Zhu, X.; Huang, Z.; Zhang, X.; Gao, W. Enhanced thermal properties of stearic acid/carbonized maize straw composite phase change material for thermal energy storage in buildings. *J. Energy Storage* **2021**, *36*, 102420. [CrossRef]
147. Shoja, M.; Mohammadi-Roshandeh, J.; Hemmati, F.; Zandi, A.; Farizeh, T. Plasticized starch-based biocomposites containing modified rice straw fillers with thermoplastic, thermoset-like and thermoset chemical structures. *Int. J. Biol. Macromol.* **2019**, *157*, 715–725. [CrossRef] [PubMed]
148. Liu, H.; Zheng, Z.; Qian, Z.; Wang, Q.; Wu, D.; Wang, X. Lamellar-structured phase change composites based on biomass-derived carbonaceous sheets and sodium acetate trihydrate for high-efficient solar photothermal energy harvest. *Sol. Energy Mater. Sol. Cell.* **2021**, *229*, 111140. [CrossRef]
149. Zhao, Y.; Min, X.; Huang, Z.; Liu, Y.; Wu, X.; Fang, M. Honeycomb-like structured biological porous carbon encapsulating PEG: A shape-stable phase change material with enhanced thermal conductivity for thermal energy storage. *Energy Build.* **2018**, *158*, 1049–1062. [CrossRef]
150. Wang, C.; Liang, W.; Yang, Y.; Liu, F.; Sun, H.; Zhu, Z.; Li, A. Biomass carbon aerogels based shape-stable phase change composites with high light-to-thermal efficiency for energy storage. *Renew. Energy* **2020**, *153*, 182–192. [CrossRef]
151. Liang, B.; Lu, X.; Li, R.; Tu, W.; Yang, Z.; Yuan, T. Solvent-free preparation of bio-based polyethylene glycol/wood flour composites as novel shape-stabilized phase change materials for solar thermal energy storage. *Sol. Energy Mater. Sol. Cell.* **2019**, *200*, 110037. [CrossRef]
152. Zhou, M.; Wang, J.; Zhao, Y.; Wang, G.; Gu, W.; Ji, G. Hierarchically porous wood-derived carbon scaffold embedded phase change materials for integrated thermal energy management, electromagnetic interference shielding and multifunctional application. *Carbon* **2021**, *183*, 515–524. [CrossRef]
153. Atinafu, D.G.; Chang, S.J.; Kim, S. Infiltration properties of n-alkanes in mesoporous biochar: The capacity of smokeless support for stability and energy storage. *J. Hazard Mater.* **2020**, *399*, 123041. [CrossRef]

154. Yang, Z.; Deng, Y.; Li, J. Preparation of porous carbonized woods impregnated with lauric acid as shape-stable composite phase change materials. *Appl. Therm. Eng.* **2019**, *150*, 967–976. [CrossRef]
155. Soares, N.; Costa, J.J.; Gaspar, A.R.; Santos, P. Review of Passive PCM Latent Heat Thermal Energy Storage Systems towards Buildings' Energy Efficiency. *Energy Build.* **2013**, *59*, 82–103. [CrossRef]
156. Kuznik, F.; Virgone, J.; Johannes, K. In-Situ Study of Thermal Comfort Enhancement in a Renovated Building Equipped with Phase Change Material Wallboard. *Renew. Energy* **2011**, *36*, 1458–1462. [CrossRef]
157. Abbas, H.M.; Jalil, J.M.; Ahmed, S.T. Experimental and Numerical Investigation of PCM Capsules as Insulation Materials Inserted into a Hollow Brick Wall. *Energy Build.* **2021**, *246*, 111127. [CrossRef]
158. Environmental Fund. Available online: https://www.fundoambiental.pt/ (accessed on 15 March 2023).
159. Ma, T.; Yang, H.; Zhang, Y.; Lu, L.; Wang, X. Using phase change materials in photovoltaic systems for thermal regulation and electrical efficiency improvement: A review and outlook. *Renew. Sustain. Energy Rev.* **2015**, *43*, 1273–1284. [CrossRef]
160. Hasan, A.; McCormack, J.S.; Huang, J.M. Evaluation of phase change material for thermal regulation enhancement of building integrated photovoltaics. *Sol. Energy* **2010**, *84*, 1601–1612. [CrossRef]
161. Waqas, A.; Jie, J.; Xu, L. Thermal behaviour of a PV panel integrated with PCM filled metallic tubes: An experimental study. *J. Renew. Sustain. Energy* **2017**, *9*, 053504. [CrossRef]
162. Kazemian, A.; Basati, Y.; Khatibi, M.; Ma, T. Performance prediction and optimization of a photovoltaic thermal system integrated with phase change material using response surface method. *J. Clean. Prod.* **2021**, *290*, 125748. [CrossRef]
163. Ma, T.; Li, Z.; Zhao, J. Photovoltaic panel integrated with phase change materials (PV-PCM): Technology overview and materials selection. *Renew. Sustain. Energy Rev.* **2019**, *116*, 109406. [CrossRef]
164. Kant, K.; Shukla, A.; Sharma, A. Ternary mixture of fatty acids as phase change materials for thermal energy storage applications. *Energy Rep.* **2016**, *2*, 274–279. [CrossRef]
165. Wheatley, G.; Rubel, R.I. Design improvement of a laboratory prototype for efficiency evaluation of solar thermal water heating system using phase change material (PCMs). *Results Eng.* **2021**, *12*, 100301. [CrossRef]
166. Bayomy, A.; Davies, S.; Saghir, Z. Domestic hot water storage tank utilizing phase change materials (PCMs): Numerical approach. *Energies* **2019**, *12*, 2170. [CrossRef]
167. Huang, K.; Feng, G.; Zhang, J. Experimental and numerical study on phase change material floor in solar water heating system with a new design. *Sol. Energy* **2014**, *105*, 126–138. [CrossRef]
168. Atalay, H. Assessment of energy and cost analysis of packed bed and phase change material thermal energy storage systems for the solar energy-assisted drying process. *Sol. Energy* **2020**, *198*, 124–138. [CrossRef]
169. Gholamibozanjani, G.; Farid, M. A comparison between passive and active PCM systems applied to buildings. *Renew. Energ.* **2020**, *162*, 112–123. [CrossRef]
170. M'hamdi, Y.; Baba, K.; Tajayouti, M.; Nounah, A. Energy, environmental, and economic analysis of different buildings envelope integrated with phase change materials in different climates. *Sol. Energy* **2022**, *243*, 91–102. [CrossRef]
171. Panayiotou, G.P.; Kalogirou, S.A.; Tassou, S.A. Evaluation of the application of Phase Change Materials (PCM) on the envelope of a typical dwelling in the Mediterranean region. *Renew. Energy* **2016**, *97*, 24–32. [CrossRef]
172. Mi, X.; Liu, R.; Cui, H.; Memon, S.A.; Xing, F.; Lo, Y. Energy and economic analysis of building integrated with PCM in different cities of China. *Appl. Energy* **2016**, *175*, 324–336. [CrossRef]
173. Sun, X.; Zhang, Q.; Medina, M.A.; Lee, K.O. Energy and economic analysis of a building enclosure outfitted with a phase change material board (PCMB). *Energy Convers. Manag.* **2014**, *83*, 73–78. [CrossRef]
174. Saffari, M.; Gracia, A.; Ushak, S.; Cabeza, L.F. Economic impact of integrating PCM as passive system in buildings using Fanger comfort model. *Energy Build.* **2016**, *112*, 159–172. [CrossRef]
175. Saafi, K.; Daouas, N. Energy and cost efficiency of phase change materials integrated in building envelopes under Tunisia Mediterranean climate. *Energy* **2019**, *187*, 115987. [CrossRef]

Disclaimer/Publisher's Note: The statements, opinions and data contained in all publications are solely those of the individual author(s) and contributor(s) and not of MDPI and/or the editor(s). MDPI and/or the editor(s) disclaim responsibility for any injury to people or property resulting from any ideas, methods, instructions or products referred to in the content.

Review

IoT-Enabled Campus Prosumer Microgrid Energy Management, Architecture, Storage Technologies, and Simulation Tools: A Comprehensive Study

Amad Ali [1], Hafiz Abdul Muqeet [2], Tahir Khan [3], Asif Hussain [4], Muhammad Waseem [5] and Kamran Ali Khan Niazi [6],*

1. Department of Electronics Engineering, Government College of Technology, Multan 60000, Pakistan
2. Electrical Engineering Technology Department, Punjab Tianjin University of Technology Lahore, Punjab 54770, Pakistan
3. College of Electrical Engineering, Zhejiang University, Hangzhou 310027, China
4. Department of Electrical Engineering, University of Management and Technology, Lahore 54782, Pakistan
5. Department of Electrical Engineering, University of Engineering and Technology, Taxila 47050, Pakistan
6. Department of Mechanical and Production Engineering—Fluids and Energy, Aarhus University, 8200 Aarhus, Denmark
* Correspondence: kkn@mpe.au.dk

Abstract: Energy is very important in daily life. The smart power system provides an energy management system using various techniques. Among other load types, campus microgrids are very important, and they consume large amounts of energy. Energy management systems in campus prosumer microgrids have been addressed in different works. A comprehensive study of previous works has not reviewed the architecture, tools, and energy storage systems of campus microgrids. In this paper, a survey of campus prosumer microgrids is presented considering their energy management schemes, optimization techniques, architectures, storage types, and design tools. The survey is comprised of one decade of past works for a true analysis. In the optimization techniques, deterministic and metaheuristic methods are reviewed considering their pros and cons. Smart grids are being installed in different campuses all over the world, and these are considered the best alternatives to conventional power systems. However, efficient energy management techniques and tools are required to make these grids more economical and stable.

Keywords: campus microgrid; prosumer market; batteries; energy management system; distributed generation; smart grid; renewable energy resources; energy storage system

1. Introduction

Energy crises have become major challenges in the economic development of a country. In this modern era, machinery is considered a more effective replacement for humans in many sectors. Smart devices are constantly being developed, which makes our routines in life much easier. In today's world, it is impossible to imagine a life without these smart devices and machinery. However, everything comes with a price, and the price of this ever-increasing dependence on machinery is the substantial consumption of energy resources [1]. These smart machines operate on electricity which is produced by the utilization of non-conventional energy resources such as coal, oil, and gas. The rising utilization of fossil fuels has result in two major environmental disorders. The first is the fast depletion of fossil fuels and the second is the production of hazardous gases and waste materials, which results in a direct increment of environmental pollution. The organization for economic cooperation and development (OECD) indicated in 2018 that the United States had the strongest gross domestic product rate [2], but British petroleum ranked the air quality index of the United States as the poorest in comparison to other countries of the world [3]. Polluted air in any country is a major cause of the demise of its people [4]. Fossil fuels are non-renewable, and so their continuous depletion results in gradually increasing energy generation prices, which increases inflation, especially in underdeveloped countries. Further, few countries have a major share of these non-renewable energy resources, which makes the ones that do powerful enough to control the economies of those countries whose electricity production largely depends upon fossil fuels. The excess utilization of nonrenewable energy resources for generation is discouraged by modern researchers [5].

Electricity produced from fossil fuels is transmitted to far-flung areas and then distributed. In addition to the energy losses during generation, these transmission and distribution phases have also several types of losses, and in some cases, these losses may rise by more than 50% [6]. An alternative approach is to eliminate the transmission phase and use distributed generation instead. In this type of generation, plants are directly located near the consumer loads. Losses can be minimized using distributed generation. These distributed power plants may use renewable energy resources such as solar [7], wind [8], and biogas [9] or nonrenewable energy resources such as geothermal energy [10], diesel generators [11], and furnace oil [12] for power generation. To minimize the environmental impacts, the usage of green energy resources is suggested in these distributed power generating stations [13]. Another benefit of these green energy resources is their renewable nature, for which they have also been termed renewable energy resources (Res). Res are environmentally friendly and renewable, but the only hurdle in the utilization of these types of resources is their intermittency, which is due to their extreme dependence upon weather conditions [14]. To reduce this problem, several techniques have been proposed in the literature, such as the incorporation of properly sized storage, architectural modifications, optimization, energy coordination schemes, etc. The generating stations operating on REs with a proper power coordination scheme and communication structure between the producers and consumers are called smart grids. Smart grids typically operate as isolated or grid-connected modes. In an isolated mode, a smart grid provides power to a connected consumer without having any connection with the main power grid, and storage then becomes necessary for these smart grids to overcome the intermittency problem of REs. In a grid-connected mode, the smart grid supports the main grid and provides ancillary services, in addition to fulfilling the consumer load requirements [15].

Small-scale smart grids which simultaneously produce and consume electrical power are termed prosumer microgrids [16]. The basic structure of a microgrid is represented in Figure 1. These microgrids can be of various types, from hospital to residential and industrial to institutional. A research institution should not depend upon the main grid for its energy requirements, especially if the main grid produces energy from conventional energy resources. A load of institutions is commercial, and these institutional microgrids are considered more important due to the research and development facilities available in an institution. These types of microgrids are also called campus microgrids [17]. The Internet of Things (IoT) is a modern technology that enables an operator to remotely monitor and control the activities of a smart grid using smart sensors [18]. The devices present in a smart grid's interface communicate bidirectionally, and they are prone to external cyber-attacks. Cyber security is also very important to secure a smart grid from external hacking attacks and to protect consumer data [19].

The organization of the paper is given here. The methodology is given in Section 2. An overview of campus energy management is presented in Section 3, different energy management schemes are discussed in Section 4, simulation tools are discussed in Section 5, IoT-enabled secured microgrids are disused in Section 6, and the conclusion is given in Section 7.

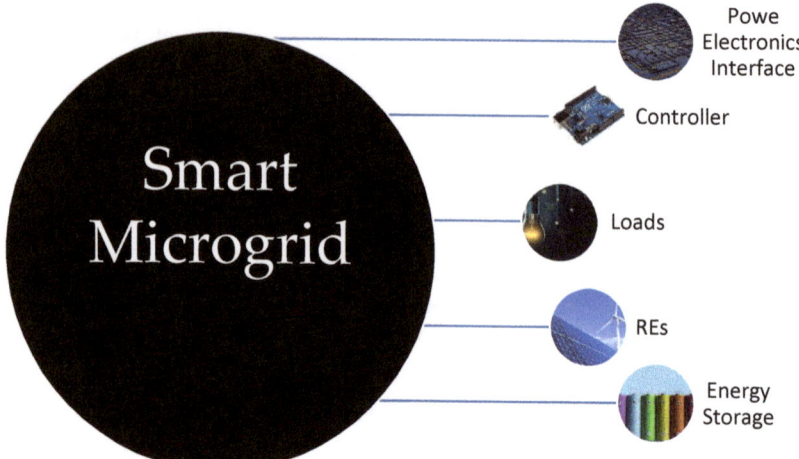

Figure 1. The basic structure of a smart microgrid.

2. Methodology

In this paper, a comprehensive review of the latest research work related to smart campus microgrid energy management is presented. The main focus of the work was to target the papers discussing real-time or simulated campus microgrids, with the restriction that each selected paper must contain at least one aspect of campus microgrids.

The methodology of the review was the same for all the selected papers as all used different energy management schemes for campus microgrid designs and optimization. The previous studies are categorized based on their architectures, storage methods, optimization techniques, simulation tools, and IOT technologies. The review study was carried out by reading the papers critically and identifying the significant points in the papers. Table 1 highlights the most important papers selected for each category.

Table 1. Criteria table of selected papers.

Sr. No	Selection Criteria	Cited Papers
1	Simulated or installed campus microgrid	[20–33]
2	Campus microgrid architecture	[34–46]
3	Storage technologies	[47–61]
4	Optimization techniques	[62–79]
5	Simulation/cost analysis tools	[80–100]
6	Internet of Things (IoT)	[101–105]

3. Overview of Campus Energy Management

Energy management is defined as a process to optimize energy production from REs and transmit this energy to consumers while cost-effectively minimizing the risk of system failure and gas emissions [106]. The concept of energy management began in the 1970s with the name of energy control centers, also known as ECCs. This concept was further expanded with the inclusion of different control schemes such as demand side management (DSM), load control (LC), demand response monitoring (DRM), etc. [107]. A simple energy management process consists of energy planning, execution, monitoring, verification, and understanding its usage. This process is represented in Figure 2.

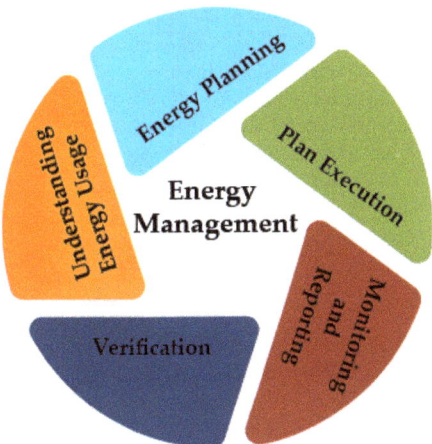

Figure 2. Simple energy management process.

The management of energy in a microgrid is critical because it is directly related to the economics of the grid. In campus prosumer microgrid energy management, the production of renewable energy resources present at a university campus is monitored, controlled, and optimized for the campus load. Worldwide, campuses of different universities are being converted to microgrids with REs as generation sources and environmentally friendly energy storage [20]. Most of the research published has contained simulated campus microgrid solutions, but practically working microgrids are also being developed at universities [21–28]. A simulation model of a campus microgrid was developed in Serbia for the University of Novi Sad in 2018, and it used photovoltaics (PV) and wind in conjunction with biomass and energy storage facilities. The concept of electric vehicles (EV) was also included in this research work, both as a consumer and a producer of energy in V2G mode [29]. In Italy, an energy-efficient microgrid solution was presented by Stefano et al. for the University of Genova. Multiple aspects of a campus microgrid were analyzed, taking into consideration the grid-connected, as well as standalone, operations [30]. However, this research lacked a feasibility analysis and the regulation problems of the proposed microgrid. Kritiawan et al. performed an in-detail feasibility analysis for a campus microgrid at Sebelas Maret University, located in Indonesia [31]. Voltage regulation is an important factor that should be taken into account when designing a campus microgrid [32]. Valentina et al. designed a microgrid for an island located in Singapore to improve the voltage regulation and power factor of the system. This design consisted of PV and diesel generators as generating sources, and it resulted in the lowest operational costs [33]. A pictorial representation of several campus microgrids installed all over the world is provided in Figure 3.

3.1. Objectives of Campus Microgrid Energy Management

The prime objective of energy management in a university campus microgrid is to optimally allocate generation and storage resources in a way that achieves the minimum per-unit cost of energy with maximum efficiency, while reducing gas emissions. Campus microgrid energy management may have single or numerous objectives such as resiliency, power quality, voltage and frequency regulation, reduced cost of energy, profit maximization, and life expectancy of transformers [108–111]. Universities can also obtain a green certificate by replacing the existing power infrastructure with a renewable energy-based microgrid [112]. Important objectives of a campus microgrid are represented in Figure 4. Campus microgrids should be efficient and reliable [113]. An energy-efficient campus microgrid solution was presented by Young et al. for the Gwanak Campus in South Korea, and it aimed to reduce the cost of energy by 21% and gas emissions by 110 TOE [114]. The economy is the most important element of a campus microgrid. Universities should be able to generate energy for the lowest possible cost. Currently, electric vehicles are becoming famous due to their environmentally friendly nature. The integration of these EVs with the lowest charging cost is another objective of campus microgrids. EV integration with a campus microgrid may cause stress on transformers [115]. Similarly, REs which are not properly sized may cause reactive power imbalances in a system, which may lead to frequent disconnections [116]. Low power outages and the continuity of supply are other objectives of campus microgrid energy management [117]. The achievement of all objectives in a single study is impossible. Most of the existing research includes a balance between economic and technical objectives.

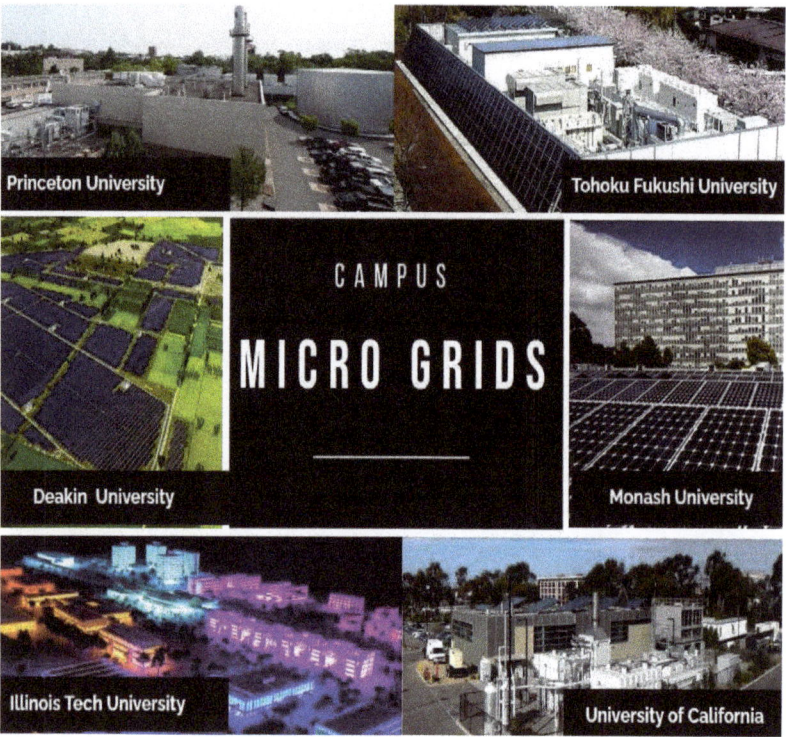

Figure 3. Campus microgrids at different universities.

3.2. Architecture of Campus Microgrids

Campus microgrids are designed to control the power production and utilization from REs and coordinate with smart metering, protective, storage, and load management devices, with the help of a control system, to achieve minimum costs and maximum efficiency. Architecture and infrastructure are two common terms used to describe the design of a campus microgrid. Microgrid infrastructure refers to all the components that a microgrid contains, such as transformers, smart metering devices, protection systems, switches, communication technologies, and cables [118]. The infrastructure of a microgrid should be resilient, which means that it should be capable enough to withstand extremely faulty conditions and recover quickly in the case of any disturbance [34]. Architecture defines how microgrid components connect to allow energy to flow and to enable storage. Both civil and electrical architecture are important factors in the design of a campus microgrid. In this paper, we will focus on the electrical architecture of different campus microgrids. A microgrid can operate in three different modes: off-grid, on-grid, and on/off-grid. Architecture generally changes slightly depending on the mode of operation. Campus microgrids typically work in the on-grid mode so that they can support the existing grid in the case of excessive supply. The common architecture of a microgrid may consist of three different bus configurations: centralized DC bus configuration, centralized AC bus configuration, and hybrid AC/DC bus configuration. The architecture of typical off-grid campus microgrids follows the DC centralized bus configuration, which is shown in Figure 5. In this architecture, the DC resources are directly connected to a centralized DC bus bar, while the AC components of the microgrid are connected via converters [35–37].

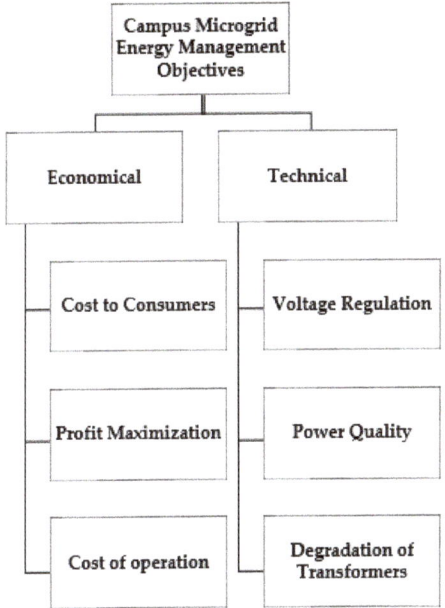

Figure 4. Important objectives of energy management in campus microgrids.

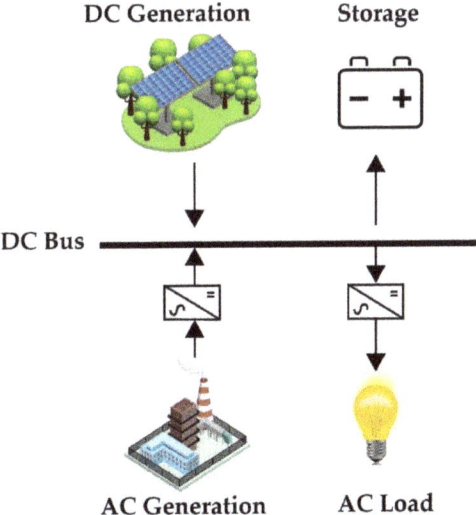

Figure 5. DC bus architecture of campus microgrids.

A typical AC bus configuration is shown in Figure 6, and it is the opposite of a DC bus configuration. This type of architecture requires higher voltage, which results in lower losses. REs largely supply direct DC, and it becomes less economical to first convert the power of REs into AC before supplying it to the central busbar; therefore, this type of configuration is found in few reach papers [38–41].

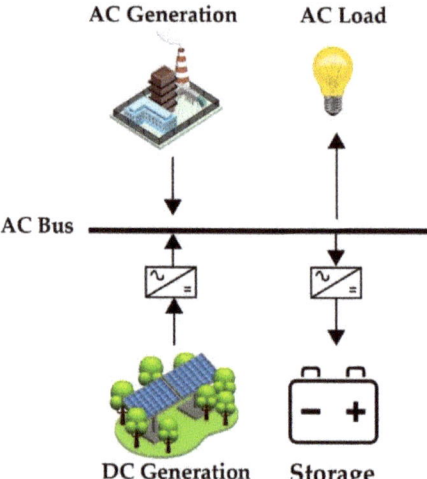

Figure 6. AC bus architecture of campus microgrids.

An advanced architecture for a campus microgrid is shown in Figure 7. It consists of both AC and DC busses, where the AC bus connects the AC components and the DC bus connects DC loads, storage, and generation. Finally, a bidirectional inverter connects these two buses [42–45]. In the literature, several studies have compared the types of architecture discussed above, and they concluded that a hybrid architecture for microgrids is more beneficial [46].

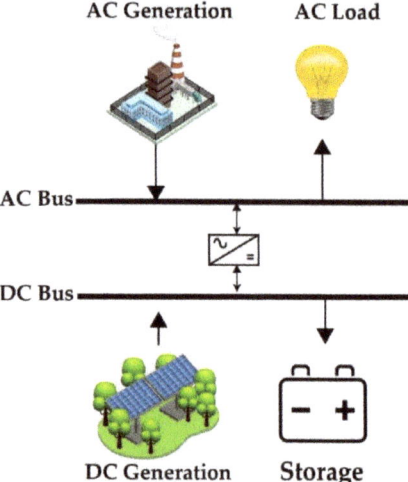

Figure 7. The hybrid bus architecture of campus microgrids.

3.3. Storage Technologies

Storage is a core component of a microgrid which serves many purposes during its operation. Besides providing power to load during the absence of generation, storage is used to provide ancillary services to smart grids, such as peak shaving [119], balancing the load profile [120], and improving system reliability [121]. In particular, a standalone campus microgrid cannot be sustained without the necessary storage facilities. A large portion of the total installation cost is attributable to the cost of storage. Storage degradation cost is another important factor that affects the overall system cost. Muqeet et.al [47] proposed a campus microgrid for the University of Engineering and Technology in Taxila, Pakistan. The authors considered the battery degradation cost in their design and compared different models, concluding that a properly scheduled storage technology results in a reduced cost of energy. There are several types of energy storage devices (ESDs) available, such as thermal [48], electrical [49], mechanical [50], electrochemical [51], and chemical [52]. With the advancements in plugin electric vehicles (PEV), these can also be used as storage devices in the vehicle-to-grid (V2G) mode of operation [53]. Storage configurations can be divided into three main categories: single storage configuration, multi-storage configuration, and swappable storage configuration [54]. In most of the existing literature, a single type of storage device is used in campus microgrids, while some advanced research designs have focused on the use of multiple types of storage technologies to improve their efficiency and useful life and reduce overall operational costs [49]. Riad Chedid et al. redesigned a campus microgrid for the American University of Beirut to reduce its dependence on a diesel generator. In the proposed design, REs replaced the fossil fuel generation with a combination of battery energy storage devices, which resulted in tremendous average annual savings of USD 1,336,000 [55]. Reyasudin Basir et al. proposed a microgrid design for the University Kuala Lumpur in Malaysia, and they proved that a grid-connected campus microgrid with battery storage was the most economical solution for this university [56]. When comparing different battery energy storage technologies, lithium-ion technology is considered the most suitable option. Yuly V. Garcia et al. designed a campus microgrid for the University of Puerto Rico in the United States. In the proposed design, solar and combined heat and power (CHP) technologies were used to reduce the fuel price. It was concluded that a combination of lithium-ion storage with solar generation and CHP provided the lowest fuel cost for a 10-year scenario [57]. A single storage device can be economical, but it always has some drawbacks which can be overcome by using multiple energy storage devices. Leskarac et al. proposed the use of PEV storage with a fuel cell to minimize the costs of operation for large commercial building microgrids [58]. A similar system was designed by Kumar et al. for the Nanyang Technological University of Singapore. It was concluded that a microgrid containing solar and natural gas as generation with PEV and fuel cells as storage could perfectly achieve the demand response targets [59]. Pedro Moura et al. practically demonstrated the use of multiple energy storage technologies for a campus microgrid at the University of Coimbra in Portugal to achieve the lowest cost of energy. The installed system contained PEV and li-ion batteries as storage systems, while grid-connected PV generation made the campus a net-zero-energy building. In [60], Hanane Dagdougui proposed the use of Li-ion batteries with a combination of a supercapacitor and hydrogen storage to improve the storage life and reduce the operating costs of the system. Rong-Jong Wai proposed the use of an ultra-capacitor and batteries as the storage medium for the economic design of the National Taiwan University of Science and Technology in Taiwan [61]. Current research is more focused on developing new storage technologies and making the existing storage technologies more compact.

4. Energy Management Schemes

The management of flowing energy between a campus microgrid, energy storage, a conventional grid, and the load is the most important element for reducing the cost of energy. Figure 8 represents the details of different optimization algorithms for the energy management of campus microgrids.

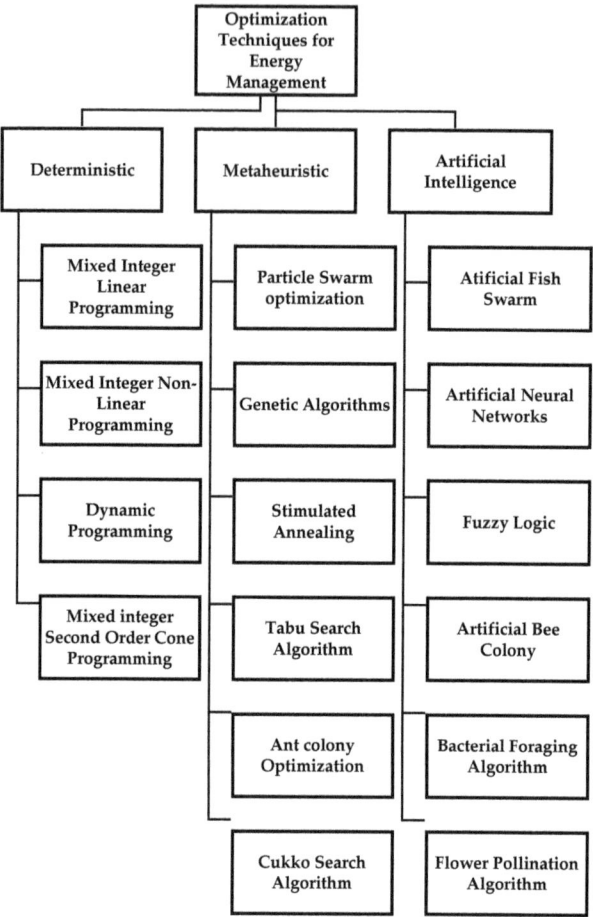

Figure 8. Classification of some important optimization algorithms for energy management.

The main objective of energy management in a microgrid is to increase efficiency by optimizing energy generation and storage systems [122]. Techniques used for optimizing campus microgrids are classified into three main categories: deterministic [62], metaheuristic [63], and artificial intelligence [64]. Each technique has unique benefits and drawbacks.

4.1. Deterministic Techniques

Deterministic techniques are primarily used to solve continuous objective functions. This technique involves the use of dynamic programming (DP), mixed integer nonlinear programming (MINLP), and linear programming (MILP) methods to solve an objective function. These methods are famous for providing precise results, but they are time-consuming and very difficult to use with distinct objective functions. Li-Bin et al. used mixed integer linear programming in MATLAB to optimize their proposed microgrid for the University of Engineering and Technology in Taxila [65]. Yeliz Yoldas et al. proposed a campus microgrid design for the Malta College of Arts, Science, and Technology. The proposed system was formulated using MILP, and compared with the stochastic approach, it was concluded that the use of MILP provided better optimal results [66]. In [67], the authors used mixed integer linear programming for the successful energy management of a campus microgrid located in Pakistan. Kayode Timothy Akindeji et al. used quadratic programming to optimize the microgrids of two campuses. The results depicted the effect of different weather conditions on campus load profiles, and it resulted in a substantial savings on fuel for the existing diesel generators by connecting them with a properly optimized microgrid [68]. For objective functions having second-order integers, the mixed integer second-order cone programming technique (MISOCP) is used [69]. In the dynamic programming technique, an objective function is divided into parts and then optimized.

4.2. Metaheuristic Techniques

Unlike deterministic techniques, metaheuristic algorithms provide an approximate solution. These are self-learning algorithms that take comparatively less time to reach a global solution. A famous example of a metaheuristic algorithm is the particle swarm algorithm. There are many metaheuristic algorithms present in the literature [70], such as Harmony Search (HS), Genetic Algorithms (GA), Particle Swarm Optimization (PSO), Ant colony Optimization (ACO), Stimulated Annealing (SA), Tabu Search (TS), Cuckoo Search (CS), Teaching Learning Based Optimization (TLBO), Jaya-Harmony search (JHS), Krill Heard (KH), Variable Neighborhood search (VNS), etc. For the campus microgrid at Yalova University in Turkey, Aykut Fatih Güven et al. proposed the use of the Jaya-Harmony search algorithm to optimize the energy management process, and they compared the results with PSO and HOMER. It was concluded that the JHS algorithm provided the most optimal component sizing [71]. To further increase the efficiency and convergence of metaheuristic algorithms, a combination of these algorithms is recommended [72]. For example, PSO provides fast convergence compared to GA [73], but it lacks flexibility, and so a combination of GA and PSO provides the best results both in terms of convergence and flexibility [74]. Mohamad Almas Prakasa and Subiyanto Subiyanto used a fusion of a genetic algorithm and a modified particle swarm optimization algorithm for the cost-optimal design of a campus microgrid located at the Universitas Negeri Semarang in Indonesia [75]. The proposed design managed to reduce operation costs by 11.9%.

4.3. Artificial Intelligence Techniques

Artificial intelligence techniques are modern optimization algorithms based on artificial intelligence. These algorithms have the benefits of quick convergence, high speed, and good precision. Machine learning involves the use of artificial intelligence algorithms. Saheed Lekan Gbadamosi and Nnamdi I. Nwulu used the machine learning Waikato Environment for Knowledge Analysis algorithm for solar and wind forecasting for a campus microgrid at the University of Johannesburg in South Africa [76]. Letícia A.L. Zaneti et al. used the rolling horizon method to reduce the charging costs of campus bus charging stations by up to 52% [77]. Jangkyum Kim et al. used IoT (Internet of Things) sensors to collect live data from the Korea Advanced Institute of Science and Technology (KAIST), Daejeon campus, and then they used this data to propose a microgrid energy management scheme. The system was optimized using an AI-based self-organizing map algorithm. Optimization by taking uncertainties into account resulted in a 3% reduction in peak power and a 2.16% reduction in the daily electricity price for the campus [101]. Deep learning techniques can be used for the prediction of energy prices and smart grid stability [78], but they are associated with some limitations. Türkücan Erdem and Süleyman Eken proposed the use of layer-wise relevance propagation to discover the relevance of each input. It was suggested that the primary input in the system was the time required for the participant response, accompanied by the pricing coefficient, and that the electricity consumption or production had a negligible effect on the stability [79]. Table 2 shows optimization techniques being utilized in the latest research.

Table 2. Some optimization techniques from the latest campus microgrid studies.

Ref	Campus Name	Country	Optimization Method	Outcome
[65]	University of Engineering and Technology, Taxila	Pakistan	MILP	• 36.6% savings for campus microgrid
[66]	Malta College of Arts, Science, and Technology	Malta	MILP	• GBP 17,226 net saving in storage • 100% optimal design
[68]	University of KwaZulu–Natal	South Africa	Quadratic Programming	Substantial savings on fuel
[71]	Yalova University	Turkey	Jaya-Harmony Search	• Better convergence. • Most optimal sizing
[75]	Universitas Negeri Semarang	Indonesia	Modified PSO and GA	• 11.99% reduction in system costs
[76]	University of Johannesburg	South Africa	Waikato Environment for Knowledge Analysis (WEKA)	• Reduced effect of RE unpredictability • Lowest system maintenance costs
[77]	University of Campinas	Brazil	Rolling Horizon	• 52% reduction in operation costs
[101]	Korea Advanced Institute of Science and Technology	Korea	Self-Organizing Map Algorithm	• 2.16% reduction in daily electricity costs • 3% reduction in peak power
[80]	Polytechnic of Porto	Portugal	Fuzzy logic	• Reliable consumption forecasting with less historical input

5. Tools Used for Energy Management of Campus Microgrids

5.1. MATLAB

MATLAB is a programming and development tool to develop several kinds of mathematical algorithms. It is widely used for campus microgrid system optimization purposes. L. Hadjidemetriou et al. used MATLAB as a tool for the optimization of their proposed microgrid for the Malta College of Arts, Science, and Technology [17]. Ali Arzani et al. proposed a campus microgrid for Clemson University, Clemson, based on the solar generation and battery energy storage that could provide power for a two-day islanding event. The authors used MATLAB for the component sizing of their proposed microgrid [101]. In Greece, Elencova designed a campus microgrid for the Democritus University of Thrace. The energy management scheme of the campus microgrid was designed using MATLAB [102].

5.2. Simulink

Simulink is a graphical programming tool for modeling work, simulating, and evaluating multidomain dynamic systems based on MATLAB. Its framework comprises a simple visual block diagramming tool and its libraries that can be modified. It is a very useful tool for constructing a simulation model of a campus microgrid. Yuly V. Garcia et al. simulated their proposed campus microgrid design containing solar, battery storage, and a CHP system for the University of Puerto Rico in the United States [57]. Moslem Uddin et al. used a Simulink (MATLAB) tool to model a campus microgrid for the Universiti Teknologi PETRONAS (UTP) in Malaysia. The proposed model was then used to perform economic and stability analyses [82].

5.3. LabVIEW

LabVIEW is also a graphical programming tool that can be considered an alternative to Simulink. This tool is widely used by designers and engineers for campus microgrid applications. Rachid Lghoul in [83] used LabVIEW and CompactRIO for data acquisition and the management of a campus microgrid at Al Akhawayn University in Morocco. Pedro Moura et al. used the LabVIEW tool for the development of monitoring software for a campus microgrid installed at the University of Coimbra in Portugal [84].

5.4. HOMER

HOMER is the abbreviation of Hybrid Optimization of Multiple Energy Resources. This tool is a favorite among researchers due to its simple interface and rich features. A complete financial analysis of a proposed microgrid can be completed using HOMER. It also provides an opportunity to compare different combinations of proposed active and passive components of a proposed microgrid to select the most economical one [85]. MD Sarwar et al. used HOMER for the design and economic analysis of a campus microgrid at Jamia Millia Islamia University in India. The results proved that the proposed system design was the most economic and environmentally friendly [86]. Ayooluwa A. Ajiboye et al. also used HOMER to identify the most economically feasible solution for Covenant University in Nigeria. The results showed that wind turbines were not suitable for this location, while a combination of solar, grid, diesel, and battery energy storage was a potential system for long-term economic benefits [87]. For green transportation at the Thiagarajar College of Engineering in India, 100 kW solar generation was recommended. The system was optimized using HOMER, and it was observed that the proposed microgrid reduced gas emissions by 49,303 kg per year [88]. Sheeraz Iqbal et al. used HOMER for the economic feasibility testing of a proposed microgrid for the King Abdullah Campus of the University of Azad Jammu and Kashmir. It was concluded that a hybrid system containing solar, battery, and grid was the best solution for this university campus [89]. T. M. I. Riayatsyah et al. performed a techno-economic analysis for Syiah Kuala University in Indonesia using HOMER, and they concluded that an RE-based system containing 62% energy from solar and 20% energy from wind could reduce the per-unit cost of campus energy utilization from $0.060 to $0.0446 per kWh [90]. Stephen Ogbikaya et al. used HOMER Pro for the optimization and economic analysis of a campus microgrid for a university located in Nigeria. The proposed optimized system resulted in 88% saving on electricity charges [91].

5.5. PVSyst

PVSyst is a powerful tool for campus microgrid design and optimization. It has many unique built-in features, such as 3D partial shading phenomena, solar system sizing [92], storage optimization, etc. However, the use of this tool is limited to the design of those microgrids which use only solar as a generation source. It is not possible to use this tool for microgrids having multiple energy generations [93]. David Morillón Gálvez et al. used PVSyst to model a grid-connected solar-based campus microgrid for the National Autonomous University of Mexico. The proposed microgrid resulted in reduced carbon footprints and had the shortest payback period (fewer than six years) [94].

5.6. CPLEX

CPLEX is a mathematical programming tool developed by IBM ILOG. This tool uses integer, mixed integer, and quadratic programming techniques to optimize microgrid design. It is compatible with several programming languages such as Java, C++, and Python [95]. Jingyun Li and Hong Zhao used a CPLEX solver to optimize a proposed campus energy management model based on an improved particle swarm optimization technique [96].

Apart from the above-mentioned softwares, there are many other useful tools for campus microgrid development and optimization, such as DER-CAM (Distributed Energy Resources Customer Adoption Model) [23], iHOGA (Hybrid Optimization by Genetic Algorithms) [97], SAM (System Advisor Model) [98], PSCAD (Power Systems Computer Aided Design) [99] and GAMS (Generic Algebraic Modelling System) [100]. Table 3 shows the energy management system and used tools in the literature.

Table 3. Tools used in the campus microgrid energy management literature.

Ref	Campus	Resources	Tool
[17]	Malta College of Arts, Science, and Technology	Solar, diesel generator, and BSS	MATLAB
[81]	Clemson University	Solar and BSS	MATLAB
[102]	Democritus University of Thrace	Solar and BSS	MATLAB
[57]	University of Puerto Rico	Solar, CHP, and BSS	Simulink
[82]	Universiti Teknologi PETRONAS	PV, gas turbine, and BSS	Simulink
[83]	Al Akhawayn University	Solar and BSS	LabVIEW
[84]	University of Coimbra	Solar, BSS, and EV	LabVIEW
[86]	Jamia Millia Islamia University	Solar, wind, and BSS	HOMER
[87]	Covenant University	Solar, diesel generator, grid, and BSS	HOMER
[88]	Thiagarajar College of Engineering	Solar	HOMER
[89]	University of Azad Jammu and Kashmir	Solar, grid, and BSS	HOMER
[90]	Syiah Kuala University	Solar and wind	HOMER
[94]	National Autonomous University of Mexico	Solar and grid	PVSyst
[123]	Seoul National University	Solar and ESS	MDStool

6. IoT Enabled Cyber-Secured Microgrid

There are several solutions available for smart campus microgrid energy management [103]. Technologies that are important for campus microgrid design, communication, and operation are RFID (radio frequency identification), cloud computing, wireless technologies, augmented reality (AR), mobile technologies, and IoT (Internet of Things) technologies [104]. Solar irradiance, humidity, and temperature sensors were used by Jangkyum Kim to acquire real-time data from a campus microgrid and then propose an effective energy management model for it [101]. M. Z. Elenkova et al. [102] produced a simulation model for the campus microgrid at the Democritus University of Thrace in Greece, and an efficient energy management scheme was introduced for operating this microgrid. This scheme used IEC 61,850 as a communication protocol. Hanaa Talei proposed the use of a cloud computing-based IoT platform to avoid unnecessary delays in campus microgrid communication systems [105].

7. Conclusions

In this paper, a comprehensive study of the various aspects of campus microgrids is presented. This paper describes the energy management of campus microgrids considering the objective, architecture, storage technologies, and different tools used. Different storage technologies are used for campus microgrids. A prosumer-based system is focused on having the characteristics of energy exchange. In modern techniques, AI is the optimal tool when compared to conventional tools. An Internet of Things (IoT) -enabled system is more advanced compared to a classical system. Therefore, IoT technologies are used for communication and signaling purposes. The softwares used for energy management systems are also studied to explore better options for simulations. Python is an advanced platform for analyses using AI tools. In the future, a technical paper will be presented focusing on advanced techniques and uncertainties of systems.

Author Contributions: A.A. proposed the idea of this research and wrote the manuscript; H.A.M. completed the literature review; T.K. prepared the figures for this manuscript; A.H., K.A.K.N. and M.W. proofread the manuscript; K.A.K.N. funded the research throughout. All authors have read and agreed to the published version of the manuscript.

Funding: This research received no external funding.

Institutional Review Board Statement: Not applicable.

Informed Consent Statement: Not applicable.

Data Availability Statement: Not applicable.

Acknowledgments: The authors would like to thank the Technical Education and Vocational Training Authority (TEVTA) and the Government College of Technology, Multan, Pakistan, for providing a formal atmosphere in which to conduct this research.

Conflicts of Interest: The authors declare no conflict of interest.

References

1. Pau, G.; Collotta, M.; Ruano, A.; Qin, J. Smart Home Energy Management. *Energies* **2017**, *10*, 382. [CrossRef]
2. Salvatore, D. Growth and Trade in the United States and the World Economy: Overview. *J. Policy Model.* **2020**, *42*, 750–759. [CrossRef] [PubMed]
3. Petroleum, B. BP Statistical Review of World Energy 2017. *Stat. Rev. World Energy* **2019**, *2019*, 65.
4. Zheng, S.; Shahzad, M.; Asif, H.M.; Gao, J.; Muqeet, H.A. Advanced Optimizer for Maximum Power Point Tracking of Photovoltaic Systems in Smart Grid: A Roadmap Towards Clean Energy Technologies. *Renew. Energy* **2023**. [CrossRef]
5. Lior, N. Energy Resources and Use: The Present Situation and Possible Paths to the Future. *Energy* **2008**, *33*, 842–857. [CrossRef]
6. Berger, L.T.; Iniewski, K. *Smart Grid Applications, Communications, and Security*; John Wiley & Sons: Hoboken, NJ, USA, 2012; ISBN 1-118-00439-6.
7. Göransson, M.; Larsson, N.; Steen, D. Cost-benefit analysis of battery storage investment for microgrid of chalmers university campus using μ-OPF framework. In Proceedings of the 2017 IEEE Manchester PowerTech, Manchester, UK, 18–22 June 2017; IEEE: New York, NY, USA, 2017; pp. 1–6.
8. Iqbal, M.M.; Waseem, M.; Manan, A.; Liaqat, R.; Muqeet, A.; Wasaya, A. IoT-Enabled Smart Home Energy Management Strategy for DR Actions in Smart Grid Paradigm. In Proceedings of the 2021 International Bhurban Conference on Applied Sciences and Technologies (IBCAST), Islamabad, Pakistan, 12–16 January 2021; pp. 352–357. [CrossRef]
9. Mazzola, S.; Astolfi, M.; Macchi, E. A Detailed Model for the Optimal Management of a Multigood Microgrid. *Appl. Energy* **2015**, *154*, 862–873. [CrossRef]
10. Raza, A.; Malik, T.N. Energy Management in Commercial Building Microgrids. *J. Renew. Sustain. Energy* **2019**, *11*, 015502. [CrossRef]
11. Muqeet, H.A.; Liaqat, R.; Jamil, M.; Khan, A.A. A State-of-the-Art Review of Smart Energy Systems and Their Management in a Smart Grid Environment. *Energies* **2023**, *16*, 472. [CrossRef]
12. Muzzammel, R.; Arshad, R. Comprehensive Analysis and Design of Furnace Oil-Based Power Station Using ETAP. *Int. J. Appl.* **2022**, *11*, 33–51. [CrossRef]
13. García Vera, Y.E.; Dufo-López, R.; Bernal-Agustín, J.L. Energy Management in Microgrids with Renewable Energy Sources: A Literature Review. *Appl. Sci.* **2019**, *9*, 3854. [CrossRef]
14. Chen, H.; Yang, C.; Deng, K.; Zhou, N.; Wu, H. Multi-Objective Optimization of the Hybrid Wind/Solar/Fuel Cell Distributed Generation System Using Hammersley Sequence Sampling. *Int. J. Hydrogen Energy* **2017**, *42*, 7836–7846. [CrossRef]
15. Machamint, V.; Oureilidis, K.; Venizelou, V.; Efthymiou, V.; Georghiou, G.E. Optimal energy storage sizing of a microgrid under different pricing schemes. In Proceedings of the 2018 IEEE 12th International Conference on Compatibility, Power Electronics and Power Engineering (CPE-POWERENG 2018), Doha, Qatar, 10–12 April 2018; IEEE: New York, NY, USA, 2018; pp. 1–6.
16. Muqeet, H.A.U.; Ahmad, A. Optimal Scheduling for Campus Prosumer Microgrid Considering Price Based Demand Response. *IEEE Access* **2020**, *8*, 71378–71394. [CrossRef]
17. Hadjidemetriou, L.; Zacharia, L.; Kyriakides, E.; Azzopardi, B.; Azzopardi, S.; Mikalauskiene, R.; Al-Agtash, S.; Al-Hashem, M.; Tsolakis, A.; Ioannidis, D. Design factors for developing a university campus microgrid. In Proceedings of the 2018 IEEE International Energy Conference (ENERGYCON), Limassol, Cyprus, 3–7 June 2018; IEEE: New York, NY, USA, 2018; pp. 1–6.
18. Al-Turjman, F.; Abujubbeh, M. IoT-Enabled Smart Grid via SM: An Overview. *Future Gener. Comput. Syst.* **2019**, *96*, 579–590. [CrossRef]
19. Gunduz, M.Z.; Das, R. Cyber-Security on Smart Grid: Threats and Potential Solutions. *Comput. Netw.* **2020**, *169*, 107094. [CrossRef]

20. Muqeet, H.A.; Munir, H.M.; Javed, H.; Shahzad, M.; Jamil, M.; Guerrero, J.M. An Energy Management System of Campus Microgrids: State-of-the-Art and Future Challenges. *Energies* **2021**, *14*, 6525. [CrossRef]
21. Hipwell, S. Developing smart campuses—A working model. In Proceedings of the 2014 International Conference on Intelligent Green Building and Smart Grid (IGBSG), Taipei, Taiwan, 23–25 April 2014; IEEE: New York, NY, USA, 2014; pp. 1–6.
22. Mitchell Finnigan, S.; Clear, A.K.; Olivier, P. SpaceBot: Towards Participatory Evaluation of Smart Buildings. In Proceedings of the Extended Abstracts of the 2018 CHI Conference on Human Factors in Computing Systems, Montreal, QC, Canada, 21–26 April 2018; pp. 1–6.
23. Jung, J.; Villaran, M. Optimal Planning and Design of Hybrid Renewable Energy Systems for Microgrids. *Renew. Sustain. Energy Rev.* **2017**, *75*, 180–191. [CrossRef]
24. Solanki, Z.; Wani, U.; Patel, J. Demand side management program for balancing load curve for CGPIT college, bardoli. In Proceedings of the 2017 International Conference on Energy, Communication, Data Analytics and Soft Computing (ICECDS), Chennai, India, 1–2 August 2017; IEEE: New York, NY, USA, 2017; pp. 769–774.
25. Mantovani, G.; Costanzo, G.T.; Marinelli, M.; Ferrarini, L. Experimental Validation of Energy Resources Integration in Microgrids via Distributed Predictive Control. *IEEE Trans. Energy Convers.* **2014**, *29*, 1018–1025. [CrossRef]
26. Tellez, S.; Alvarez, D.; Montano, W.; Vargas, C.; Cespedes, R.; Parra, E.; Rosero, J. National laboratory of smart grids (LAB+ i) at the National University of Colombia-Bogota Campus. In Proceedings of the 2014 IEEE PES Transmission & Distribution Conference and Exposition-Latin America (PES T&D-LA), Medellin, Colombia, 10–13 September 2014; IEEE: New York, NY, USA, 2014; pp. 1–6.
27. Xu, P.; Jin, Z.; Zhao, Y.; Wang, X.; Sun, H. Design and Operation Experience of Zero-Carbon Campus. *EDP Sci.* **2018**, *48*, 03004. [CrossRef]
28. Tatro, R.; Vadhva, S.; Kaur, P.; Shahpatel, N.; Dixon, J.; Alzanoon, K. Building to Grid (B2G) at the California smart grid center. In Proceedings of the 2010 IEEE International Conference on Information Reuse & Integration, Las Vegas, NV, USA, 4–6 August 2010; IEEE: New York, NY, USA, 2010; pp. 382–387.
29. Savić, N.S.; Katić, V.A.; Katić, N.A.; Dumnić, B.; Milićević, D.; Čorba, Z. Techno-Economic and environmental analysis of a microgrid concept in the University Campus. In Proceedings of the 2018 International Symposium on Industrial Electronics (INDEL), Banja Luka, Bosnia and Herzegovina, 1–3 November 2018; pp. 1–6.
30. Bracco, S.; Delfino, F.; Laiolo, P.; Rossi, M. The smart city energy infrastructures at the savona campus of the university of genoa. In Proceedings of the 2016 AEIT International Annual Conference (AEIT), Capri, Italy, 5–7 October 2016; pp. 1–6.
31. Kristiawan, R.B.; Widiastuti, I.; Suharno, S. Technical and Economical Feasibility Analysis of Photovoltaic Power Installation on a University Campus in Indonesia. *MATEC Web Conf.* **2018**, *197*, 08012. [CrossRef]
32. Morales González, R.; van Goch, T.A.J.; Aslam, M.F.; Blanch, A.; Ribeiro, P.F. Microgrid design considerations for a smart-energy university campus. In Proceedings of the IEEE PES Innovative Smart Grid Technologies, Europe, Istanbul, Turkey, 12–15 October 2014; pp. 1–6.
33. Bertolotti, V.; Procopio, R.; Rosini, A.; Bracco, S.; Delfino, F.; Soh, C.B.; Cao, S.; Wei, F. Energy management system for pulau ubin islanded microgrid test-bed in Singapore. In Proceedings of the 2020 IEEE International Conference on Environment and Electrical Engineering and 2020 IEEE Industrial and Commercial Power Systems Europe (EEEIC / I CPS Europe), Madrid, Spain, 9–12 June 2020; pp. 1–6.
34. Ibrahim, M.; Alkhraibat, A. Resiliency Assessment of Microgrid Systems. *Appl. Sci.* **2020**, *10*, 1824. [CrossRef]
35. Wang, Z.; Gu, C.; Li, F. Flexible Operation of Shared Energy Storage at Households to Facilitate PV Penetration. *Renew. Energy* **2018**, *116*, 438–446. [CrossRef]
36. Morin, D.; Stevenin, Y.; Grolleau, C.; Brault, P. Evaluation of Performance Improvement by Model Predictive Control in a Renewable Energy System with Hydrogen Storage. *Int. J. Hydrogen Energy* **2018**, *43*, 21017–21029. [CrossRef]
37. Huang, W.; Fu, Z.; Hua, L. Research on optimal capacity configuration for distributed generation of island micro-grid with wind/solar/battery/diesel engine. In Proceedings of the 2018 2nd IEEE Conference on Energy Internet and Energy System Integration (EI2), Beijing, China, 20–22 October 2018; IEEE: New York, NY, USA, 2018; pp. 1–6.
38. Eseye, A.T.; Zheng, D.; Zhang, J.; Wei, D. Optimal Energy Management Strategy for an Isolated Industrial Microgrid Using a Modified Particle Swarm Optimization. In Proceedings of the 2016 IEEE International Conference on Power and Renewable Energy (ICPRE), Shanghai, China, 21–23 October 2016; pp. 494–498.
39. Madiba, T.; Bansal, R.; Justo, J.; Kusakana, K. Optimal control system of under frequency load shedding in microgrid system with renewable energy resources. In *Smart Energy Grid Design for Island Countries*; Springer: Berlin/Heidelberg, Germany, 2017; pp. 71–96.
40. Alvarez, S.R.; Ruiz, A.M.; Oviedo, J.E. Optimal design of a diesel-PV-wind system with batteries and hydro pumped storage in a Colombian community. In Proceedings of the 2017 IEEE 6th International Conference on Renewable Energy Research and Applications (ICRERA), San Diego, CA, USA, 5–8 November 2017; IEEE: New York, NY, USA, 2017; pp. 234–239.
41. An, L.N.; Dung, T.T.M.; Quoc-Tuan, T. Optimal energy management for an on-grid microgrid by using branch and bound method. In Proceedings of the 2018 IEEE International Conference on Environment and Electrical Engineering and 2018 IEEE Industrial and Commercial Power Systems Europe (EEEIC/I&CPS Europe), Palermo, Italy, 12–15 June 2018; IEEE: New York, NY, USA, 2018; pp. 1–5.
42. Wang, H.; Huang, J. Joint Investment and Operation of Microgrid. *IEEE Trans. Smart Grid* **2015**, *8*, 833–845. [CrossRef]

43. Devi, V.K.; Premkumar, K.; Beevi, A.B. Energy Management Using Battery Intervention Power Supply Integrated with Single Phase Solar Roof Top Installations. *Energy* **2018**, *163*, 229–244. [CrossRef]
44. Ramli, M.A.; Bouchekara, H.; Alghamdi, A.S. Optimal Sizing of PV/Wind/Diesel Hybrid Microgrid System Using Multi-Objective Self-Adaptive Differential Evolution Algorithm. *Renew. Energy* **2018**, *121*, 400–411. [CrossRef]
45. Tayab, U.B.; Yang, F.; El-Hendawi, M.; Lu, J. Energy Management system for a grid-connected microgrid with photovoltaic and battery energy storage system. In Proceedings of the 2018 Australian & New Zealand Control Conference (ANZCC), Melbourne, VIC, Australia, 7–8 December 2018; IEEE: New York, NY, USA, 2018; pp. 141–144.
46. Solorzano del Moral, J.; Egido, M.Á. Simulation of AC, DC, and AC-DC Coupled mini-grids. In search of the most efficient system. In Proceedings of the 6th European Conference on PV hybrids an Mini Grids, Chambery, France, 25–27 April 2012.
47. Abd ul Muqeet, H.; Munir, H.M.; Ahmad, A.; Sajjad, I.A.; Jiang, G.-J.; Chen, H.-X. Optimal Operation of the Campus Microgrid Considering the Resource Uncertainty and Demand Response Schemes. *Math. Probl. Eng.* **2021**, *2021*, 5569701. [CrossRef]
48. Ibrahim, H.; Belmokhtar, K.; Ghandour, M. Investigation of Usage of Compressed Air Energy Storage for Power Generation System Improving-Application in a Microgrid Integrating Wind Energy. *Energy Procedia* **2015**, *73*, 305–316. [CrossRef]
49. Jing, W.; Lai, C.H.; Wong, W.S.H.; Wong, M.L.D. Dynamic Power Allocation of Battery-Supercapacitor Hybrid Energy Storage for Standalone PV Microgrid Applications. *Sustain. Energy Technol. Assess.* **2017**, *22*, 55–64. [CrossRef]
50. Arani, A.K.; Karami, H.; Gharehpetian, G.; Hejazi, M. Review of Flywheel Energy Storage Systems Structures and Applications in Power Systems and Microgrids. *Renew. Sustain. Energy Rev.* **2017**, *69*, 9–18. [CrossRef]
51. Konstantinopoulos, S.A.; Anastasiadis, A.G.; Vokas, G.A.; Kondylis, G.P.; Polyzakis, A. Optimal Management of Hydrogen Storage in Stochastic Smart Microgrid Operation. *Int. J. Hydrogen Energy* **2018**, *43*, 490–499. [CrossRef]
52. Alsaidan, I.; Khodaei, A.; Gao, W. A Comprehensive Battery Energy Storage Optimal Sizing Model for Microgrid Applications. *IEEE Trans. Power Syst.* **2017**, *33*, 3968–3980. [CrossRef]
53. Liu, Z.; Chen, Y.; Zhuo, R.; Jia, H. Energy Storage Capacity Optimization for Autonomy Microgrid Considering CHP and EV Scheduling. *Appl. Energy* **2018**, *210*, 1113–1125. [CrossRef]
54. Ali, A.; Shakoor, R.; Raheem, A.; Awais, Q.; Khan, A.A.; Jamil, M. Latest Energy Storage Trends in Multi-Energy Standalone Electric Vehicle Charging Stations: A Comprehensive Study. *Energies* **2022**, *15*, 4727. [CrossRef]
55. Chedid, R.; Sawwas, A.; Fares, D. Optimal Design of a University Campus Micro-Grid Operating under Unreliable Grid Considering PV and Battery Storage. *Energy* **2020**, *200*, 117510. [CrossRef]
56. Khan, M.R.B.; Pasupuleti, J.; Al-Fattah, J.; Tahmasebi, M. Optimal Grid-Connected PV System for a Campus Microgrid. *Indones. J. Electr. Eng. Comput. Sci.* **2018**, *12*, 899–906.
57. Garcia, Y.V.; Garzon, O.; Andrade, F.; Irizarry, A.; Rodriguez-Martinez, O.F. Methodology to Implement a Microgrid in a University Campus. *Appl. Sci.* **2022**, *12*, 4563. [CrossRef]
58. Leskarac, D.; Moghimi, M.; Liu, J.; Water, W.; Lu, J.; Stegen, S. Hybrid AC/DC Microgrid Testing Facility for Energy Management in Commercial Buildings. *Energy Build.* **2018**, *174*, 563–578. [CrossRef]
59. Kumar, K.P.; Saravanan, B. Real Time Optimal Scheduling of Generation and Storage Sources in Intermittent Microgrid to Reduce Grid Dependency. *Indian J. Sci. Technol.* **2016**, *9*, 1–4.
60. Dagdougui, H.; Dessaint, L.; Gagnon, G.; Al-Haddad, K. Modeling and optimal operation of a university campus microgrid. In Proceedings of the 2016 IEEE Power and Energy Society General Meeting (PESGM), Boston, MA, USA, 17–21 June 2016; IEEE: New York, NY, USA, 2016; pp. 1–5.
61. Wai, R.-J. Systematic Design of Energy-Saving Action Plans for Taiwan Campus by Considering Economic Benefits and Actual Demands. *Energies* **2022**, *15*, 6530. [CrossRef]
62. Gao, H.-C.; Choi, J.-H.; Yun, S.-Y.; Lee, H.-J.; Ahn, S.-J. Optimal Scheduling and Real-Time Control Schemes of Battery Energy Storage System for Microgrids Considering Contract Demand and Forecast Uncertainty. *Energies* **2018**, *11*, 1371. [CrossRef]
63. Panda, S.; Yegireddy, N.K. Multi-Input Single Output SSSC Based Damping Controller Design by a Hybrid Improved Differential Evolution-Pattern Search Approach. *ISA Trans.* **2015**, *58*, 173–185. [CrossRef]
64. Husein, M.; Chung, I.-Y. Day-Ahead Solar Irradiance Forecasting for Microgrids Using a Long Short-Term Memory Recurrent Neural Network: A Deep Learning Approach. *Energies* **2019**, *12*, 1856. [CrossRef]
65. Bin, L.; Shahzad, M.; Javed, H.; Muqeet, H.A.; Akhter, M.N.; Liaqat, R.; Hussain, M.M. Scheduling and Sizing of Campus Microgrid Considering Demand Response and Economic Analysis. *Sensors* **2022**, *22*, 6150. [CrossRef] [PubMed]
66. Yoldas, Y.; Goren, S.; Onen, A.; Ustun, T.S. Dynamic Rolling Horizon Control Approach for a University Campus. *Energy Rep.* **2022**, *8*, 1154–1162. [CrossRef]
67. Muqeet, H.A.; Ahmad, A.; Sajjad, I.A.; Liaqat, R.; Raza, A.; Iqbal, M.M. Benefits of distributed energy and storage system in prosumer based electricity market. In Proceedings of the 2019 IEEE International Conference on Environment and Electrical Engineering and 2019 IEEE Industrial and Commercial Power Systems Europe (EEEIC/I CPS Europe), Genova, Italy, 11–14 June 2019; pp. 1–6.
68. Akindeji, K.T.; Tiako, R.; Davidson, I. Optimization of University Campus Microgrid for Cost Reduction: A Case Study. *Trans. Tech. Publ.* **2022**, *45*, 77–96. [CrossRef]
69. Esmaeili, S.; Anvari-Moghaddam, A.; Jadid, S. Optimal Operation Scheduling of a Microgrid Incorporating Battery Swapping Stations. *IEEE Trans. Power Syst.* **2019**, *34*, 5063–5072. [CrossRef]

70. Abdel-Basset, M.; Abdel-Fatah, L.; Sangaiah, A. Chapter 10-Metaheuristic algorithms: A comprehensive review. In *Computational Intelligence for Multimedia Big Data on the Cloud with Engineering Applications*; Elsevier: Amsterdam, The Netherlands, 2018.
71. Güven, A.F.; Yörükeren, N.; Samy, M.M. Design Optimization of a Stand-Alone Green Energy System of University Campus Based on Jaya-Harmony Search and Ant Colony Optimization Algorithms Approaches. *Energy* **2022**, *253*, 124089. [CrossRef]
72. Suresh, M.; Meenakumari, R. Optimum Utilization of Grid Connected Hybrid Renewable Energy Sources Using Hybrid Algorithm. *Trans. Inst. Meas. Control* **2021**, *43*, 21–33. [CrossRef]
73. Twaha, S.; Ramli, M.A. A Review of Optimization Approaches for Hybrid Distributed Energy Generation Systems: Off-Grid and Grid-Connected Systems. *Sustain. Cities Soc.* **2018**, *41*, 320–331. [CrossRef]
74. Ali, A.F.; Tawhid, M.A. A Hybrid Particle Swarm Optimization and Genetic Algorithm with Population Partitioning for Large Scale Optimization Problems. *Ain Shams Eng. J.* **2017**, *8*, 191–206. [CrossRef]
75. Almas Prakasa, M.; Subiyanto, S. Optimal Cost and Feasible Design for Grid-Connected Microgrid on Campus Area Using the Robust-Intelligence Method. *Clean Energy* **2022**, *6*, 823–840. [CrossRef]
76. Gbadamosi, S.L.; Nwulu, N.I. Optimal Microgrid Sizing Incorporating Machine Learning Forecasting. In Proceedings of the International Conference on Industrial Engineering and Operations Management, Toronto, ON, Canada, 23–25 October 2019; pp. 1637–1642.
77. Zaneti, L.A.; Arias, N.B.; de Almeida, M.C.; Rider, M.J. Sustainable Charging Schedule of Electric Buses in a University Campus: A Rolling Horizon Approach. *Renew. Sustain. Energy Rev.* **2022**, *161*, 112276. [CrossRef]
78. Breviglieri, P.; Erdem, T.; Eken, S. Predicting Smart Grid Stability with Optimized Deep Models. *SN Comput. Sci.* **2021**, *2*, 1–12. [CrossRef]
79. Erdem, T.; Eken, S. *Layer-Wise Relevance Propagation for Smart-Grid Stability Prediction*; Springer: Berlin/Heidelberg, Germany, 2022; pp. 315–328.
80. Jozi, A.; Pinto, T.; Praça, I.; Vale, Z.; Soares, J. Day Ahead Electricity Consumption Forecasting with MOGUL Learning Model. In Proceedings of the 2018 International Joint Conference on Neural Networks (IJCNN), Rio de Janeiro, Brazil, 8–13 July 2018; IEEE: New York, NY, USA, 2018; pp. 1–6.
81. Arzani, A.; Boshoff, S.; Arunagirinathan, P.; Enslin, J.H. System design, economic analysis and operation strategy of a campus microgrid. In Proceedings of the 2018 International Joint Conference on Neural Networks (IJCNN), Rio de Janeiro, Brazil, 8–13 July 2018; IEEE: New York, NY, USA, 2018; pp. 1–7.
82. Uddin, M.; Romlie, M.; Abdullah, M.; Hassan, K.M.; Tan, C.; Bakar, A. Modeling of campus microgrid for off-grid application. In Proceedings of the 5th IET International Conference on Clean Energy and Technology (CEAT2018), Kuala Lumpur, Malaysia, 5–6 September 2018; pp. 1–5.
83. Lghoul, R.; Abid, M.R.; Khallaayoun, A.; Bourhnane, S.; Zine-Dine, K.; Elkamoun, N.; Khaidar, M.; Bakhouya, M.; Benhaddou, D. Towards a real-world university campus micro-grid. In Proceedings of the 2018 International Conference on Smart Energy Systems and Technologies (SEST), Seville, Spain, 10–12 September 2018; IEEE: New York, NY, USA, 2018; pp. 1–6.
84. Moura, P.; Correia, A.; Delgado, J.; Fonseca, P.; de Almeida, A. University Campus microgrid for supporting sustainable energy systems operation. In Proceedings of the 2020 IEEE/IAS 56th Industrial and Commercial Power Systems Technical Conference (I&CPS), Las Vegas, NV, USA, 29 June–29 July 2008; IEEE: New York, NY, USA, 2020; pp. 1–7.
85. Zhang, G.; Xiao, C.; Razmjooy, N. Optimal Operational Strategy of Hybrid PV/Wind Renewable Energy System Using Homer: A Case Study. *Int. J. Ambient Energy* **2022**, *43*, 3953–3966. [CrossRef]
86. Sarwar, M.; Warsi, N.A.; Siddiqui, A.S.; Kirmani, S. Optimal Selection of Renewable Energy–Based Microgrid for Sustainable Energy Supply. *Int. J. Energy Res.* **2022**, *46*, 5828–5846. [CrossRef]
87. Ajiboye, A.A.; Popoola, S.I.; Adewuyi, O.B.; Atayero, A.A.; Adebisi, B. Data-Driven Optimal Planning for Hybrid Renewable Energy System Management in Smart Campus: A Case Study. *Sustain. Energy Technol. Assess.* **2022**, *52*, 102189. [CrossRef]
88. Praveen, T.; Nishanthy, J. optimization of solar energy system for charging of ev at an institution campus. In Proceedings of the 2022 International Virtual Conference on Power Engineering Computing and Control: Developments in Electric Vehicles and Energy Sector for Sustainable Future (PECCON), Chennai, India, 5–6 May 2022; IEEE: New York, NY, USA, 2022; pp. 1–5.
89. Iqbal, S.; Jan, M.U.; Rehman, A.U.; Shafiq, A.; Rehman, H.U.; Aurangzeb, M. Feasibility Study and Deployment of Solar Photovoltaic System to Enhance Energy Economics of King Abdullah Campus, University of Azad Jammu and Kashmir Muzaffarabad, AJK Pakistan. *IEEE Access* **2022**, *10*, 5440–5455. [CrossRef]
90. Riayatsyah, T.; Geumpana, T.; Fattah, I.R.; Rizal, S.; Mahlia, T.I. Techno-Economic Analysis and Optimisation of Campus Grid-Connected Hybrid Renewable Energy System Using HOMER Grid. *Sustainability* **2022**, *14*, 7735. [CrossRef]
91. Ogbikaya, S.; Iqbal, M.T. Design and Sizing of a Microgrid System for a University Community in Nigeria. In Proceedings of the 2022 IEEE 12th Annual Computing and Communication Workshop and Conference (CCWC), Las Vegas, NV, USA, 26–29 January 2022; IEEE: New York, NY, USA, 2022; pp. 1049–1054.
92. Salmi, M.; Baci, A.B.; Inc, M.; Menni, Y.; Lorenzini, G.; Al-Douri, Y. Desing and Simulation of an Autonomous 12.6 KW Solar Plant in the Algeria's M'sila Region Using PVsyst Software. *Optik* **2022**, *262*, 169294. [CrossRef]
93. Mohamed, N.; Sulaiman, S.; Rahim, S. Design of Ground-Mounted Grid-Connected Photovoltaic System with Bifacial Modules Using PVsyst Software. *J. Phys. Conf. Ser.* **2022**, *2312*, 012058. [CrossRef]
94. Gálvez, D.M.; Kerdan, I.G.; Carmona-Paredes, G. Assessing the Potential of Implementing a Solar-Based Distributed Energy System for a University Using the Campus Bus Stops. *Energies* **2022**, *15*, 3660. [CrossRef]

95. Duarte, R.V.; Lata-García, J. Optimization of the Economic Dispatch of a Hybrid Renewable Energy System Using CPLEX. In *Communication, Smart Technologies and Innovation for Society*; Springer: Berlin/Heidelberg, Germany, 2022; pp. 623–633.
96. Li, J.; Zhao, H. Construction of an Optimal Scheduling Method for Campus Energy Systems Based on Deep Learning Models. *Math. Probl. Eng.* **2022**, *2022*, 5350786. [CrossRef]
97. Recioui, A.; Benaissa, N.; Dekhandji, F.Z. Hybrid Renewable Energy System Optimization Using IHOGA. *Alger. J. Signals Syst.* **2022**, *7*, 99–108. [CrossRef]
98. Gul, E.; Baldinelli, G.; Bartocci, P.; Bianchi, F.; Domenghini, P.; Cotana, F.; Wang, J. A Techno-Economic Analysis of a Solar PV and DC Battery Storage System for a Community Energy Sharing. *Energy* **2022**, *244*, 123191. [CrossRef]
99. Pradhan, J.D.; Hadpe, S.S.; Shriwastava, R.G. Analysis and Design of Overcurrent Protection for Grid-Connected Microgrid with PV Generation. *Glob. Transit. Proc.* **2022**, *3*, 349–358. [CrossRef]
100. Dashtdar, M.; Bajaj, M.; Hosseinimoghadam, S.M.S. Design of Optimal Energy Management System in a Residential Microgrid Based on Smart Control. *Smart Sci.* **2022**, *10*, 25–39. [CrossRef]
101. Kim, J.; Oh, H.; Choi, J.K. Learning Based Cost Optimal Energy Management Model for Campus Microgrid Systems. *Appl. Energy* **2022**, *311*, 118630. [CrossRef]
102. Elenkova, M.; Papadopoulos, T.; Psarra, A.; Chatzimichail, A. A Simulation platform for smart microgrids in university campuses. In Proceedings of the 2017 52nd international Universities Power Engineering Conference (UPEC), Heraklion, Greece, 28–31 August 2017; IEEE: New York, NY, USA, 2017; pp. 1–6.
103. Muqeet, H.A.; Javed, H.; Akhter, M.N.; Shahzad, M.; Munir, H.M.; Nadeem, M.U.; Bukhari, S.S.H.; Huba, M. Sustainable Solutions for Advanced Energy Management System of Campus Microgrids: Model Opportunities and Future Challenges. *Sensors* **2022**, *22*, 2345. [CrossRef]
104. Kourgiozou, V.; Commin, A.; Dowson, M.; Rovas, D.; Mumovic, D. Scalable Pathways to Net Zero Carbon in the UK Higher Education Sector: A Systematic Review of Smart Energy Systems in University Campuses. *Renew. Sustain. Energy Rev.* **2021**, *147*, 111234. [CrossRef]
105. Talei, H.; Essaaidi, M.; Benhaddou, D. Smart Campus Energy Management System: Advantages, Architectures, and the Impact of Using Cloud Computing. In Proceedings of the 2017 International Conference on Smart Digital Environment, Rabat, Morocco, 21–23 July 2017; pp. 1–7.
106. Hussain, A.; Bui, V.-H.; Kim, H.-M. A Resilient and Privacy-Preserving Energy Management Strategy for Networked Microgrids. *IEEE Trans. Smart Grid* **2016**, *9*, 2127–2139. [CrossRef]
107. Handschin, E.; Petroianu, A. *Energy Management Systems: Operation and Control of Electric Energy Transmission Systems*; Springer Science & Business Media: Berlin/Heidelberg, Germany, 2012; ISBN 3-642-84041-8.
108. Sexauer, J.M.; McBee, K.D.; Bloch, K.A. Applications of Probability Model to Analyze the Effects of Electric Vehicle Chargers on Distribution Transformers. *IEEE Trans. Power Syst.* **2012**, *28*, 847–854. [CrossRef]
109. Lunz, B.; Yan, Z.; Gerschler, J.B.; Sauer, D.U. Influence of Plug-in Hybrid Electric Vehicle Charging Strategies on Charging and Battery Degradation Costs. *Energy Policy* **2012**, *46*, 511–519. [CrossRef]
110. Mahmood, A.; Amjad, M.; Malik, M.; Ali, A.; Muhammad, A. Reactive Power Control of A 220kv Transmission Line Using PWM Based Statcom with Real Time Data Implementation. *Univ. Eng. Technol. Taxila Tech. J.* **2016**, *21*, 43.
111. Ali, A.; Amjad, M.; Mehmood, A.; Asim, U.; Abid, A. Cost Effective Power Generation Using Renewable Energy Based Hybrid System for Chakwal, Pakistan. *Sci. Int.* **2015**, *27*, 6017–6022.
112. Lazaroiu, G.C.; Dumbrava, V.; Costoiu, M.; Teliceanu, M.; Roscia, M. Smart campus-an energy integrated approach. In Proceedings of the 2015 International Conference on Renewable Energy Research and Applications (ICRERA), Palermo, Italy, 22–25 November 2015; IEEE: New York, NY, USA, 2015; pp. 1497–1501.
113. Xie, Y.; Lin, S.; Liang, W.; Yang, Y.; Liu, M. An Interval Probabilistic Energy Flow Calculation Method for CCHP Campus Microgrids. *IEEE Syst. J.* **2022**, *16*, 6219–6230. [CrossRef]
114. Campus Microgrid Project by LSIS at SNU—News—Newsroom—SNU NOW. Available online: https://en.snu.ac.kr/snunow/snu_media/news?md=v&bbsidx=126250 (accessed on 1 September 2022).
115. Kavousi-Fard, A.; Abunasri, A.; Zare, A.; Hoseinzadeh, R. Impact of Plug-in Hybrid Electric Vehicles Charging Demand on the Optimal Energy Management of Renewable Micro-Grids. *Energy* **2014**, *78*, 904–915. [CrossRef]
116. Ovalle, A.; Hably, A.; Bacha, S.; Ahmed, M. Voltage Support by optimal integration of plug-in hybrid electric vehicles to a residential grid. In Proceedings of the IECON 2014—40th Annual Conference of the IEEE Industrial Electronics Society, Dallas, TX, USA, 29 October–1 November 2014; IEEE: New York, NY, USA, 2014; pp. 4430–4436.
117. Aram, A. Microgrid Market in the USA. *Glob. Innov. Rep.* **2017**, *2017*, 2630.
118. Dehkordi, N.M.; Baghaee, H.R.; Sadati, N.; Guerrero, J.M. Distributed Noise-Resilient Secondary Voltage and Frequency Control for Islanded Microgrids. *IEEE Trans. Smart Grid* **2018**, *10*, 3780–3790. [CrossRef]
119. Levron, Y.; Shmilovitz, D. Power Systems' Optimal Peak-Shaving Applying Secondary Storage. *Electr. Power Syst. Res.* **2012**, *89*, 80–84. [CrossRef]
120. Díaz-González, F.; Sumper, A.; Gomis-Bellmunt, O.; Villafáfila-Robles, R. A Review of Energy Storage Technologies for Wind Power Applications. *Renew. Sustain. Energy Rev.* **2012**, *16*, 2154–2171. [CrossRef]
121. Subburaj, A.S.; Pushpakaran, B.N.; Bayne, S.B. Overview of Grid Connected Renewable Energy Based Battery Projects in USA. *Renew. Sustain. Energy Rev.* **2015**, *45*, 219–234. [CrossRef]

122. Cui, S.; Wang, Y.-W.; Xiao, J.-W.; Liu, N. A Two-Stage Robust Energy Sharing Management for Prosumer Microgrid. *IEEE Trans. Ind. Inform.* **2018**, *15*, 2741–2752. [CrossRef]
123. Husein, M.; Chung, I.-Y. Optimal Design and Financial Feasibility of a University Campus Microgrid Considering Renewable Energy Incentives. *Appl. Energy* **2018**, *225*, 273–289. [CrossRef]

Disclaimer/Publisher's Note: The statements, opinions and data contained in all publications are solely those of the individual author(s) and contributor(s) and not of MDPI and/or the editor(s). MDPI and/or the editor(s) disclaim responsibility for any injury to people or property resulting from any ideas, methods, instructions or products referred to in the content.

Review

Digitalization, Industry 4.0, Data, KPIs, Modelization and Forecast for Energy Production in Hydroelectric Power Plants: A Review

Crescenzo Pepe * and Silvia Maria Zanoli *

Dipartimento di Ingegneria dell'Informazione, Università Politecnica delle Marche, Via Brecce Bianche 12, 60131 Ancona, Italy
* Correspondence: c.pepe@univpm.it (C.P.); s.zanoli@univpm.it (S.M.Z.)

Abstract: Intelligent water usage is required in order to target the challenging goals for 2030 and 2050. Hydroelectric power plants represent processes wherein water is exploited as a renewable resource and a source for energy production. Hydroelectric power plants usually include reservoirs, valves, gates, and energy production devices, e.g., turbines. In this context, monitoring and maintenance policies together with control and optimization strategies, at the different levels of the automation hierarchy, may represent strategic tools and drivers for energy efficiency improvement. Nowadays, these strategies rely on different basic concepts and elements, which must be assessed and investigated in order to provide a reliable background. This paper focuses on a review of the state of the art associated with these basic concepts and elements, i.e., digitalization, Industry 4.0, data, KPIs, modelization, and forecast.

Keywords: hydroelectric power plant; digitalization; Industry 4.0; data; KPIs; modelization; forecast

1. Introduction

The efficiency of large production facilities, in terms of energy and more generally in terms of the use of the resources exploited, is of primary importance in a world that aims to respect nature and that desires to limit air pollution and the waste of environmental resources. Increasing alarm over current climate change and the future of the planet is leading to a shift away from nonrenewable resources (e.g., fossil fuels). Renewable energy source (RES) exploitation is encouraged, and this encouragement contributes to decarbonization. Projects on energy sustainability are shifting political–economic focus from fossil fuel consumption to the use of alternative and less polluting energies, such as wind, hydropower or solar energy [1–3].

Water represents a key RES, and its rational usage represents a crucial milestone when targeting the challenging goals for 2030 and 2050 [4,5]. Water resource systems require advanced and innovative solutions to guarantee the needed performance required to focus on, assess and approach these objectives. Water is exploited in many sectors, e.g., water distribution networks (WDNs) [6,7] and renewable energy production. Among the various types of renewable energy that exploit water, hydroelectricity is certainly of fundamental importance. Hydropower is a type of renewable energy generation based on the production of electricity via exploitation of the movement of large masses of water. Electricity is generated thanks to the action and to the force of these masses, moved by gravity within penstocks. In fact, moving or falling water generates kinetic energy and this kinetic energy is converted into electricity by plants equipped with turbines, generators and transformers. It is indeed essential to make full use of available resources, e.g., water from river courses, in order to avoid waste and maximize plants' efficiency [8].

Hydroelectric power plants are experiencing strong growth in global energy production, partly due to the exploitation of innovative technologies that have a focus on electricity

Citation: Pepe, C.; Zanoli, S.M. Digitalization, Industry 4.0, Data, KPIs, Modelization and Forecast for Energy Production in Hydroelectric Power Plants: A Review. *Energies* **2024**, *17*, 941. https://doi.org/10.3390/en17040941

Academic Editors: Chiara Martini and Claudia Toro

Received: 8 December 2023
Revised: 10 February 2024
Accepted: 13 February 2024
Published: 17 February 2024

Copyright: © 2024 by the authors. Licensee MDPI, Basel, Switzerland. This article is an open access article distributed under the terms and conditions of the Creative Commons Attribution (CC BY) license (https://creativecommons.org/licenses/by/4.0/).

production monitoring and the proper use of water resources [9,10]. Hydropower plants can take the form of a reservoir (water is managed through one or more reservoirs) or run-of-river (placed directly on the course of rivers). Within reservoir hydropower plants, a particular subcategory is that represented by pumped storage stations. In this type of plant, reservoirs are located at significantly different vertical levels and water is brought back to the upper reservoirs through pumps. The output of a plant, on the other hand, depends basically on two factors, i.e., flow rate and head. The mass of water flowing through a point in the unit of time is named as flow rate and the head is the difference in height that exists between the elevation at which the water resource is available and the level at which it is returned, after passing through the turbine. Hydroelectric power plants are characterized by electrical, hydraulic and mechanical coupling effects and their management can also involve market conditions and incentives [11–13]. Different components can be included in hydroelectric power plants, including rivers, intakes, regulation gates, water collection reservoirs, sand traps, turbines, floodgates, and dams. Figure 1 depicts some typical elements of a hydroelectric power plant. Figure 2 depicts a general scheme of a hydroelectric power plant.

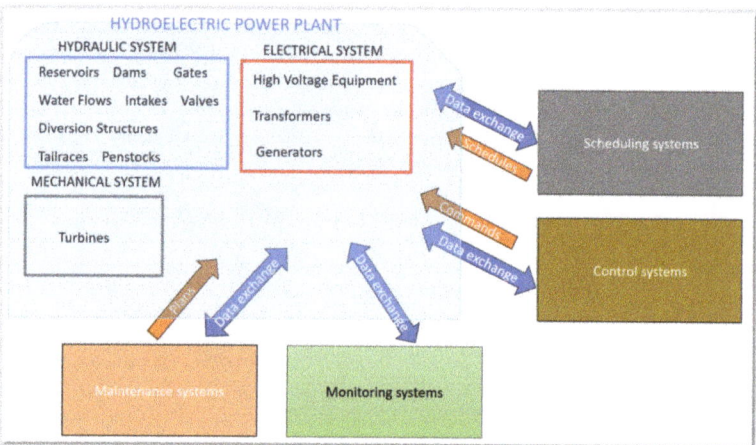

Figure 1. Overview of the general features of a hydroelectric power plant and of the different systems that can be installed on a hydropower plant.

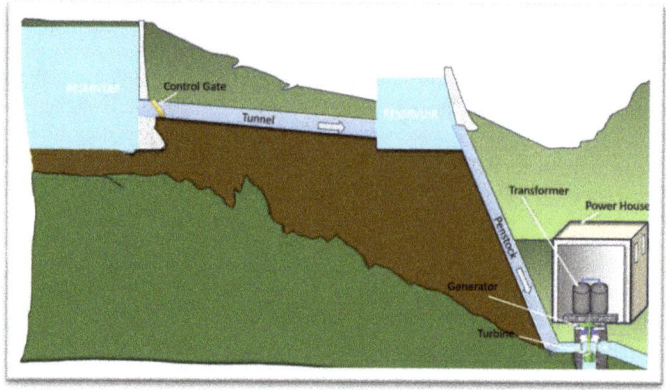

Figure 2. Example of a hydroelectric power plant.

In the last decades, equipment in hydropower plants has been enhanced in order to target high levels of availability, performance, and flexibility, and these results have been contributing to the energy transition. The efficiency of hydroelectric power plants has been improved thanks to the key role of Industry 4.0 and digitalization in monitoring, maintenance, control, and optimization applications [14–18]. Tailored cross-fertilization procedures can be applied in order to exploit the available knowledge on these topics in other fields, thus customizing them for the hydropower sector. Monitoring, maintenance, control and optimization systems are included in Figure 1, together with some typical information exchanged with the hydroelectric power plant.

The design, manufacture and installation of hydropower plants require a large initial investment. Moreover, in many cases, the revenue that can be obtained depends on the weather; periods of water scarcity, resulting in low inflows from rivers, could limit the production of this type of plant. In this sense, and similar to other types of renewable energy production, effective production is not always guaranteed for hydropower and, as a consequence, management of the plants is not a trivial task [19]. Energy efficiency is a crucial need in the management of hydropower plants. Energy efficiency can be associated with the single components of the plants and/or subparts of the plants; tailored key performance indicators (KPIs) must be defined and exploited for efficiency certification. In order to optimize the plants' management with regard to different aspects, tailored monitoring, maintenance, control, and optimization procedures must be applied.

Different review papers are present in the current literature of hydroelectric power plants. Some of the proposed topics are:

- Assessment of innovative technological aspects for hydropower plants, focusing on research and development (R&D) [9];
- Climate change mitigation and adaptation for the hydropower sector [20,21];
- Energy revolution for the hydropower sector [22];
- Analysis of the greenhouse gas (GHG) emissions in the hydropower sector [23];
- Operation strategies for hydropower plants [24–29];
- Methods for the solution of scheduling problems in hydroelectric power plants [30–35];
- Modelization and control in hydroelectric power plants [26,27,36–38];
- Hydropower case study collection [39].

Table 1 reports the topic and the main findings of the previously mentioned review papers.

Table 1. Main features of some review papers for the hydropower sector.

Topic	Main Findings	Ref.
Analysis of emerging technologies in the hydropower sector.	Review of R&D aspects in the hydroelectric power sector, including digitalization and operation.	[9]
Climate change mitigation and adaptation.	Significant role of hydropower technology in the mitigation of climate change and its crucial role in the adaptation of the availability of water resources to climate change.	[20]
Climate change mitigation and adaptation.	Assessment of the relationship between climate change and the hydropower sector. Identification of methods for improvement on the analysis of the net pros of hydropower technology subject to climate change.	[21]
Energy revolution.	Revealing the as-yet untapped potential of the hydropower sector in most areas of Europe and of its contribution to future energy needs.	[22]

Table 1. *Cont.*

Topic	Main Findings	Ref.
GHG emissions.	Discussion on GHG emissions from hydroelectric reservoirs, mitigation techniques, methodological know-how, and relationship between the parameters affecting GHG emissions. Review of crucial approaches and methods for the prediction of GHG emissions associated with reservoirs, investigating also life cycle assessment, uncertainty sources, and knowledge gaps.	[23]
Operation strategies.	Review of the optimization of operation in hydropower plants with a focus on minimizing the costs, minimizing the environmental impact, and maximizing energy generation.	[24]
Operation strategies.	Evaluation of revenue maximization focused on head impact and price values. Comparison between mean yearly energy generation and mean yearly revenue. Identification of benefits provided exploiting various operation methodologies.	[25]
Operation strategies, modelization and control.	Investigation of applications, control, operation, modeling and environmental impacts of hydroelectric power with a focus on power systems.	[26]
Operation strategies, modelization and control.	Focus on the priorities of the European Union (EU) with regard to applications, case studies, challenges, limitations, benefits, technology readiness level (TRL), and transversal pros associated with predictive operation and maintenance (O&M) and energy generation	[27]
Operation strategies.	Review of impact and role of distributed flexible AC transmission system (D-FACTS) devices in the function of microgrids.	[28]
Operation strategies.	Review of retrofit and upgrade of hydro power plants through the analysis of research papers, reports, guidelines and standards.	[29]
Scheduling problem.	Survey on optimal scheduling generation methodologies with a focus on meta-heuristic optimization methods.	[30]
Scheduling problem.	Review of machine learning (ML) for short-term hydropower scheduling, considering the cyber–physical systems (CPSs) paradigm.	[31]
Scheduling problem.	Review of mathematical programming methods for the solution of the short-term scheduling problem on single hydropower units. Classification of different techniques for the modelling of constraints and objectives.	[32]
Scheduling problem.	Review of the hydropower scheduling problem with focus on ML and artificial intelligence (AI) applications for optimization, prediction, and scheduling.	[33]
Scheduling problem.	Review of ML techniques for hydropower operation optimization, with a focus on optimal dispatch of hydropower plants in the context of energy production enhancement.	[34]
Scheduling problem.	Review of hydropower optimization R&D activities using a metaheuristic approach, also considering scheduling problems.	[35]
Modelization and control.	Review of modelling and control systems development of the hydro turbine.	[36]
Modelization and control.	Review of models, stability analysis and control methods for hydro turbine governing systems.	[37]
Modelization and control.	Review of data-driven methods for modeling plant operations.	[38]
Hydropower case study collection.	Discussion of significant case studies on hydropower installations of different companies, highlighting decarbonization and ecosystem protection aspects.	[39]

The present paper aims to supply a comprehensive review of the existing literature on the basic concepts and elements needed for the design of monitoring, maintenance, control and optimization strategies with a focus on energy production. Some basic concepts and elements, if in-depth approached, may provide the ingredients for the construction and implementation of these strategies. Based on the authors' knowledge and based on the literature analysis of review papers reported in Table 1, a thorough overview on the proposed topics is not present on the literature. Figure 3 reports a word cloud containing the most important words of the paper.

Figure 3. Word cloud containing the most important words of the review paper.

The organization of the paper is as follows: Section 1.1 details the methodology used for the conduction of the literature analysis. Section 2 focuses on digitalization and Industry 4.0 for the hydropower sector. Section 3 analyzes data and KPIs while modelization and forecast are detailed in Section 4. Section 5 reports some discussion while conclusions and future research directions are reported in Section 6.

1.1. Methodology for the Conduction of the Literature Analysis

For the identification of the documents to be analyzed in the proposed review, different databases have been explored, i.e., IEEE Xplore, Scopus, Web of Science, Google Scholar, Springer Link, and MDPI. In addition, Google search has been exploited.

Different search strings were used to identify significant documents. An example of one of these search strings is "hydropower" AND "word". "word" was replaced based on the effective topic to be investigated, e.g., "Industry 4.0". Starting from the available documents, a selection based on the impact and on the relevance to the considered topics was performed. A total of 159 references was considered.

Figure 4 reports the number of reviewed and investigated documents associated with previous review works in the hydropower sector (see Table 1), based on the year of publication. The total number of documents is 21. Among the 159 references, 115 documents on hydropower plants were exploited for the technical core of the paper (Sections 2–4). Figure 5 reports the number of reviewed and investigated documents on hydropower plants that were exploited for the technical core of the paper (Sections 2–4), based on the year of publication. Figure 6 reports the number of reviewed and investigated documents on hydropower plants that were exploited for the technical core of the paper (Sections 2–4), based on the topic. Some documents were considered for more than one topic, so these documents are considered more than once in Figure 6. In addition, all of the references considered for each topic are summarized in Table 2.

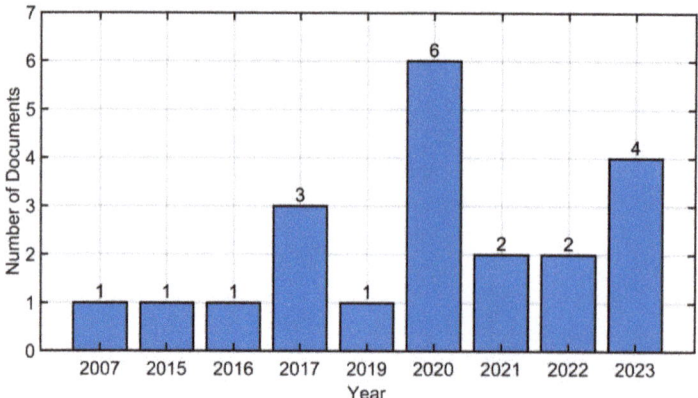

Figure 4. Bar graph showing the number of reviewed and investigated documents (previous review works) based on the year of publication.

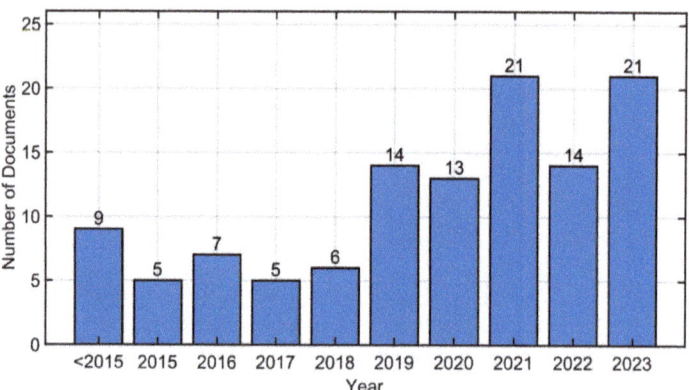

Figure 5. Bar graph showing the number of reviewed and investigated documents (hydropower plants) based on the year of publication.

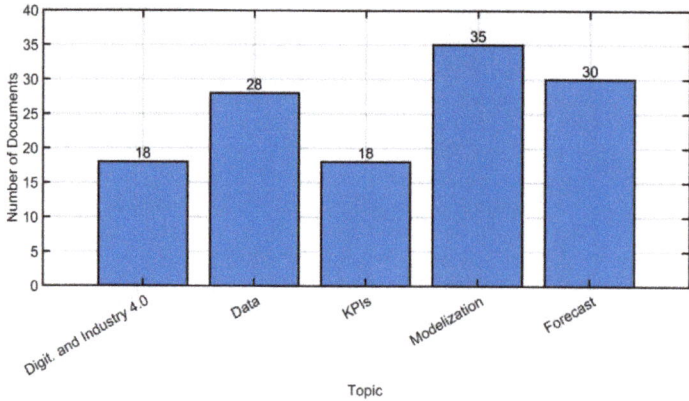

Figure 6. Bar graph showing the number of reviewed and investigated documents (hydropower plants) based on the topic.

Table 2. References associated with each topic.

Topic	Associated References
Digitalization and Industry 4.0	[9,15,16,27,39–55]
Data	[9,31,41,47,53,56–80]
KPIs	[6,7,75,81–101]
Modelization	[57,75,94–96,102–136]
Forecast	[75,94–96,117,127,130–132,137–157]

2. Role of Digitalization and Industry 4.0 in the Hydropower Sector

In this section, the role of digitalization and Industry 4.0 in the hydropower sector is analyzed. In particular, some research works about these themes are reported, with a focus on the main topic, on the scope and on the main findings.

The concept of Industry 4.0 was created as part of the fourth industrial revolution. Industry 4.0 represents the way to digitalization and it is focused on the encapsulation of the "digital industry" concept in different industrial areas. Applying cross-fertilization procedures to the sectors that exploit RESs, the Industry 4.0 paradigm can allow conventional plants to evolve into smart plants. In particular, Industry 4.0 technologies, e.g., the internet of things (IoT), simulation, cloud computing, augmented reality (AR), big data analytics, CPSs, cybersecurity, blockchain, AI and ML, can contribute to the improvement of the energy processes [15]. In the hydropower sector, digitalization can allow remote monitoring that is oriented to diagnostic actions to be executed for sustainable operation. This digital transformation can be coupled to modernization actions in order to support energy flexibility [16]. This potential could be effectively verified if the field implementation of digitalization takes place in the hydropower sector. In this context, a revolution is occurring because a large number of hydropower plants were designed a long time ago and their design working conditions consequently differ from current scenarios [9]. Practical projects can be implemented in order to exploit Industry 4.0 and digitalization to maintain and monitor hydropower plants. Some examples of applications are condition monitoring, fault detection, and operations' optimization aimed at efficiency improvement [39,40]. These projects could lead to the implementation of Industry 4.0 platforms, which can then contribute to risk reduction and to a sustainable energy supply through the facilitation of data consultation and analysis [41]. In addition, digital technology platforms may represent a key tool in the proposal of pioneering business models [42].

Digitalization in the hydropower sector includes digital technologies, e.g., systems, infrastructures, networks, sensors, and connected devices [43–45]. The digital technologies aim to improve efficiency and/or to increase the competitiveness of the business models. Digitalization provides access to new sources of data and communications, while exploiting decision support systems (DSSs), automation, and control. Digitalization tools, aimed at the convergence between information technology (IT) and operation technology (OT), can be exploited for design, construction, investment decisions, O&M, rehabilitation, and modernization. Some fields of interest are safety, sustainability, and commerce [46]. In [47], five key themes are claimed as drivers of the digital transition of hydropower plants: production management and optimization, asset analytics, process improvement and automation, connected workers, and cybersecurity. Digitalization activates an improved reliability of forecasting associated with internal factors (e.g., reservoir level), external factors (e.g., water inflow and weather), and market conditions (e.g., price and demand). Predictions that are more reliable can play a key role in the automation, control, optimization, and maintenance fields [48]. Power production, spillage minimization and energy sources integration represent some of the benefits that could be obtained. In addition, the selection, acquisition, storage, analysis and visualization of data on O&M of assets represent drivers for the transition from periodic maintenance to predictive maintenance. Furthermore, IoT can contribute to process improvement and automation, enabling remote operations which can in turn improve safety and reduce costs [49]. In this context, the connected worker is a fundamental requirement. As a main drawback, data protection can be subjected to threats and cybersecurity approaches are needed [47]. In [50], digitalization is defined as a modernization technology for hydropower plants' upgrade. In particular, a hydropower gain can be obtained with regard to the provided power and to the optimization of storage. In [51], the digitalization process in hydropower plants is defined as a driver for asset performance management, equipment condition monitoring, outage management, supervisory control and data acquisition (SCADA) systems, IoT, AI, and ML. In addition, the concept of "Hydropower 4.0" is mentioned. All of these features allow one to obtain an enhanced flexibility of operations in hydropower stations and an enhancement of the digital tools in order to fully exploit the resources and to provide upgraded digital security levels [52]. Industry 4.0-enabling technologies, e.g., IoT, ML and big data analytics, can be successfully combined with tools provided by other areas, e.g., photogrammetry [53]. Connecting and contaminating different research areas, it can be observed that the digital and the green transitions are interconnected challenges. In this context, digitalization plays a key role in the path toward these transitions. Environmental and energy benefits can be obtained through this process in the hydropower sector, e.g., hydropeaking mitigation, and water management improvement [27]. Tailored methodologies can be used to evaluate the potential economic benefits provided by digitalization in hydropower generation and operation. Increase of efficiency, improvement of the reservoir operation and damage prevention can be taken into account together with the fact that the hydropower sector can be associated with other sectors in a water–energy–food ecosystem [54]. Optimal trade-offs between empowering renewable energy and the reduction of the costs in hydropower plants must be ensured;, hydropower trading and predictive maintenance on hydro plants are methodologies that can be accelerated by digital transformation [55].

Table 3 reports some research work about digitalization and Industry 4.0 in the hydropower sector, with a focus on the main topic, on the scope and on the main findings.

Table 3. Main features of the selected papers regarding digitalization and Industry 4.0 in the hydropower sector.

Main Topic	Scope	Main Findings	Ref.
Optimization opportunities in the digital revolution for renewable energy sector.	Evaluation of the impact of the Industry 4.0 paradigm within RESs energy generation sectors.	Industry 4.0 can enhance the existing plants and offer some opportunities to increase efficiency, e.g., sustainable circular economy aimed at waste management.	[15]
Digitalization and modernization of hydropower operating facilities.	Define the actions for the extension of the lifetime of a hydropower plant.	Evaluation of the residual life of generators and retrofit of components can help in the modernization of hydropower plants. Digitalization can allow remote monitoring oriented to diagnostic actions to be executed during sustainable operation.	[16]
Analysis of emerging technologies in the hydropower sector.	Collection of information on the challenges, innovation trends and emerging hydropower technologies.	There is an emerging need to digitalize hydropower design and operation. In this sense, a revolution in hydropower plants is expected.	[9]
Hydropower case study collection.	Discussion regarding a number of representative examples of hydropower installations from different companies, highlighting decarbonization and ecosystem protection aspects.	Digitalization can improve efficiency in hydropower plants through different ways, e.g., preventing failures and optimizing their operation.	[39]

Table 3. *Cont.*

Main Topic	Scope	Main Findings	Ref.
Report on an Industry 4.0 and digitalization research project applied to maintenance and fault detection of hydropower plants.	Provision of a set of methods for the implementation of Industry 4.0 and digitalization on hydropower plants	Proof of the potential of Industry 4.0 and digitalization methods for the optimization of maintenance through condition monitoring and fault detection.	[40]
Case report on a developed commercial digital transformation process for a hydropower plant.	Implementation of a modern platform based on Industry 4.0 tools in order to facilitate consultation and analysis of data for the management of applications of hydrological, operational, and market information, and commercial information systems and platforms.	The developed activities contribute to risk reduction and to a sustainable energy supply.	[41]
Report on the exploitation of digital technology platforms in an RES context.	Investigation of the impact of digital technology platforms on the implementation of innovative business models in the RES sector.	Digital technology platforms play a key role in the development of innovative business models.	[42]
Report on the digital revolution of hydropower in Latin American countries.	Investigation of the digitalization status in hydropower plants of Latin American countries.	Assessment of the level of digitalization in hydropower sector in Latin American countries through the detection of barriers and challenges of incorporating digitalization.	[46]
Report on the opportunity to transform the future of hydropower sector through digital generation.	Investigation of challenges of digitalization and its risks.	Assessment of the current challenges toward a digitally enabled and data-driven operation of hydropower plants. Assessment of the risks to be taken into account and to be mitigated within the digital transition.	[47]
Report on the technical route, maturity evaluation and content planning associated with the digital transformation of hydropower plants	Study and analysis of the technical route, the maturity evaluation and construction planning of hydropower digital transformation.	Assessment of the problems encountered within the digital transformation process. Definition of a digital maturity evaluation model for hydropower systems. Assessment of digital transformation planning, including top-level design, information and communication standards, and intelligent application.	[48]
Report on digital technologies for hydropower plant operation.	Study on the impact of digitalization, contemporary technologies and optimization on hydropower plants operations.	Enhanced flexibility, better safety, larger energy production and lower costs of maintenance can be targeted through the implementation of digital technologies.	[49]
Report on the energy potential of modernizing the European hydropower fleet.	Evaluation of different options for upgrading existing facilities, excluding measures that are expected to increase hydromorphological pressure on water bodies.	Indicative estimation of the additional annual generation and power that could be obtained compared with the current condition if modernization techniques are implemented.	[50]

Table 3. Cont.

Main Topic	Scope	Main Findings	Ref.
Report on state-of-the-art hydropower technologies.	Provision of an overview of the state of the art of hydropower technologies and techniques, in order to set a baseline reference to identify and prioritize future R&D actions.	Process analysis and assessment of recent technologies for hydropower plants aimed at enhancing flexibility, improving efficiency and resilience, and optimizing O&M. Discussion of current techniques and technologies by which to mitigate the environmental impact of hydropower projects.	[51]
Report on the most recent advances in hydropower technology.	Presentation of a set of research activities for hydropower plants.	Focus on technological projects. Focus on studies aimed at the improvement of the simulations of the hydrological cycle and climate change associated with hydropower operation.	[52]
Case study for the design of a small hydropower plant.	Exploitation of IoT, photogrammetry, ML and big data analytics as support tools toward the digital transition.	Evaluation and application of the proposed design method based on data analysis, cost estimation, and revenues.	[53]
Digitalization and real-time control of rivers for the mitigation of environmental impacts.	Discussion on the development of different technologies for the hydropower sector.	Focus on the priorities of the EU with regard to challenges, case studies, applications, limitations, benefits, and TRL.	[27]
Digitalization of the European water sector to nudge the digital and green transitions.	Investigation of advantages provided by digital solutions in the hydropower sector in terms of economic benefits for operation and generation.	Computation of the benefits for EU states and for UK (excluding environmental and social benefits).	[54]
Digital transformation in renewable energy.	Provision of significant use cases and experiences with regard to a Nordic power producer.	The digital transformation can act for the empowering the value of renewable energy and to the reduction of the costs. Examples of significant use cases are hydropower trading and predictive maintenance on hydro plants.	[55]

3. Role of Data and KPIs in the Hydropower Sector

In this section, the role of data and KPIs in the hydropower sector is analyzed. In particular, some research works about these themes are reported, with focuses on the main topic, on the scope and on the main findings.

Thanks to Industry 4.0 and digitalization, data selection, acquisition, and storage is acquiring high relevance in the hydropower sector. The overall hydropower fleet can benefit from data gathering and from data analysis [9]. The development of information technology in the hydropower sector has resulted in abundant data resources, thus introducing the hydropower big data era [56]. In this context, big data analytics can play a crucial role at all O&M and business levels through the harnessing of data for proactive decision-making [9]. Cross-fertilization procedures can be applied from industrial sectors in order to define efficient data analysis methodologies [57]. The collection and analysis of data associated with the overall equipment of hydropower plants represents a challenging objective. These steps can provide huge advantages thanks to the information contained in the data [31]. Consultation and analysis of data through specific platforms is a real added value for hydropower plants [41]. Decision-making could benefit from these platforms thanks to the integration of big data collection, storage, and analysis for monitoring purposes [58]. Data-driven O&M and business management may represent a nudge for energy efficiency improvement and O&M cost reduction [47]. In order to acquire a large amount of data, different technologies can be used, e.g., IoT, photogrammetry, and satellite [53,59,60]. Big data can be exploited for different types of analysis, e.g., the investigation of potentially untapped hydropower exploitation based on the exploration of the energy markets [61]. A strategic combination of cloud computing, big data and IoT can support an intelligent hydropower enterprise construction process. Efficient and scalable unified data management and analysis platforms can guarantee intelligent, safe, efficient and economic operation and can support command and decision-making [62]. In order to exploit all of the needed data for a defined scope, extensive databases may be needed. In these databases, data associated with the crucial general information and process variables in terms of plant operation are stored. For example, spatial distribution and georeferenced data may represent important sources of information for preliminary analysis [63,64], while rainfall data can support the decisions [65]. On the other hand, extensive datasets may represent a driver for hydropower system modelling purposes [66]. Data on full operating ranges may represent an added value because they may be able to cover a large number of working conditions [67]. In this context, data mining can play a significant role. It would introduce big data characteristics, connotation and origin, allowing in-depth analysis to be performed and providing a basis for the principles of large data associated with hydropower science [68]. Data mining techniques can be integrated in a SCADA system in order to speed up the data processing and allow an acceleration of the decision processes to the decision makers [69]. In fact, decision makers are not usually able to work with large databases or to extract information in a short time horizon. The extracted information can be used for different purposes, e.g., scheduling [70]. An issue to be tackled with big data is the analysis of the data quality. Data quality can be subjected to different issues, e.g., failures of sensors, actuators and data transmission systems, and defects of data storage systems. Data reliability is a fundamental requirement when extracting the needed information from the available data. In this context, data cleaning represents a strategy by which to detect abnormal conditions within data and to correct them [71]. In order to fully exploit data potential, data classification and classification management aspects are important steps. These phases can be based on databases and aim to provide support for data analysis [72]. In addition, in the exploration of massive data, data island and data heterogeneity issues can arise, and customized platforms are thus required [73].

The added value provided by data can also be appreciated in terms of the long-term analysis associated with market and business development for hydropower plants. For example, a recent contaminated combination between the data of different research areas, e.g., market and policy, showed an expected slowdown of the growth of the hydropower sector in the decade 2021–2030 [74].

Data analysis, selection, storage, and acquisition can also assume a crucial role for the feasibility study of an advanced process control (APC) project in order to perform an in-depth plant inspection with the help of information related to that plant's devices [75]. Another field of application of data analysis in hydropower plants can be the evaluation of silt erosion in order to intelligently plan maintenance scheduling activities. In this context, clustering and ML can be exploited [76]. Massive data information can also be used for the development of digital twins aimed at fault diagnosis in hydropower plants [77]. For the creation of advanced systems like an APC system and digital twins in a hydropower CPS, networks of connected subsystems are needed. For example, digital twins need to simulate the real world in cyber space. Real-time connections between physical and digital systems are required and cybersecurity represents a challenging objective [78]. Cyber-attacks and cyber threats must be managed through multilevel protection architectures, which must be characterized by encryption, access control, and secure virtual private networks [79]. Suitable standards, e.g., the IEC 62443 standard, can be exploited for the development of cyber security systems in hydropower plants with different generation units [80].

Table 4 reports some research works about data in the hydropower sector, with focuses on the main topic, on the scope and on the main findings.

Table 4. Main features of the selected papers focused on data for the hydropower sector.

Main Topic	Scope	Main Findings	Ref.
Analysis of emerging technologies in the hydropower sector.	Collection of information on the challenges, innovation trends and emerging hydropower technologies.	The overall hydropower fleet can benefit from data gathering and from data analysis. In this context, big data analytics can play a crucial role at all O&M and business levels through harnessing data for proactive decision-making.	[9]
Big data assessment for the hydropower sector.	Exploration for promoting the exploitation of big data and the associated resources.	The exploration, analysis and application of big data will retain an important role in hydropower facilities, thus providing new opportunities and challenges.	[56]
Review of ML techniques for hydropower scheduling.	In-depth assessment of the state-of-the-art of ML for scheduling in hydropower plants.	The collection and analysis of data associated with the overall equipment of hydropower plants represents a challenging objective. These steps can provide huge advantages thanks to the information contained in the data.	[31]
Report on a developed commercial digital transformation process for a hydropower plant.	Implementation of a modern Industry 4.0 platform in order to facilitate consultation and analysis of data for the management of applications of hydrological, operational, and market information, and commercial information systems and platforms.	The developed activities contribute to risks' reduction and to a sustainable energy supply.	[41]
Design of university hydropower intelligent decision service platform.	Implementation of a big data-driven platform for decision-making support based on data collection, storage, and analysis.	Design of a platform architecture with explicit references to technologies and tools.	[58]
Report on the opportunity to transform the future of hydropower sector through digital generation.	Investigation of challenges of digitalization and its risks.	Assessment of the current challenges toward a digitally enabled and data-driven operation of hydropower plants. Assessment of the risks to be taken into account and to be mitigated within the digital transition.	[47]
Case study for the design of a small hydropower plant.	Exploitation of IoT, photogrammetry, ML and big data Analytics as support tools toward the digital transition.	Evaluation and put into practice of the proposed design method based on data analysis, cost estimation, and revenues.	[53]
Assessment of the reliability of the hydropower sector in Malawi.	Design and validation of an energy-climate-water system associating remotely sensed data from multiple satellite missions and instruments and field observations.	A framework for modelling exploiting open-access data from satellites and based on ML algorithms and regression analysis can provide data and enhance vulnerabilities explanation.	[60]
Big data analysis for the hydropower sector.	Investigation of the hydropower development potential within ASEAN-8.	Existence of an untapped potential for hydropower development. The potential development must assume an international power exchange.	[61]
Research on intelligent construction of hydropower enterprises.	Provision of a strategic framework and scheme of intelligent construction of hydropower enterprises.	Cloud computing, big data and IoT can reinforce operation, command and decision-making in hydropower enterprises.	[62]

Table 4. *Cont.*

Main Topic	Scope	Main Findings	Ref.
Spatial distribution and key parameters of hydropower plants.	Investigation of spatial distribution and key parameters of bio- and river hydro powerplants in Germany.	Provision of a dataset that includes the spatial configuration, capacity, and year of commissioning of river- and bio hydropower plants in Germany.	[63]
Enhance the consideration of data for decision support tools.	Investigation of hydropower plants in Africa.	Georeferenced database on existing and proposed plants in Africa.	[64]
Quantitative precipitation estimates.	Investigation of quantitative precipitation estimates and the needed checks for ensure good data quality.	Provision of a dataset that contains precipitation statistics calculated by German Weather Service and their combination with rainfall statistics through rain gauge data.	[65]
Enhance the consideration of data for power system modelling purposes.	Creation of a dataset for European hydropower plants modelling.	Provision of a dataset that collects some information on the European hydro-power plants.	[66]
Digitalization in hydropower generation.	Development of a model-based Smart Power Plant Supervisor.	Digital tools can play a crucial role with regard to the optimization of O&M to enhance the supply of auxiliary services to the electric system.	[67]
Big data and data mining in the hydropower sector.	Investigation of the hydropower big data science and technology based on data mining.	Provision of a value analysis application prospect of hydropower science large data.	[68]
Enhance operation of hydropower plants through data mining.	Empowering of a SCADA system through the inclusion of data mining algorithms.	Implementation of a SCADA system with the integration of data mining techniques to support decision makers.	[69]
Enhance the scheduling operations in the hydropower sector through data mining.	Investigation of the application of data mining and clustering techniques for scheduling pattern identification.	Implementation of a rule-based procedure aimed at daily generation scheduling enhancement.	[70]
Research on data cleaning method for hydropower plants.	Investigation of data quality enhancement through data cleaning on dispatch and operation datasets.	Design and implementation of a data cleaning framework.	[71]
Big data management for hydropower enterprises.	Investigation of classification issues for big data.	Development of a data classification and classification management framework for hydropower enterprises big data.	[72]
Massive data value exploitation in hydropower stations.	Investigation of the construction of a platform for the development of a smart hydropower station.	Assessment of the origin, definition, and development of data middle platform for smart hydropower station. Test of the designed system on practical application scenarios.	[73]
Exploration of the data related to market for hydropower plants.	Data-driven investigation of past, present and future of hydropower sector, with outlook to 2030.	Based on today's policy settings, the growth of hydropower globally is set to slow in the decade 2021–2030.	[74]
Reservoir advanced process control (APC) for hydroelectric power production	Design and implementation of an APC system for water management in a hydroelectric power plant.	Definition of data-driven methods for in-depth plant study before the design of an APC system. The selection, acquisition, storage, and analysis of data cover a crucial role in the plant study, together with an in-depth analysis of the devices.	[75]
Application of ML for silt data analysis in hydropower plants.	Exploitation of ML for the reduction of the risk caused by the silt erosion.	Based on silt data analysis, maintenance planning of hydropower plants can be obtained.	[76]

Table 4. *Cont.*

Main Topic	Scope	Main Findings	Ref.
Digital twin for hydropower plants.	Construction and application of a data-driven model for different purposes.	The model is investigated with a hydropower fault diagnosis application.	[77]
Digital twins for energy systems and smart cities.	Review of literature and practices of digital twin in energy systems and smart cities.	Assessment of cybersecurity, efficiency, sustainability and reliability through digital twin.	[78]
Security in automated control systems of hydropower engineering facilities.	Assessment of the concept of CPS to manage computer security of automated process control systems at hydropower engineering facilities.	Security systems can be integrated in an efficient manner with automated process control systems of hydropower engineering facilities.	[79]
Cybersecurity in hydropower plants.	Assessment of cybersecurity in hydropower plants.	Exploitation of the standard IEC 62443-2-1 for the implementation of a cybersecurity management system on a hydropower plant with two generation units.	[80]

In the context of Industry 4.0 and digitalization, thanks to the added value provided by massive data, KPIs represent a fundamental tool in the hydropower sector. Tailored KPIs must be defined based on the studied context. KPIs can be associated with different objectives, e.g., business, optimization, control, and maintenance. Generally, KPIs can be exploited to evaluate whether a project really provides benefits in terms of efficiency and optimal performance. As mentioned above, a project can be referred to different areas. KPIs represent a metric by which to evaluate performance. Examples of general project KPIs in hydroelectric power plants are availability, revenue, return on investment, efficiency, energy output, capacity factor, and generation capacity. KPIs can support a wide variety of economic, ecological, cultural, and social objectives [81].

In order to enhance a system's availability, i.e., the percentage of current uptime considering the total service hours, tailored plant models can be obtained using the reliability block Diagram and of the stochastic block diagram. This approach can support decision makers in the optimization of maintenance management [82]. In addition, intelligent mitigation measures can be adopted in order to manage the trade-off between economic constraints, climatic variability, and increased demand. These measures can be associated with tailored indicators that explain the production reliability of hydropower plants [83]. Plant modelization can be used also to obtain tailored sustainability KPIs in order to assess the sustainability of hydropower plants [84]. Generally, KPIs can be formulated based on tailored process models in order to achieve a global evaluation of the performance of hydropower plants [85]. For example, transfer function modelling can be exploited and models can be obtained based on input/output data and the statistic features of the computed indicators can be analyzed in order to compare different production periods [86].

KPIs can also be exploited for companies' sustainability analyses based on key environmental, technical and social aspects [87]. For example, relationships between water supply, biodiversity and ecosystem service losses can be investigated [88]. Furthermore, electricity consumption of hydropower plants, economic growth and a power system's losses can be related in order to support decision-making regarding electric energy policies [89]. In addition, KPIs can be used as proof of the feasibility of hydropower projects thanks to their capability to take into account financial and non-financial aspects [90].

Tailored KPIs can also be exploited for the design of condition monitoring, early diagnostics and predictive maintenance systems aimed at supporting and enhancing the O&M of hydropower plants [91]. In-depth analysis is required in order to formulate peculiar indicators [92]. Poor maintenance can cause process failures that can represent threats to revenue, operation feasibility and people's health. For this reason, the performance of maintenance must be monitored and analyzed through specific indicators based on tailored regulations [93].

In order to formulate the mathematical optimization problems and to verify the performance of APC systems in hydropower plants, tailored KPIs can be formulated, monitored and analyzed [75,94]. An example of a KPI to be verified is the service factor, i.e., the time percentage associated with full service of the APC system [6,7]. The service factor can also represent a significant performance metric for the achievement of plant operators' feedback when they are required to use APC systems based on different methodologies and techniques, e.g., model predictive control (MPC) [95–99]. The formulation of mathematical optimization problems associated with control and optimization systems may depend on tailored KPIs, represented by the cost functions to be minimized/maximized. These cost functions can be formulated taking into account different objectives, e.g., the maximization of hydropower production, water supply reliability, spill prevention, and revenue income [100,101].

Table 5 reports some research works about KPIs in the hydropower sector, with focuses on the main topic, on the scope and on the main findings.

Table 5. Main features of the selected papers focused on KPIs for the hydropower sector.

Main Topic	Scope	Main Findings	Ref.
System availability improvement.	Maintenance management optimization aimed at system availability improvement.	Design of a method based on a stochastic block diagram adopting a hydropower plant as case study.	[82]
Reliability of performance in hydropower plants.	Selection of features able to provide a reliability analysis associated with the performance of hydropower plants.	Adoption of multi-criteria decision-making methods for the identification of the indicators.	[83]
Assessment of the sustainability in the hydropower sector	Development of a sustainability framework.	Design of a sustainability index based on plant mathematical modelization.	[84]
Evaluation of the performance of Brazilian hydropower plants.	Investigation of the performance of the largest Brazilian hydropower plants based on O&M indicators and quality of service.	Design of an approach for the construction of composite indicators based on modelization.	[85]
Hydropower generation system performance evaluation.	Development of a method for the evaluation of the hydropower generation facilities for performance improvement.	Exploitation of transfer function modelling to obtain performance indicators.	[86]
Environmental reporting associated with some Italian hydropower companies.	Analysis on the evaluation of sustainability data by the companies.	Reporting on the key environmental, technical, and social aspects (significance, indicators, and improvement objectives).	[87]
Integrated energy and economic evaluation for hydropower plants.	Assessment of the impact of the donor-side and the user-side on ecosystem services.	Design of an integrated evaluation framework able to evaluate impacts on water supply, biodiversity and ecosystem service losses.	[88]
Losses (power), consumption, performance (economic) of hydroelectricity in Indonesia.	Investigation of causality relationship between electricity consumption of hydropower plants, losses of the power system and economic growth.	Proposal of a statistical method for the evaluation of the considered causality relationships and exploitation of the results for the decision of electric energy policy.	[89]
Small-scale hydropower project attractiveness analysis.	Design of tools for the assessment of small-scale hydropower plants performance in the pre-implementation phase.	Exploitation of balanced scorecard as evaluation tool in the feasibility studies on plant implementation.	[90]

Table 5. Cont.

Main Topic	Scope	Main Findings	Ref.
Condition monitoring and predictive maintenance methodologies for hydropower equipment.	Development of a system able to support O&M through condition monitoring, early diagnostics, and predictive maintenance.	Formulation of a KPI for operating hydropower plants in order to identify faults and to support O&M tasks.	[91]
Performance analysis of hydropower units.	Condition monitoring of hydropower generation units.	Formulation a correlation transmissibility indicator to analyze degradation performance.	[92]
Maintenance monitoring and analysis in hydropower plants.	Determination of maintenance performance indicators for asset management.	Formulation of tailored indicators associated with different evaluation levels and test of the proposed methodology on a Brazilian hydroelectric power plant.	[93]
Reservoir advanced process control (APC) for hydroelectric power production.	Design and implementation of an APC system for water management in a hydroelectric power plant.	Use of tailored KPIs for the mathematical formulation of the control problem. The satisfactory performance of the developed APC system was proven by the high service factor achieved.	[75]
APC for water resources systems.	Presentation and analysis of two case studies where APC systems were successfully applied.	Use of tailored KPIs for the mathematical formulation of the control problem. Use of the service factor as support KPI for APC system's performance evaluation.	[94]
Production plan satisfaction for hydropower production.	Design and implementation of an MPC strategy for the satisfaction of the production plan in a hydroelectric power plant.	Use of tailored KPIs for the mathematical formulation of the control problem. The satisfactory performance of the developed MPC strategy was proven by the high service factor achieved.	[95]
Reservoir water management for hydropower production.	Design and implementation of an MPC strategy for reservoir water management.	Use of tailored KPIs for the mathematical formulation of the control problem. The satisfactory performance of the developed MPC strategy was proven by the high service factor achieved.	[96]
Performance appraisal of agricultural–water systems.	Assessment of the performance of agricultural–water systems based on comparative performance indicators.	Formulation of a high-level optimization problem to maximize monthly irrigation and hydropower releases.	[100]
Performance evaluation for multipurpose multi-reservoir system operation.	Development of a performance evaluation model.	Formulation of a performance index that takes into account spill prevention, reliability of water supply, revenue income, and energy production.	[101]

4. Role of Modelization and Forecast in the Hydropower Sector

In this section, the role of modelization and forecast in the hydropower sector is analyzed. In particular, some research works about these themes are reported, with focuses on the main topic, on the scope and on the main findings.

Modelization can play a crucial role in the hydropower sector. Modelization practices represent an important aspect in both design and in the O&M phases of the hydropower plants. Data support the modelization phase, especially in the O&M phase. Reliable data allows one to obtain robust models, which are crucial for different applications [57]. Design could also benefit from reliable data. For example, data from already commissioned hydropower plants could be exploited to improve the design of future plants.

Another area that can significantly affect the design and O&M of hydropower plants is associated with the capability to forecast. Forecasts can be exploited for different purposes and in different contexts.

4.1. Modelization for Design

Effective design procedures require reliability assessment. In [102], a mixed integer non-linear program (MINLP) is proposed as an optimization problem by which to guarantee the maximization of the hydropower production while considering the reliability level. The model-based optimal design is applied to a real plant. Reliability assessment can represent a crucial aspect in large-scale hydropower plants, e.g., hydropower plants characterized by multiple interconnected reservoirs. In order to guarantee an optimal design for the energy generation, cellular–automata-based approaches can be exploited for the modelization of reliability, taking into account the interdependency between O&M and design [103]. Reliability could also be referred to the energy yield in multi-reservoir systems. For this purpose, a reliability-based simulation model is needed in order to formulate objective functions, taking into account reservoir releases and/or reservoir storages [104].

Accurate modelization procedures could also be exploited for design changes on already installed plants. For example, multi-objective optimization models can be used to analyze economic–ecological tradeoffs in order to support decision making regarding dam removal [105]. Additionally, ex-post assessments can be performed in order to discover trade-offs, dependencies and robustness in hydropower plants, e.g., in dam design and operation. For this purpose, design problems associated with operation and dam sizing must be formulated and solved [106]. In addition, sensitivity analysis could support plants' modifications; an example is represented by a dam that supplies drinking water, where, thanks to sensitivity analysis, some margins for the installation of a hydropower unit can be detected [107]. Furthermore, different tools can be used for the computation of the potential associated with a design change. For example, geographic information system (GIS) tools can be exploited for the evaluation of the benefits that can be achieved through the transformation of conventional hydropower schemes into pumped hydropower schemes [108].

O&M and design challenges may result in a highly non-convex optimization problem, so tailored methods for the achievement of a feasible solution must be applied, e.g., honey bee mating optimization (HBMO) [109]. In this context, multi-objective evolutionary algorithms can reduce the computational cost [110]. Specific software and routines are needed to solve the obtained optimization problems. An example of such software is Water Evaluation and Planning (WEAP), which can be exploited as simulation software for water resources planning. The invasive weed optimization (IWO) algorithm represents another specific tool [111].

Effective design procedures require reliable simulation frameworks. In this way, some crucial features of the hydropower plants to be constructed can be simulated, e.g., the capacity of installation (design discharge) [112]. In order to provide holistic simulation frameworks, environmental impact evaluation must be incorporated into optimization and decision making, wherein different scenarios are needed for the optimization of the design of the facilities [113]. Various factors must be included in the simulation frameworks, e.g., market price. These factors could support the assessment of the economic feasibility of a hydropower project. In this context, design parameters associated with plant sizing cover a crucial role and the computation of their optimal values can be performed through effective simulation frameworks based on reliable models and efficient optimizers [114]. Hydropower projects can affect the ecosystem and a proper planning of hydropower projects must include an efficient cost–benefit analysis framework. In this way, decision-making can also take into account ecosystem services [115]. For the mitigation of the huge budget required for projects of the hydropower field, the design phase must optimize both the design and the future operation. An example is represented by reservoir plants equipped with pump stations, because these systems are energy producers characterized by cost effectiveness; an optimization model can be formulated in order to maximize the net benefit [116].

The potential of AI can be exploited for planning and design in dam engineering [117]. In addition, the visualization of a construction process could represent an added value for the design of hydropower plants. This involves a modelization approach that is able to provide real-time and interactive 3D scenes through virtual reality (VR) [118]. In this context, realistic terrain simulation could also aid in the design through dynamic visual simulation based on VR [119].

Table 6 reports some research works about modelization for design in the hydropower sector, with focuses on the main topic, on the scope and on the main findings.

Table 6. Main features of the selected papers focused on modelization (design) for the hydropower sector.

Main Topic	Scope	Main Findings	Ref.
Optimization of the design and of the operation of hydropower plants.	Formulate an optimization problem aimed at the maximization of the produced energy and at controlling the reliability level.	The developed system is exploited for the capacity optimization of the plants to be designed.	[102]
Hydropower plants design.	Incorporate reliability models in the design procedures.	Cellular automata are exploited for reliability model formulation.	[103]
Analysis of the integrated operation of multi-reservoir hydropower systems.	Reliability assessment as a support for the design.	The developed model allows one to exploit tailored objective functions able to take into account both reservoir release and storage.	[104]
Proof-of-concept demonstration on the effects of dam's removal on a plant.	Exploit modelization for decision-making support.	The developed optimization problem allows one to take into account both economic and ecological objectives.	[105]
Ex-post assessment of the design of hydropower plants.	Solve design problems that include operation and sizing of dams.	The proposed method allows one to reduce capital costs.	[106]
Sensitivity analysis.	Detect potential margins for the installation of a hydropower plant on an already existing dam.	The proposed method revealed that a small-scale hydropower plant can be installed on the already existing plant.	[107]
Calculation of the potential that can be achieved from hydropower plants' design change.	Development of a GIS-based model.	The developed model was tested on some case studies and revealed its applicability.	[108]
Design-operation of multi-hydropower reservoirs.	Illustrate and test the option to use the HBMO algorithm in a highly non-convex optimization problem.	The HBMO algorithm outperforms other existing algorithms.	[109]
Design of cascade hydropower reservoir systems.	Formulation of a multi-objective optimization model for the determination of the design parameters.	Multi-objective differential evolution can be exploited to solve hydropower design optimization problems.	[110]
Optimal design and operation of hydropower plants.	Formulate an optimization problem aimed at generated energy maximization and at flood damage minimization.	WEAP software and IWO algorithm can be exploited for the optimal design and operation of a hydropower plant.	[111]
Simulation of hydropower plants aimed at design optimization.	Formulation of a simulation model.	The developed model allows for the simulation of hydropower plants' main features, e.g., capacity of installation.	[112]
Sustainable planning of multipurpose hydropower reservoirs.	Provide a simulation–optimization framework for design optimization based on different scenarios.	Environmental economics can significantly affect the project results.	[113]
Optimal design of hydropower projects.	Provide a simulation-optimization model for multi-reservoir systems design.	The proposed framework exploits WEAP software and adds a hydropower computation module; the optimization problem is solved through PSO and takes into account different economic factors.	[114]

Table 6. Cont.

Main Topic	Scope	Main Findings	Ref.
Optimal development of hydropower projects.	Inclusion of ecosystem services within the optimal development of hydropower projects.	The formulated cost–benefit analysis framework allows incorporation of the services of the ecosystem into decision-making and into models for the optimization of the projects.	[115]
Plan a design phase equipped with simultaneous design-operation optimization for pumping systems in reservoir power plants.	Maximization of the net benefit of installations in reservoir systems.	Pumped-storage systems have better outcomes than individual hydropower systems, taking into account benefits and efficiency.	[116]
AI and digital technologies in dam engineering.	Review of AI and digital technologies application in dam engineering.	Assessment of the role of AI and digital technologies in dam engineering with regard to design and planning.	[117]
Exploitation of VR for the construction process of dams in hydropower plants.	Development of a system able to achieve a realistic, 3D real-time and interactive virtual scene.	The developed system can represent a support analytical tool to optimize the construction management of hydropower plants.	[118]
Dynamic visual simulation of hydropower construction project.	Development of a system that uses VR.	The developed system could provide a design and management environment for hydropower projects.	[119]

4.2. Modelization for O&M

O&M activities can be supported by modelization and simulation tools. For example, deep learning, support vector regression and artificial neural network (ANN) techniques can be exploited for reservoir operation modeling and simulation in order to support decision making [120]. In addition, data-driven AI can also be exploited for reservoir operation policy computation; in this context, pattern recognition and metaheuristic optimization are useful concepts [121]. Dam operation can also benefit from deep learning in order to build models able to provide key insights through the simulation of different scenarios [122]. In the field of dam engineering, AI and digital technologies could provide predictive modeling [117]. Dam–reservoir system modelization can be exploited for the computation of a balance between environmental management and hydropower generation [123]. Cascade hydropower systems can be considered as complex systems and cellular automata methodologies can be exploited for modelization and optimization in order to maximize energy production [124]. In order to solve reservoir operation optimization problems, animal-inspired evolutionary algorithms can be applied, e.g., particle swarm optimization (PSO), ant colony optimization (ACO), artificial bee colony (ABC), shuffled frog leaping algorithm (SFLA), firefly algorithm (FA), HBMO, bat algorithm (BA), and cuckoo search (CS) [125]. In addition, surrogate modeling could contribute to release decisions at the desired timescale [126].

Different variables and indexes can be modeled for O&M activities. A significant index that can be modeled is the reliability level of hydro-energy production. The inclusion of this index in the optimization problem associated with optimal design and operation strategies can represent an overall enhancement of the considered plant [102]. Among the variables that can be taken into account are design and operation variables. These variables can be affected by different factors and they can retain different impacts on the overall system performance [111]. A crucial process operation variable that can be taken into account is the dammed water level. In order to modelize it, different approaches can be exploited, e.g., neural networks, support vector regression, and Gaussian processes [127].

Techno-ecological synergy frameworks may be needed in order to evaluate the sustainability of hydropower plant operation at local and regional levels. Life-cycle assessment can be exploited for this purpose [128]. In addition, externality theory models can be used for the evaluation of the impact of a hydropower project [129].

O&M of hydropower plants could benefit from modeling and simulation combined with forecasting. Hydropower allocation can be guided by forecast-informed reservoir operations. Hydroclimatic predictions can be converted into actionable information for the enhancement of the management of hydropower plants [130]. Another context of application of the combination between modeling, simulation and forecasting is short-term hydropower operations planning. In this context, a short-term model is required in order to dispatch the available water to the turbines on a daily basis. The model can be characterized by a stochastic nature in order to include uncertainties, e.g., inflow uncertainty [131]. The scheduling task can also be performed through multi-objective models that are particularly recommended for large-scale systems [132].

An additional field of application of the combination between modeling, simulation, and forecasting is reservoir level control. Different APC techniques can be applied to solve this problem, e.g., MPC [133–136]. Among the features needed to apply MPC, model and forecasts reliability cover a crucial role. A reliable model of the plant allows the provision of a tool with which to simulate and predict the effect of the manipulated and unmanipulated inputs on the outputs; on the other hand, forecast reliability can be referred to the availability of predictions associated with the unmanipulated inputs in order to include them in the control law computation [75]. Unmanipulated inputs can be represented by the water discharged toward the turbines, which is connected to the energy that has to be generated. The relationship between the hydroelectric energy and the water that has to be provided to the turbines can be obtained by exploiting linear regression models [94]. The reliability of these models can impact the reliability of the models of the upstream devices, e.g., reservoirs. The modelization of reservoirs' levels can be performed using different types of model, e.g., first-principles models [95]. In order to mitigate model uncertainties, suitable model mismatch compensation strategies must be used [96]. Table 7 reports some research works about modelization for O&M in the hydropower sector, with focuses on the main topic, on the scope and on the main findings.

Table 7. Main features of the selected papers focused on modelization (O&M) for the hydropower sector.

Main Topic	Scope	Main Findings	Ref.
Modeling and simulation of reservoir operation.	Development of models for decision-making support.	ANN, support vector regression and deep learning techniques are exploited to build models for simulation of reservoir operation.	[120]
Hydropower reservoir operation policy computation.	Exploitation of AI for modelization.	The developed models are included in a metaheuristic optimizer and satisfactory operation results are provided.	[121]
Modeling and simulation of dam operation.	Development of deep learning models for decision-making support.	Assessment of the explainability of the developed deep learning model on dam operation.	[122]
Exploration of digital technologies and AI for dams.	Review of digital technologies and AI application in dam engineering.	Assessment of the role of AI and digital technologies with regard to predictive modeling.	[117]
Dam–reservoir system management.	Mathematical modelling and computation of a balance between hydropower generation and environmental management.	The balanced operation policy is obtained through the solution of a stochastic control problem.	[123]

Table 7. Cont.

Main Topic	Scope	Main Findings	Ref.
Energy production maximization in complex plants.	Development of a modelization and optimization method for energy production maximization.	Cellular automata can be used for modelization, simulation and optimization of complex hydropower plants.	[124]
Modelization of the operation of reservoirs.	Review of applications of animal-inspired evolutionary algorithms in reservoir operation modelling.	PSO, ACO, ABC, SFLA, HBMO, CS, FA and BA algorithms can be exploited for reservoir operation modelling.	[125]
Hydropower optimization.	Exploitation of ANNs surrogate models and water quality models.	The proposed approach provides high-fidelity modelization and allows one to support decisions at the desired timescale.	[126]
Optimal operation and design of hydropower plants.	Formulate an optimization problem aimed at produced energy maximization and at controlling the reliability level.	The developed system is exploited for the capacity optimization of the plants to be designed.	[102]
Optimal design and operation of hydropower plants.	Formulate an optimization problem aimed at generated energy maximization and at flood damage minimization.	The optimization of design variables can affect the system performance in a different manner with respect to the optimization of operation variables.	[111]
Dammed water level analysis and prediction in a hydropower reservoir.	Short- and long-term analyses.	Machine learning could represent a significant driver.	[127]
GHG mitigation in hydropower plants.	Investigation of techno-ecological synergies of hydropower plants.	The developed techno-ecological framework, based on life cycle assessment, can be exploited for the evaluation of the supply and of the demand of hydropower plants in order to evaluate their sustainability level.	[128]
Evaluation of hydropower projects.	Investigation of externality evaluation models for hydropower projects.	Externality theory can be exploited for hydropower projects evaluation.	[129]
Guide hydropower allocation.	Provision of forecast-informed reservoir operations.	The developed model allows one to convert hydroclimatic predictions into actionable information for the enhancement of plants' management.	[130]
Short-term hydropower operations planning.	Exploitation of modeling, simulation and forecasting for the optimal dispatch of the water to the turbines.	Stochastic short-term models can be exploited due to their ability to deal with uncertainty.	[131]
Short-term hydropower scheduling.	Development of a multi-objective model for peak shaving.	The proposed approach is applicable, tractable and robust enough to obtain near optimal results efficiently.	[132]
Reservoir advanced process control (APC) for hydroelectric power production.	Design and implementation of an APC system for water management in a hydroelectric power plant.	The achievement of a robust plant model and of reliable forecasts on the production plan represent fundamental requirements for the application of MPC technique.	[75]
APC for water resources systems.	Presentation and analysis of two case studies where APC systems were successfully applied.	The relationship between the hydroelectric energy and the water that has to be provided to the turbines can be obtained by exploiting linear regression models.	[94]

Table 7. *Cont.*

Main Topic	Scope	Main Findings	Ref.
Production plan satisfaction for hydropower production.	Design and implementation of an MPC strategy for the satisfaction of a production plan.	Different types of models may be formulated and integrated in an MPC strategy for production plan satisfaction.	[95]
Reservoir water management for hydropower production.	Design and implementation of an MPC strategy for reservoir water management.	The design of ad hoc model mismatch compensation strategies can mitigate the model uncertainties.	[96]

4.3. Forecast

Dam engineering could benefit from predictive modeling for the forecasts needed in order to make the due decisions at the right times [117]. Crucial process variables need to be predicted in order to support the management of areas with hydrology stress; in this context, machine learning could represent a significant driver [127]. In addition, hydroclimatic predictions can support the decision-making process in order to optimally allocate hydropower energy [130]. The hydropower generation represents a crucial process variable and its prediction is a very challenging task. Tailored algorithms can be used for this purpose, e.g., the developed wildebeest herd optimization (DWHO) algorithm; the convergence speed and the required time to provide a solution represent significant factors to be considered in the selection of an algorithm [137]. Combinations of different techniques can be exploited for hydropower generation prediction, e.g., grey wolf optimization (GWO), and the adaptive neuro-fuzzy inference system (ANFIS) [138]. Additionally, ANNs can be exploited for the prediction of power production. Good performance can be obtained if large amounts of significant data are available [139]. The water inflow into reservoirs represents another crucial process variable to be predicted. For this purpose, historical data can be exploited together with dynamic non-linear auto-regressive (NAR) and non-linear auto-regressive with eXogenous input (NARX) models [140]. Additionally, ensemble forecasts can be used for the verification of the inflow into hydropower reservoirs, taking into account different meteorological models. These forecasts can provide enhancement in different aspects, e.g., benefits and flood control [141].

With regard to streamflow, the quality and the value of the forecast are crucial aspects to be taken into account. The quality refers to the accuracy of predictions while the value refers to the impact of the decisions' predictions on the results of the operations [142]. Machine learning could be very useful in the modelization of complex, non-stationary, and non-linear time series aimed at streamflow forecasting. In particular, data preprocessing can be combined with modelization and an optimizer can be selected for the achievement of a feasible parameter configuration [143]. Middle and long-term streamflow forecasts can also be exploited for hydropower maximization; in order to deal with the stochastic nature of streamflow, the combination of AI, adaptation and parameter optimization could be needed [144]. Another paradigm that can be used for streamflow forecasts is that of the ensemble forecasts used in order to deal with uncertainties [145]. Biases on the ensemble streamflow predictions could impact the electricity production in the hydropower reservoir management. Methods for streamflow prediction with a poor robustness to uncertainties could cause a high risk for the level of reservoirs in the pursuit of maximized short-time profit; the introduction of suitable constraints can mitigate this aspect [146]. With regard to short-term streamflow prediction, the combination of stochastic weather generation and ensemble weather forecasts can help in the extension of the forecasting horizon [147].

The optimal dispatch of the available water to the turbines in a hydroelectric power plant can benefit from reliable predictions. In particular, stochastic short-term models can be exploited in order to provide reliable one day ahead forecasts [131]. In addition, multi-objective short term scheduling models can be exploited for this purpose. The main challenges are represented by large scale systems, numerous power-receiving grids, complex constraints, and cascaded systems [132]. In this context, the impact of hydrological forecasts on the revenue and on the management of reservoirs can be obtained through sensitivity analysis. In addition, the forecast quality affects the stock evolution, the spillage, the production rates, and the production hours [148]. Different performances can be provided by probabilistic and deterministic forecasts in the optimization (short-term) of the reservoirs. Ensemble forecasts (probabilistic) and multi-stage optimization (stochastic) techniques could provide a higher level of flood protection while guaranteeing an acceptable energy production [149]. Another problem that can benefit from reliable forecasts is the monthly runoff prediction in hydropower plants. For this purpose, for the enhancement of the accuracy of the predictions, hybrid prediction models can be formulated through the combination of empirical mode decomposition (EMD), time varying filtering (TVF), extreme learning machines (ELM), and salp swarm algorithm (SSA) [150]. ELM algorithms can be integrated with the Monte Carlo method in order to provide reliable predictions of hydropower production and energy saving [151].

The outputs of the optimal dispatch problems are usually represented by the hydroelectric energy production plan. This plan is connected with the water that has to be sent to the turbines by the upstream devices. A reliable model between the hydropower generation and the provided water can provide reliable forecasts of the water that has to be discharged from the upstream devices [75]. Reliable forecasts of the water that has to be discharged from the upstream devices can be obtained through the application of different algorithms, e.g., ANFIS and the cooperative search algorithm (CSA) [152]. Reliable forecasts of the water that has to be discharged from the upstream devices allows one to obtain reliable predictions of the process variables associated with these devices, e.g., the reservoir level [94]. These predictions also depend on the inflows to the upstream devices; reliability on these forecasts is required in order to accurately predict the needed process variables, e.g., the reservoir level [95,96]. Inflow prediction is significantly influenced by precipitation forecasts. Uncertainties and probabilities of the precipitation must be taken into account [153]. Inflows can be characterized by different lead times. For this purpose, short-, long-, and medium-term forecasts on inflow can be computed [154]. As mentioned above, a strategy that can be implemented based on predictions is MPC. In this regard, the prediction horizon must be selected based on the effective reliability window of the computed forecasts [155].

Reliable forecasting methods could also be useful in the evaluation of the impact of hydropower facilities on sustainable development. For example, the long-term prediction of GHG risk is needed in order to assess life cycle emissions [156]. GHG risk could be detected through CO_2 and CH_4 fluxes prediction. Specific tools can be exploited for the evaluation of GHG emissions vulnerability; in-depth analysis must be conducted in order to evaluate the reliability of the predictions [157].

Table 8 reports some research works about forecast in the hydropower sector, with focuses on the main topic, on the scope and on the main findings.

Table 8. Main features of the selected papers focused on forecast for the hydropower sector.

Main Topic	Scope	Main Findings	Ref.
AI and digital technologies in dam engineering.	Review of AI and digital technologies application in dam engineering.	Assessment of the role of AI and digital technologies in dam engineering with regard to predictive modeling.	[117]
Analysis and prediction of dammed water level in a hydropower reservoir.	Long- and short-term predictions of dammed water level in a hydropower reservoir.	Machine learning could represent a significant tool.	[127]
Guide hydropower allocation.	Provision of forecast-informed reservoir operations.	Hydroclimatic predictions can be converted into actionable information in order to actively support the decision-making process.	[130]

Table 8. *Cont.*

Main Topic	Scope	Main Findings	Ref.
Prediction of hydropower generation.	Investigation of the algorithms that can be used.	DWHO algorithm represents a suitable solution due to its convergence speed and to the required time to provide the solution.	[137]
Prediction of hydropower generation.	Investigation of the algorithms that can be used.	GWO and ANFIS algorithms can be combined in order to provide accurate predictions.	[138]
Prediction of hydropower generation.	Investigation of the algorithms that can be used.	ANNs algorithms can represent a suitable solution.	[139]
Water inflow prediction for dam reservoirs.	Development of a prediction method.	NAR and NARX models can be exploited for the prediction.	[140]
Verification of inflow into hydropower reservoirs.	Exploitation of ensemble forecasts and TIGGE database.	The obtained forecasts can provide enhancement in terms of benefit and flood control.	[141]
Subseasonal hydrometeorological forecasts for hydropower operations.	Assessment of the quality and of value of subseasonal hydrometeorological forecasts for hydropower operations.	The improvement of forecast quality and value is strictly related to the used preprocessing techniques. The forecast value–quality relationship is complex and it is related to many factors.	[142]
Streamflow time series forecasting of hydropower reservoir.	Improvement of the conventional hydrological forecasting models.	Machine learning and tailored optimizers could represent significant tools.	[143]
Middle and long-term streamflow forecasts.	Deal with the stochastic nature of streamflow in order to maximize the hydropower generation.	The combination of AI, adaptation and parameters' optimization could represent a significant tool.	[144]
Streamflow predictions on short and long ranges.	Deal with the uncertainties that characterize the streamflows.	Ensemble forecasts can successfully predict the streamflow.	[145]
Ensemble prediction on streamflow.	Evaluation of the impact of prediction bias on electric energy production.	Exploiting algorithms with a poor robustness to uncertainty, the reservoir levels can be calibrated to configurations characterized by high risk in order to maximize short-term profit. Constraints can mitigate this aspect.	[146]
Short-term streamflow prediction.	Development of a method that combines stochastic weather generation and ensemble weather forecasts.	The proposed strategy could help in the extension of the forecasting horizon.	[147]
Short-term hydropower operations planning.	Exploitation of modeling, simulation and forecasting for the optimal dispatch of water to the turbines.	Stochastic short-term models can be exploited due to their ability to deal with uncertainty.	[131]
Short term hydropower scheduling.	Development of a multi-objective model for peak shaving.	The proposed approach is applicable, tractable and robust enough to obtain near optimal results efficiently.	[132]
Evaluation of the impact of the hydrological forecast quality on operation of reservoirs.	Development of a conceptual approach for the evaluation of the hydrological forecast quality impact on management and revenue.	The developed approach allows one to evaluate the impact of the revenue and forecasts, stock evolution, spillage, production rates, and production hours.	[148]
Short-term optimization of hydropower reservoirs.	Evaluation of the performance of deterministic and probabilistic forecasts.	The exploitation of probabilistic forecasts is more convenient due to its major robustness with respect to flood control while guaranteeing an acceptable level of energy production.	[149]

Table 8. Cont.

Main Topic	Scope	Main Findings	Ref.
Monthly runoff prediction in hydropower plants.	Formulated an enhanced prediction formulation.	The combination of different techniques, i.e., TVF, EMD, SSA and ELM, can provide significant enhancements.	[150]
Production capacity prediction of hydropower industries for energy optimization.	Investigation of a method for the prediction of hydropower production and energy saving.	ELM algorithms can be successfully integrated with the Monte Carlo method.	[151]
Reservoir advanced process control (APC) for hydroelectric power production.	Design and implementation of an APC system for water management in a hydroelectric power plant.	The availability of reliable forecasts on the hydroelectric power production plan and the formulation of a reliable model between hydropower generation and discharged water represent fundamental ingredients of an MPC strategy.	[75]
Prediction of discharge time series under hydropower reservoir operation.	Investigation of the algorithms that can be used.	ANFIS and CSA algorithms can be combined in order to achieve better performance with respect ANFIS algorithm.	[152]
APC for water resources systems.	Presentation and analysis of two case studies where APC systems were successfully applied.	Reliable predictions of the water discharged toward the turbines allow one to obtain reliable predictions of crucial process variables, e.g., reservoir levels.	[94]
Production plan satisfaction for hydropower production.	Design and implementation of an MPC strategy for the satisfaction of the production plan in a hydroelectric power plant.	Reliable predictions of the water supplied to the reservoirs contribute to the computation of reliable predictions on some reservoirs' crucial process variables, e.g., reservoir level.	[95]
Reservoir water management for hydropower production.	Design and implementation of an MPC strategy for reservoir water management.	Reliable predictions of the water supplied to the reservoirs contribute to the computation of reliable predictions on some reservoirs' crucial process variables, e.g., reservoir level.	[96]
Real-time decision-making in hydropower operations.	Development of a decision-making strategy that includes precipitation forecasts.	Uncertainty of precipitation, operation policies and a risk evaluation model are integrated into the strategy.	[153]
Optimal operation model of hydropower stations.	Development of a model that includes inflow forecasts with different lead-times.	The developed reservoir model can exploit the short-, long- and medium-term forecasts on inflow.	[154]
Optimization of the generated hydroelectric energy.	Evaluation of forecast and decision horizons assuming medium-range precipitation forecasts.	The efficiency and reliability are improved with a shortened effective decision horizon.	[155]
Long-term prediction of greenhouse gas risk in hydropower reservoirs.	Evaluation of the impact on the sustainable development of electricity production by hydropower reservoirs.	The proposed method predicts long-term GHG risk and the associated life cycle emissions.	[156]
Assessment of risk of GHG emissions in hydropower reservoirs.	Development of a tool for the prediction of CO_2 and CH_4 fluxes.	The developed tool allows one to evaluate the selected fluxes taking into account potential prediction errors.	[157]

5. Discussion

Decarbonization represents a primary objective and the use of RESs can act as a driver in this context. Water is an RES and its rational exploitation is a fundamental requirement. Renewable energy production can exploit water, such as in hydroelectric power plants. Hydropower production is a very complex sector due to process, environmental and economic reasons. With regard to the process, hydraulic, mechanical and electrical aspects interact and the optimal management of these interactions may have to take into account conditions and incentives associated with the market. The economic area is also involved in hydropower plants management for this reason.

Customization of monitoring, maintenance, control and optimization strategies for the hydropower sector is not a trivial task. This task can be tackled by highlighting the basic elements and concepts that can represent the basis for the construction of these strategies. These strategies, if effectively approached, could represent strategic items for the mitigation of the huge costs that characterize the setup and the management of a hydropower plant. The design of proficient approaches for the development and implementation of monitoring, maintenance, control and optimization strategies strictly depends on the level of robustness associated with some basic elements and concepts, i.e., digitalization, Industry 4.0, data, KPIs, modelization and forecast.

Digitalization and Industry 4.0 play key roles and will retain these roles in the hydropower sector. Industry 4.0 is a driver for digitalization and suitable cross-fertilization procedures are being applied for its adaptation to non-industrial sectors like hydropower. Digitalization, through its capability to merge IT and OT and through the connection and contamination between different research areas, can speed up the green and digital transitions.

Energy production and efficiency can benefit from the tools provided by Industry 4.0. Industry 4.0 technologies, e.g., the internet of things (IoT), simulation, cloud computing, augmented reality (AR), big data analytics, CPSs, cybersecurity, blockchain, AI and ML, can speed up the evolution of conventional plants into smart plants. This evolution can massively support the digital and energy transition.

Digitalization and Industry 4.0 highlight the importance of data from different points of view: selection, acquisition, storage, analysis and visualization. Hydropower 4.0 and hydropower CPSs provide hydropower big data. Design, O&M and business levels can benefit from an effective exploitation of the information provided by data. In this context, DSSs can be designed and implemented in order to enhance command and decision making at all levels. In this way, data-driven policies can be conceived and implemented thanks to the shrinking of the time horizon required for making decisions. Databases, data classification, data mining, data quality and data reliability represent strategic features to be implemented in order to totally exploit data potential.

To compute, evaluate, assess, analyze and process information about the efficiency of energy production in hydropower plants, tailored KPIs are needed. These KPIS could be associated with availability, revenue, return on investment, efficiency, energy output, capacity factor, and generation capacity. The automatic evaluation of these parameters represent an additional powerful tool with which to optimize the time horizon for making decisions, as well as the optimality of those decisions.

Figure 7 reports a summary of the key results that digitalization, Industry 4.0, data and KPIs can obtain on hydropower plants.

All of the previously cited elements and concepts, together with the knowledge provided by different research areas, can be exploited for modelization and forecast. Design, O&M and business can benefit from reliable modelization and forecast. Data, process knowledge and methodology represent the two main elements needed for the development of robust modelization and forecast frameworks. Modelization could be referred to design and O&M. Robust models, together with different Industry 4.0 tools, e.g., AR and VR, are innovating the design field. Meanwhile, effective and optimized O&M solutions can massively benefit from the information provided by robust models. In this context, robust models implemented on the field and run with reliable data have become the digital twins of their physical counterparts. Figure 8 reports a summary of the role of modelization and forecast in hydropower plants.

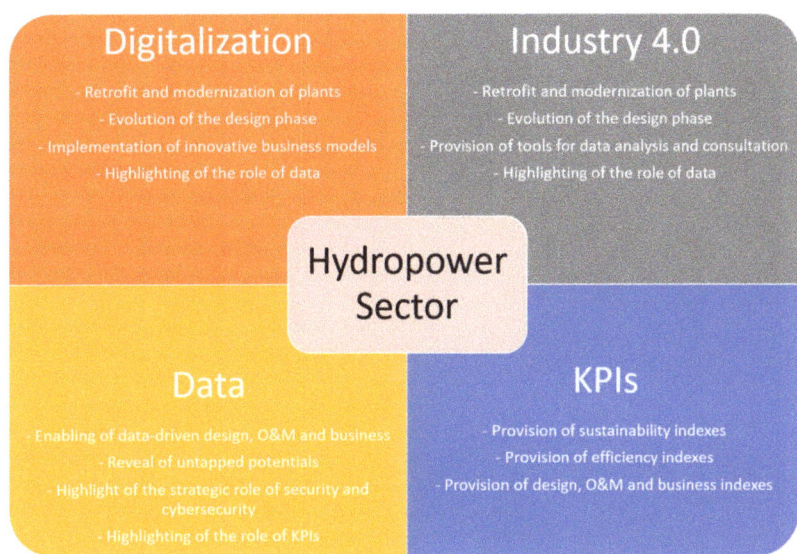

Figure 7. Summary of the key results that digitalization, Industry 4.0, data and KPIs can obtain on hydropower plants.

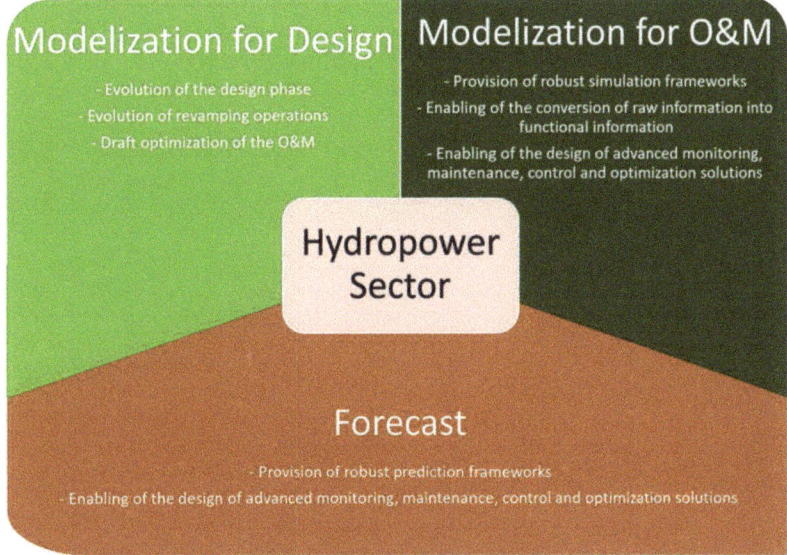

Figure 8. Summary of the role of modelization and forecast in hydropower plants.

The presented discussion highlights the potential of each basic element and concept addressed in the present review paper but at the same time highlights the interactions that can occur between them.

The described concepts represent a solid background for many applications associated with monitoring, maintenance, control and optimization strategies. In the authors' opinion, in order to further assess and enhance the potential of digitalization, Industry 4.0, data,

KPIs, modelization and forecast in hydropower plants, the following principles must be applied:

- Continue to create multidisciplinary teams for the development of specific projects: due to the extreme complexity of the hydropower CPSs, many competences must contaminate each other and a fusion of knowledge is required.
- Reduce the gap between university and facilities, the theoretical and scientific approach provided by university can represent a win-win solution only and only if it is accompanied by the field experience and knowledge provided by facilities. The reduction of the gap between university and facilities can speed up the digital and energy transition in the hydropower sector. In fact, the combination between the innovative methodologies proposed by the university research groups and the field experience and knowledge retained by plant managers, engineers, practitioners and operators represents a strategy with huge potential.
- Increase the small-scale laboratories: small-scale laboratories can support the design and the O&M of real hydropower plants through the small-scale implementation of subparts of the real plants. Small-scale implementation can support the design and the O&M thanks to the fact that it is more convenient and practical to perform modifications and tests in laboratories instead of on the real plants. These modifications and tests can enhance both the methodological and practical aspects associated with design and O&M.
- Increase the open access datasets: data represent the main source of information and the availability of open access datasets reporting, for example, issues and problems, can significantly promote the cross-fertilization of already existing algorithms to hydropower sector and the conceiving of new strategies tailored for this sector. This cross-fertilization can represent a powerful method by which to import into the hydropower sector effective solutions for the speed-up of the energy and digital transition.
- Exploit the analyzed basic elements and concepts, i.e., digitalization, Industry 4.0, data, KPIs, modelization and forecast, for the development of advanced monitoring, maintenance, control and optimization solutions. Based on the provided analysis, a clear and straight connection was created between the basic elements/concepts and these strategies.
- Exploit the analyzed basic elements and concepts for the enhancement of the control and monitoring rooms (onsite and remote) of the hydropower plants.
- Exploit the analyzed basic elements and concepts for the search of the best decision in terms of design, retrofit and O&M on hydropower plants.

6. Conclusions and Future Research Directions

A comprehensive literature review of hydropower plant technology with a focus on the basic elements and concepts needed for monitoring, maintenance, control and optimization tasks has been proposed in the present paper. The authors agree that digitalization, Industry 4.0, data, KPIs, modelization and forecast represent milestones for the design of complex tools; for this reason, an assessment and an outline of the existing state of the art associated with these basic concepts and elements has been proposed in this paper. In addition, some insights associated with methods and concepts that can also be applied for their further assessment and enhancement in hydropower research area have been proposed.

Future research directions should be associated with the following:

- Assessment of Industry 5.0 for the hydropower sector: the "technology-driven" concept promoted by Industry 4.0 (born in 2011) has to be adapted to a new "value-driven" concept (born in 2017). In this context, a hydropower sector investigation of the co-existence between Industry 4.0 and Industry 5.0 and on the benefits each can derive from the other must be conducted [158]. Human-centricity, resiliency and sustainability concepts must be assessed for the hydropower sector. Based on the authors' knowledge, an assessment of resiliency and sustainability concepts was begun from a methodological point of view, but it still needs massive field implementation. For example, resiliency can be referred to the idea of implementing flexibility into hydropower plants with respect to market conditions in order to maximize the desired KPIs. On the other hand, sustainability can be considered a rational usage of the water RES. With regard to human-centricity, for example, advanced control and monitoring rooms (onsite and remote) can represent a driver. Operators can be placed out of the lower-level loops in order to gain a crucial supervisory role.

- Assessment of Industry 6.0 for the hydropower sector: the AR/VR and digital twin concepts must acquire major importance within the hydropower sector; in addition, accurate analysis on the possible utilization of quantum computing can be performed [159]. Based on the authors' knowledge, AR/VR and digital twin concepts are gaining the attention of researchers, practitioners, engineers and managers in the hydropower field. For example, the design of new hydropower plants and the proof of their features can be massively supported by AR/VR. In addition, digital twins can support the road to sustainability. With regard to quantum computing, this can support the computation at different automation and business levels, especially with very complex CPSs.
- Reduce the gap between simulation and field implementation: projects that examine hydropower sectors characterized by lasting field implementation are not widespread. The real implementation of a system in the field requires additional robustness and reliability assessment with respect to the requirements of a system tested through virtual environment simulations.
- Further assessment of KPIs aimed at further emphasizing the potential impact of innovative technologies in the hydropower sector.

Author Contributions: Conceptualization, C.P. and S.M.Z.; formal analysis, C.P. and S.M.Z.; investigation, C.P. and S.M.Z.; methodology, C.P. and S.M.Z.; validation, C.P. and S.M.Z.; visualization, C.P. and S.M.Z.; writing—original draft, C.P. and S.M.Z.; writing—review and editing, C.P. and S.M.Z. All authors have read and agreed to the published version of the manuscript.

Funding: This research was funded by Green Hope s.r.l.

Data Availability Statement: Not applicable.

Conflicts of Interest: The authors declare no conflict of interest.

References

1. Razmjoo, A.; Gakenia Kaigutha, L.; Vaziri Rad, M.A.; Marzband, M.; Davarpanah, A.; Denai, M. A Technical analysis investigating energy sustainability utilizing reliable renewable energy sources to reduce CO_2 emissions in a high potential area. *Renew. Energy* **2021**, *164*, 46–57. [CrossRef]
2. Muh, E.; Tabet, F. Comparative analysis of hybrid renewable energy systems for off-grid applications in Southern Cameroons. *Renew. Energy* **2019**, *135*, 41–54. [CrossRef]
3. Hosseini, S.E.; Andwari, A.M.; Wahid, M.A.; Bagheri, G. A review on green energy potentials in Iran. *Renew. Sustain. Energy Rev.* **2013**, *27*, 533–545. [CrossRef]
4. Agenda 2030. Available online: https://unric.org/it/agenda-2030/ (accessed on 31 August 2023).
5. Ramos, H.M.; Carravetta, A.; Nabola, A.M. New Challenges in Water Systems. *Water* **2020**, *12*, 2340. [CrossRef]
6. Zanoli, S.M.; Pepe, C.; Astolfi, G.; Orlietti, L. Applications of Advanced Process Control Techniques to an Italian Water Distribution Network. *IEEE Trans. Control Netw. Syst.* **2022**, *9*, 1767–1779. [CrossRef]
7. Zanoli, S.M.; Astolfi, G.; Orlietti, L.; Frisinghelli, M.; Pepe, C. Water Distribution Networks Optimization: A real case study. *IFAC-PapersOnLine* **2020**, *53*, 16644–16650. [CrossRef]
8. Hydropower Europe. Available online: https://hydropower-europe.eu/ (accessed on 31 August 2023).
9. Kougias, I.; Aggidis, G.; Avellan, F.; Deniz, S.; Lundin, U.; Moro, A.; Muntean, S.; Novara, D.; Pérez-Díaz, J.I.; Quaranta, E.; et al. Analysis of emerging technologies in the hydropower sector. *Renew. Sustain. Energy Rev.* **2019**, *113*, 109257. [CrossRef]
10. Kougias, I. *Hydropower: Technology Development Report*, EUR 29912 EN; Publications Office of the European Union: Luxembourg, 2019; ISBN 978-92-76-12437-5. [CrossRef]
11. Yang, W. *Hydropower Plants and Power Systems—Dynamic Processes and Control for Stable and Efficient Operation*; Springer: Cham, Switzerland, 2019. [CrossRef]
12. Munoz-Hernandez, G.A.; Mansoor, S.P.; Jones, D.I. *Modelling and Controlling Hydropower Plants*; Springer: London, UK, 2013. Available online: https://link.springer.com/book/10.1007/978-1-4471-2291-3 (accessed on 11 September 2023).

13. Bogardi, J.J.; Gupta, J.; Nandalal, K.W.; Salamé, L.; van Nooijen, R.R.; Kumar, N.; Tingsanchali, T.; Bhaduri, A.; Kolechkina, A.G. *Handbook of Water Resources Management: Discourses, Concepts and Examples*; Springer: Cham, Switzerland, 2021. Available online: https://link.springer.com/book/10.1007/978-3-030-60147-8 (accessed on 11 September 2023).
14. Bundesministerium für Wirtschaft und Klimaschutz. Available online: https://www.plattform-i40.de/ (accessed on 31 August 2023).
15. Pandey, V.; Sircar, A.; Bist, N.; Solanki, K.; Yadav, K. Accelerating the renewable energy sector through Industry 4.0: Optimization opportunities in the digital revolution. *Int. J. Innov. Stud.* **2023**, *7*, 171–188. [CrossRef]
16. Leguizamon-Perilla, A.; Rodriguez-Bernal, J.S.; Moralez-Cruz, L.; Farfán-Martinez, N.I.; Nieto-Londoño, C.; Vásquez, R.E.; Escudero-Atehortua, A. Digitalisation and Modernisation of Hydropower Operating Facilities to Support the Colombian Energy Mix Flexibility. *Energies* **2023**, *16*, 3161. [CrossRef]
17. Ristić, B.; Božić, I. A short overview on Industry 4.0 in maintenance of hydropower plants. In Proceedings of the 8th International Conference on Industrial Engineering, Belgrade, Serbia, 29–30 September 2022. Available online: https://machinery.mas.bg.ac.rs/bitstream/id/14021/bitstream_14021.pdf (accessed on 11 September 2023).
18. IEEE. *IEC/IEEE Guide for Computer-Based Control for Hydroelectric Power Plant Automation*; IEEE: New York, NY, USA, 2013; pp. 1–83. [CrossRef]
19. Pandey, R.; Shrestha, R.; Bhattarai, N.; Dhakal, R. Problems identification and performance analysis in small hydropower plants in Nepal. *Int. J. Low-Carbon Technol.* **2023**, *18*, 561–569. [CrossRef]
20. Berga, L. The Role of Hydropower in Climate Change Mitigation and Adaptation: A Review. *Engineering* **2016**, *2*, 313–318. [CrossRef]
21. Wasti, A.; Ray, P.; Wi, S.; Folch, C.; Ubierna, M.; Karki, P. Climate change and the hydropower sector: A global review. *Wiley Interdiscip. Rev. Clim. Change* **2022**, *13*, e757. [CrossRef]
22. Manzano-Agugliaro, F.; Taher, M.; Zapata-Sierra, A.; Juaidi, A.; Montoya, F.G. An overview of research and energy evolution for small hydropower in Europe. *Renew. Sustain. Energy Rev.* **2017**, *75*, 476–489. [CrossRef]
23. Kumar, A.; Kumar, A.; Chaturvedi, A.K.; Joshi, N.; Mondal, R.; Malyan, S.K. Greenhouse gas emissions from hydroelectric reservoirs: Mechanistic understanding of influencing factors and future prospect. *Environ. Sci. Pollut. Res.* **2023**. [CrossRef]
24. Singh, V.K.; Singal, S.K. Operation of hydro power plants-a review. *Renew. Sustain. Energy Rev.* **2017**, *69*, 610–619. [CrossRef]
25. Ak, M.; Kentel, E.; Savasaneril, S. Operating policies for energy generation and revenue management in single-reservoir hydropower systems. *Renew. Sustain. Energy Rev.* **2017**, *78*, 1253–1261. [CrossRef]
26. Shahgholian, G. An Overview of Hydroelectric Power Plant: Operation, Modeling, and Control. *J. Renew. Energy Environ.* **2020**, *7*, 14–28. [CrossRef]
27. Quaranta, E.; Bejarano, M.D.; Comoglio, C.; Fuentes-Pérez, J.F.; Pérez-Díaz, J.I.; Sanz-Ronda, F.J.; Schletterer, M.; Szabo-Meszaros, M.; Tuhtan, J.A. Digitalization and real-time control to mitigate environmental impacts along rivers: Focus on artificial barriers, hydropower systems and European priorities. *Sci. Total Environ.* **2023**, *875*, 162489. [CrossRef]
28. Shahgholian, G. A Brief Overview of Microgrid Performance Improvements Using Distributed FACTS Devices. *J. Renew. Energy Environ.* **2023**, *10*, 43–58. [CrossRef]
29. Rahi, O.P.; Chandel, A.K. Refurbishment and uprating of hydro power plants—A literature review. *Renew. Sustain. Energy Rev.* **2015**, *48*, 726–737. [CrossRef]
30. Thaeer Hammid, A.; Awad, O.I.; Sulaiman, M.H.; Gunasekaran, S.S.; Mostafa, S.A.; Manoj Kumar, N.; Khalaf, B.A.; Al-Jawhar, Y.A.; Abdulhasan, R.A. A Review of Optimization Algorithms in Solving Hydro Generation Scheduling Problems. *Energies* **2020**, *13*, 2787. [CrossRef]
31. Bordin, C.; Skjelbred, H.I.; Kong, J.; Yang, Z. Machine Learning for Hydropower Scheduling: State of the Art and Future Research Directions. *Procedia Comput. Sci.* **2020**, *176*, 1659–1668. [CrossRef]
32. Kong, J.; Skjelbred, H.I.; Fosso, O.B. An overview on formulations and optimization methods for the unit-based short-term hydro scheduling problem. *Electr. Power Syst. Res.* **2020**, *178*, 106027. [CrossRef]
33. Villeneuve, Y.; Séguin, S.; Chehri, A. AI-Based Scheduling Models, Optimization, and Prediction for Hydropower Generation: Opportunities, Issues, and Future Directions. *Energies* **2023**, *16*, 3335. [CrossRef]
34. Bernardes, J., Jr.; Santos, M.; Abreu, T.; Prado, L., Jr.; Miranda, D.; Julio, R.; Viana, P.; Fonseca, M.; Bortoni, E.; Bastos, G.S. Hydropower Operation Optimization Using Machine Learning: A Systematic Review. *AI* **2022**, *3*, 78–99. [CrossRef]
35. Azad, A.S.; Rahaman, M.S.A.; Watada, J.; Vasant, P.; Gamez Vintaned, J.A. Optimization of the hydropower energy generation using Meta-Heuristic approaches: A review. *Energy Rep.* **2020**, *6*, 2230–2248. [CrossRef]
36. Kishor, N.; Saini, R.P.; Singh, S.P. A review on hydropower plant models and control. *Renew. Sustain. Energy Rev.* **2007**, *11*, 776–796. [CrossRef]
37. Xu, B.; Zhang, J.; Egusquiza, M.; Chen, D.; Li, F.; Behrens, P.; Egusquiza, E. A review of dynamic models and stability analysis for a hydro-turbine governing system. *Renew. Sustain. Energy Rev.* **2021**, *144*, 110880. [CrossRef]
38. Hunter-Rinderle, R.; Sioshansi, R. Data-Driven Modeling of Operating Characteristics of Hydroelectric Generating Units. *Curr. Sustain. Renew. Energy Rep.* **2021**, *8*, 199–206. [CrossRef]
39. Quaranta, E.; Bonjean, M.; Cuvato, D.; Nicolet, C.; Dreyer, M.; Gaspoz, A.; Rey-Mermet, S.; Boulicaut, B.; Pratalata, L.; Pinelli, M.; et al. Hydropower Case Study Collection: Innovative Low Head and Ecologically Improved Turbines, Hydropower in Existing Infrastructures, Hydropeaking Reduction, Digitalization and Governing Systems. *Sustainability* **2020**, *12*, 8873. [CrossRef]

40. Welte, T.; Foros, J.; Nielsen, M.; Adsten, M. MonitorX—Experience from a Norwegian-Swedish research project on industry 4.0 and digitalization applied to fault detection and maintenance of hydropower plants. In Proceedings of the Hydro 2018—Progress through Partnership, Gdansk, Poland, 15–17 October 2018. Available online: https://hdl.handle.net/11250/2576645 (accessed on 11 September 2023).
41. Giraldo, S.; la Rotta, D.; Nieto-Londoño, C.; Vásquez, R.E.; Escudero-Atehortúa, A. Digital Transformation of Energy Companies: A Colombian Case Study. *Energies* 2021, *14*, 2523. [CrossRef]
42. Bartczak, K. Digital Technology Platforms as an Innovative Tool for the Implementation of Renewable Energy Sources. *Energies* 2021, *14*, 7877. [CrossRef]
43. Pierleoni, P.; Marzorati, S.; Ladina, C.; Raggiunto, S.; Belli, A.; Palma, L.; Cattaneo, M.; Valenti, S. Performance Evaluation of a Low-Cost Sensing Unit for Seismic Applications: Field Testing During Seismic Events of 2016–2017 in Central Italy. *IEEE Sens. J.* 2018, *18*, 6644–6659. [CrossRef]
44. Pierleoni, P.; Belli, A.; Palma, L.; Pernini, L.; Valenti, S. An accurate device for real-time altitude estimation using data fusion algorithms. In Proceedings of the 2014 IEEE/ASME 10th International Conference on Mechatronic and Embedded Systems and Applications (MESA), Senigallia, Italy, 10–12 September 2014. [CrossRef]
45. Palma, L.; Pernini, L.; Belli, A.; Valenti, S.; Maurizi, L.; Pierleoni, P. IPv6 WSN solution for integration and interoperation between smart home and AAL systems. In Proceedings of the 2016 IEEE Sensors Applications Symposium (SAS), Catania, Italy, 20–22 April 2016. [CrossRef]
46. Alarcon, A.D. (Ed.) *The Digital Revolution of Hydropower in Latin American Countries*; Inter-American Development Bank: Washington, DC, USA, 2019.
47. Agostini, M.; Corbetti, C.; Ogbonna, D.; Stark, M. *Hydro'a Digital Generation. Transforming for the Future of Hydropower*; Accenture: Dublin, Ireland, 2020.
48. Ren, J.; Zhang, L.; Jin, L.; He, J.; Gao, Y. Digital Transformation of Hydropower Stations: Technical Route, Maturity Evaluation and Content Planning. In Proceedings of the 2022 IEEE 5th International Electrical and Energy Conference (CIEEC), Nangjing, China, 27–29 May 2022. [CrossRef]
49. Ristić, B.; Bozic, I. Digital technologies emergence in the contemporary hydropower plants operation. In Proceedings of the International Conference Power Plants, Belgrade, Serbia, 16–17 December 2021. Available online: https://machinery.mas.bg.ac.rs/handle/123456789/5970 (accessed on 30 August 2023).
50. Quaranta, E.; Aggidis, G.; Boes, R.M.; Comoglio, C.; De Michele, C.; Patro, E.R.; Georgievskaia, E.; Harby, A.; Kougias, I.; Muntean, S.; et al. Assessing the energy potential of modernizing the European hydropower fleet. *Energy Convers. Manag.* 2021, *246*, 114655. [CrossRef]
51. Corà, E.; Fry, J.J.; Bachhiesl, M.; Schleiss, A. *Hydropower Technologies: The State-of-the-Art*; Hydropower Europe: Brussels, Belgium, 2020. Available online: https://hydropower-europe.eu/uploads/news/media/The%20state%20of%20the%20art%20of%20hydropower%20industry-1600164483.pdf (accessed on 3 October 2023).
52. Kougias, I. *Hydropower Technology Development Report 2020*; EUR 30510 EN; Publications Office of the European Union: Luxembourg, 2020; ISBN 978-92-76-27285-4. [CrossRef]
53. Ramos, H.M.; Coronado-Hernández, O.E. IoT, machine learning and photogrammetry in small hydropower towards energy and digital transition: Potential energy and viability analyses. *J. Appl. Res. Technol. Eng.* 2023, *4*, 69–86. [CrossRef]
54. Quaranta, E.; Ramos, H.M.; Stein, U. Digitalisation of the European Water Sector to Foster the Green and Digital Transitions. *Water* 2023, *15*, 2785. [CrossRef]
55. Xing, L.; Sizov, G.; Gundersen, O.E. Digital Transformation in Renewable Energy: Use Cases and Experiences from a Nordic Power Producer. In *Digital Transformation in Norwegian Enterprises*; Mikalef, P., Parmiggiani, E., Eds.; Springer: Cham, Switzerland, 2022. [CrossRef]
56. Xing, W.; Tian, W. Research on the Key Technology of Hydropower's Large Data and Its Resources. In Proceedings of the 2015 Seventh International Conference on Measuring Technology and Mechatronics Automation, Nanchang, China, 13–14 June 2015. [CrossRef]
57. Zanoli, S.M.; Pepe, C.; Moscoloni, E.; Astolfi, G. Data Analysis and Modelling of Billets Features in Steel Industry. *Sensors* 2022, *22*, 7333. [CrossRef]
58. Chen, K.; He, J. Big-Data-Based Research on the Architecture Design of University Hydropower Intelligent Decision Service Platform. In Proceedings of the 2021 9th International Conference on Communications and Broadband Networking (ICCBN '21), Shanghai, China, 25–27 February 2021. [CrossRef]
59. Di Stefano, F.; Sanità, M.; Malinverni, E.S.; Doti, G. GEOMATIC TECHNOLOGIES TO VALORIZE HISTORICAL WATERMILLS. *Int. Arch. Photogramm. Remote Sens. Spatial Inf. Sci* 2023, *XLVIII-M-2-2023*, 511–518. [CrossRef]
60. Falchetta, G.; Kasamba, C.; Parkinson, S.C. Monitoring hydropower reliability in Malawi with satellite data and machine learning. *Environ. Res. Lett.* 2020, *15*, 014011. [CrossRef]
61. Hu, Y.; Jin, X.; Guo, Y. Big data analysis for the hydropower development potential of ASEAN-8 based on the hydropower digital planning model. *J. Renew. Sustain. Energy* 2018, *10*, 034502. [CrossRef]
62. Zhang, W.; Ding, Z.; Yang, G.; Xiong, Z. Research on intelligent construction of Hydropower Enterprises. *E3S Web Conf.* 2021, *276*, 01005. [CrossRef]

63. Eichhorn, M.; Scheftelowitz, M.; Reichmuth, M.; Lorenz, C.; Louca, K.; Schiffler, A.; Keuneke, R.; Bauschmann, M.; Ponitka, J.; Manske, D.; et al. Spatial Distribution of Wind Turbines, Photovoltaic Field Systems, Bioenergy, and River Hydro Power Plants in Germany. *Data* **2019**, *4*, 29. [CrossRef]
64. Peters, R.; Berlekamp, J.; Tockner, K.; Zarfl, C. RePP Africa—A georeferenced and curated database on existing and proposed wind, solar, and hydropower plants. *Sci. Data* **2023**, *10*, 16. [CrossRef]
65. Kreklow, J.; Tetzlaff, B.; Kuhnt, G.; Burkhard, B. A Rainfall Data Intercomparison Dataset of RADKLIM, RADOLAN, and Rain Gauge Data for Germany. *Data* **2019**, *4*, 118. [CrossRef]
66. JRC Hydro-Power Plants Database. Available online: https://github.com/energy-modelling-toolkit/hydro-power-database (accessed on 31 August 2023).
67. Xing, W.; Hongfu, T. Research on the Hydropower Science and Technology in the Era of Big Data Based on Data Mining. In Proceedings of the 2016 International Conference on Smart Grid and Electrical Automation (ICSGEA), Zhangjiajie, China, 11–12 August 2016. [CrossRef]
68. Garbea, R.; Scarlatache, F.; Grigoras, G.; Neagu, B.-C. Integration of Data Mining Techniques in SCADA System for Optimal Operation of Hydropower Plants. In Proceedings of the 2021 13th International Conference on Electronics, Computers and Artificial Intelligence (ECAI), Pitesti, Romania, 1–3 July 2021. [CrossRef]
69. Garbea, R.; Grigoras, G. Clustering-Using Data Mining-based Application to Identify the Hourly Loading Patterns of the Generation Units from the Hydropower Plants. In Proceedings of the 2022 International Conference and Exposition on Electrical and Power Engineering (EPE), Iasi, Romania, 20–22 October 2022. [CrossRef]
70. Lin, H.; Xie, S.; Tang, Z.; Xu, Y.; Wang, Y. Research on Data Cleaning Method for Dispatching and Operation of Cascade Hydropower Stations. In *WRE 2022: Proceedings of the 8th International Conference on Water Resource and Environment*; Weng, C.H., Ed.; Lecture Notes in Civil Engineering; Springer: Singapore, 2023; Volume 341, p. 341. [CrossRef]
71. Luo, W.; Xu, J.; Zhou, Z. Design of Data Classification and Classification Management System for Big Data of Hydropower Enterprises Based on Data Standards. *Mob. Inf. Syst.* **2022**, *2022*, 8103897. [CrossRef]
72. Lyu, Y.; Luo, Y.; Fei, W.; Zheng, B. Research on the Construction of Data Middle Platform for Smart Hydropower Station. In Proceedings of the 2021 2nd International Conference on Computer Engineering and Intelligent Control (ICCEIC), Chongqing, China, 12–14 November 2021. [CrossRef]
73. IEA. *Hydropower Data Explorer*; IEA: Paris, France, 2021. Available online: https://www.iea.org/data-and-statistics/data-tools/hydropower-data-explorer (accessed on 30 August 2023).
74. Vagnoni, E.; Gerini, F.; Cherkaoui, R.; Paolone, M. Digitalization in hydropower generation: Development and numerical validation of a model-based Smart Power Plant Supervisor. *IOP Conf. Ser. Earth Environ. Sci.* **2021**, *774*, 012107. [CrossRef]
75. Zanoli, S.M.; Pepe, C.; Astolfi, G.; Luzi, F. Reservoir Advanced Process Control for Hydroelectric Power Production. *Processes* **2023**, *11*, 300. [CrossRef]
76. Kumar, K.; Saini, R.P. Application of machine learning for hydropower plant silt data analysis. *Mater. Today Proc.* **2021**, *46*, 5575–5579. [CrossRef]
77. Zhao, Z.; Li, D.; She, J.; Zhang, H.; Zhou, Y.; Zhao, L. Construction and Application of Digital Twin Model of Hydropower Plant Based on Data-driven. In Proceedings of the 2021 3rd International Workshop on Artificial Intelligence and Education (WAIE), Xi'an, China, 19–21 November 2021. [CrossRef]
78. Cali, U.; Dimd, B.D.; Hajialigol, P.; Moazami, A.; Gourisetti, S.N.G.; Lobaccaro, G.; Aghaei, M. Digital Twins: Shaping the Future of Energy Systems and Smart Cities through Cybersecurity, Efficiency, and Sustainability. In Proceedings of the 2023 International Conference on Future Energy Solutions (FES), Vaasa, Finland, 12–14 June 2023. [CrossRef]
79. Vasiliev, Y.S.; Zegzhda, P.D.; Zegzhda, D.P. Providing security for automated process control systems at hydropower engineering facilities. *Therm. Eng.* **2016**, *63*, 948–956. [CrossRef]
80. Heluany, J.B.; Galvão, R. IEC 62443 Standard for Hydro Power Plants. *Energies* **2023**, *16*, 1452. [CrossRef]
81. Fekete, B.M.; Stakhiv, E.Z. Performance Indicators in the Water Resources Management Sector. In *The Global Water System in the Anthropocene*; Bhaduri, A., Bogardi, J., Leentvaar, J., Marx, S., Eds.; Springer Water: Cham, Switzerland, 2014. [CrossRef]
82. Murad, C.A.; Henrique de Andrade Melani, A.; de Carvalho Michalski, M.A.; Francisco Martha de Souza, G. Maintenance Management Optimization to Improve System Availability Based on Stochastic Block Diagram. In Proceedings of the 2021 Annual Reliability and Maintainability Symposium (RAMS), Orlando, FL, USA, 24–27 May 2021. [CrossRef]
83. Majumder, P.; Majumder, M.; Saha, A.K.; Nath, S. Selection of features for analysis of reliability of performance in hydropower plants: A multi-criteria decision making approach. *Environ. Dev. Sustain.* **2020**, *22*, 3239–3265. [CrossRef]
84. Sahimi, N.S.; Turan, F.M.; Johan, K. Development of Sustainability Assessment Framework in Hydropower sector. *IOP Conf. Ser. Mater. Sci. Eng* **2017**, *226*, 012048. [CrossRef]
85. Calabria, F.A.; Camanho, A.S.; Zanella, A. The use of composite indicators to evaluate the performance of Brazilian hydropower plants. *Int. Trans. Oper. Res.* **2016**, *25*, 1323–1343. [CrossRef]
86. Nwobi-Okoye, C.C.; Igboanugo, A.C. Performance evaluation of hydropower generation system using transfer function modelling. *Electr. Power Energy Syst.* **2012**, *43*, 245–254. [CrossRef]
87. Comoglio, C.; Castelluccio, S.; Fiore, S. Environmental reporting in the hydropower sector: Analysis of EMAS registered hydropower companies in Italy. *Front. Environ. Sci.* **2023**, *11*, 1178037. [CrossRef]

88. Liu, X.; Pan, H.; Zheng, X.; Zhang, X.; Lyu, Y.; Deng, S.; Guo, X. Integrated emergy and economic evaluation of 8 hydropower plants in Zagunao Basin, Southwest of China. *J. Clean. Prod.* **2022**, *353*, 131665. [CrossRef]
89. Yusri Syam Akil, Y.S.; Lateko, A.A.H.; Rahim, A. Hydroelectricity consumption, power losses and economic performance in Indonesia. *AIP Conf. Proc.* **2019**, *2097*, 030024. [CrossRef]
90. De Souza Machado, A.C.C.; Filho, G.L.T.; de Abreu, T.M.; Facchini, F.; da Silva, R.F.; Pinto, L.F.R. Use of Balanced Scorecard (BSC) Performance Indicators for Small-Scale Hydropower Project Attractiveness Analysis. *Energies* **2023**, *16*, 6615. [CrossRef]
91. Betti, A.; Crisostomi, E.; Paolinelli, G.; Piazzi, A.; Ruffini, F.; Tucci, M. Condition monitoring and predictive maintenance methodologies for hydropower plants equipment. *Renew. Energy* **2021**, *171*, 246–253. [CrossRef]
92. Tong, K.; Mao, H.; Wu, R.; Zhong, J.; Mao, H.; Huang, Z.; Li, X. Correlation transmissibility damage indicator for deterioration performance analysis of hydropower generator unit. *J. Vib. Control* **2023**, 10775463231154665. [CrossRef]
93. Da Silva, R.F.; de Andrade Melani, A.H.; de Carvalho Michalski, M.A.; Martha de Souza, G.F.; Nabeta, S.I.; Hiroyuki Hamaji, F. Defining Maintenance Performance Indicators for Asset Management Based on ISO 55000 and Balanced Scorecard: A Hydropower Plant Case Study. In Proceedings of the 30th European Safety and Reliability Conference and the 15th Probabilistic Safety Assessment and Management Conference, Venice, Italy, 1–5 November 2020. [CrossRef]
94. Zanoli, S.M.; Pepe, C.; Astolfi, G. Advanced Process Control Applications to Water Resources Systems: Two Industrial Case Studies. *IFAC-PapersOnLine* **2022**, *55*, 99–104. [CrossRef]
95. Zanoli, S.M.; Pepe, C.; Astolfi, G.; Cervigni, I. Model Predictive Control aimed at satisfying the production plan of a hydroelectric plant. In Proceedings of the 2022 IEEE 17th International Conference on Control & Automation (ICCA), Naples, Italy, 27–30 June 2022. [CrossRef]
96. Zanoli, S.M.; Pepe, C.; Astolfi, G.; Luzi, F. Model Predictive Control for Hydroelectric Power Plant Reservoirs. In Proceedings of the 2022 23rd International Carpathian Control Conference (ICCC), Sinaia, Romania, 29 May–1 June 2022. [CrossRef]
97. Maciejowski, J.M. *Predictive Control with Constraints*; Prentice-Hall, Pearson Education Limited: Harlow, UK, 2002.
98. Bemporad, A.; Morari, M.; Ricker, N.L. *Model Predictive Control Toolbox User's Guide*; MathWorks: Natick, MA, USA, 2015.
99. Rawlings, J.B.; Mayne, D.Q.; Diehl, M.M. *Model Predictive Control: Theory and Design*; Nob Hill Publishing: Madison, WI, USA, 2020. Available online: http://www.nobhillpublishing.com/mpc-paperback/index-mpc.html (accessed on 10 August 2023).
100. Chanda, N.; Chintalacheruvu, M.R.; Choudhary, K.A. Performance Appraisal of Ravi Shankar Sagar Project Using Comparative Indicators. In *Recent Advances in Civil Engineering. ICSTE 2023. Lecture Notes in Civil Engineering*; Swain, B.P., Dixit, U.S., Eds.; Springer: Singapore, 2024; Volume 431. [CrossRef]
101. Joshi, G.S.; Gupta, K. Performance Evaluation Model for Multipurpose Multireservoir System Operation. *Water Resour. Manag.* **2010**, *24*, 3051–3063. [CrossRef]
102. Afsharian Zadeh, N.; Mousavi, S.J.; Jahani, E.; Kim, J.H. Optimal Design and Operation of Hydraulically Coupled Hydropower Reservoirs System. *Procedia Eng.* **2016**, *154*, 1393–1400. [CrossRef]
103. Azizipour, M.; Sattari, A.; Afshar, M.H.; Goharian, E. Incorporating reliability into the optimal design of multi-hydropower systems: A cellular automata-based approach. *J. Hydrol.* **2022**, *604*, 127227. [CrossRef]
104. Afzali, R.; Mousavi, S.J.; Ghaheri, A. Reliability-Based Simulation-Optimization Model for Multireservoir Hydropower Systems Operations: Khersan Experience. *J. Water Resour. Plan. Manag.* **2008**, *134*, 24–33. [CrossRef]
105. Kuby, M.J.; Fagan, W.F.; ReVelle, C.S.; Graf, W.L. A multiobjective optimization model for dam removal: An example trading off salmon passage with hydropower and water storage in the Willamette basin. *Adv. Water Resour.* **2005**, *28*, 845–855. [CrossRef]
106. Bertoni, F.; Castelletti, A.; Giuliani, M.; Reed, P.M. Discovering Dependencies, Trade-Offs, and Robustness in Joint Dam Design and Operation: An Ex-Post Assessment of the Kariba Dam. *Earth's Future* **2019**, *7*, 1367–1390. [CrossRef]
107. Aslan, Y.; Arslan, O.; Yasar, C. A sensitivity analysis for the design of small-scale hydropower plant: Kayabogazi case study. *Renew. Energy* **2008**, *33*, 791–801. [CrossRef]
108. Fitzgerald, N.; Lacal Arántegui, R.; McKeogh, E.; Leahy, P. A GIS-based model to calculate the potential for transforming conventional hydropower schemes and non-hydro reservoirs to pumped hydropower scheme. *Energy* **2012**, *41*, 483–490. [CrossRef]
109. Bozorg Haddad, O.; Afshar, A.; Mariño, M.A. Design-Operation of Multi-Hydropower Reservoirs: HBMO Approach. *Water Resour. Manag.* **2008**, *22*, 1709–1722. [CrossRef]
110. Yazdi, J.; Moridi, A. Multi-Objective Differential Evolution for Design of Cascade Hydropower Reservoir Systems. *Water Resour. Manag.* **2018**, *32*, 4779–4791. [CrossRef]
111. Hatamkhani, A.; Shourian, M.; Moridi, A. Optimal Design and Operation of a Hydropower Reservoir Plant Using a WEAP-Based Simulation–Optimization Approach. *Water Resour. Manag.* **2021**, *35*, 1637–1652. [CrossRef]
112. Zahedi, R.; Eskandarpanah, R.; Akbari, M.; Rezaei, N.; Mazloumin, P.; Farahani, O.N. Development of a New Simulation Model for the Reservoir Hydropower Generation. *Water Resour Manag.* **2022**, *36*, 2241–2256. [CrossRef]
113. Hatamkhani, A.; Moridi, A.; Randhir, T.O. Sustainable planning of multipurpose hydropower reservoirs with environmental impacts in a simulation–optimization framework. *Hydrol. Res.* **2023**, *54*, 31–48. [CrossRef]
114. Hatamkhani, A.; Moridi, A.; Yazdi, J. A simulation—Optimization models for multi-reservoir hydropower systems design at watershed scale. *Renew. Energy* **2020**, *149*, 253–263. [CrossRef]
115. Hatamkhani, A.; Moridi, A.; Haghighi, A.T. Incorporating ecosystem services value into the optimal development of hydropower projects. *Renew. Energy* **2023**, *203*, 495–505. [CrossRef]

116. Haddad, O.B.; Ashofteh, P.S.; Rasoulzadeh-Gharibdousti, S.; Mariño, M.A. Optimization Model for Design-Operation of Pumped-Storage and Hydropower Systems. *J. Energy Eng.* **2013**, *140*, 04013016. [CrossRef]
117. Hariri-Ardebili, M.A.; Mahdavi, G.; Nuss, L.K.; Lall, U. The role of artificial intelligence and digital technologies in dam engineering: Narrative review and outlook. *Eng. Appl. Artif. Intell.* **2023**, *126*, 106813. [CrossRef]
118. Zhao, C.; Dong, J.; Zhou, Y.; Wu, H.; Hu, C. Dynamic Visualization of Dam Construction Process Based on Virtual Reality. In Proceedings of the 2009 International Conference on Information Technology and Computer Science, Kiev, Ukraine, 25–26 July 2009. [CrossRef]
119. Wang, L. Research on Dynamic Visual Simulation of Hydropower Project Construction Based on Virtual Reality. *Int. J. Sci. Eng. Appl.* **2023**, *12*, 91–93. [CrossRef]
120. Zhang, D.; Lin, J.; Peng, Q.; Wang, D.; Yang, T.; Sorooshian, S.; Liu, X.; Zhuang, J. Modeling and simulating of reservoir operation using the artificial neural network, support vector regression, deep learning algorithm. *J. Hydrol.* **2018**, *565*, 720–736. [CrossRef]
121. Feng, Z.; Niu, W.; Zhang, T.; Wang, M.; Yang, T. Deriving hydropower reservoir operation policy using data-driven artificial intelligence model based on pattern recognition and metaheuristic optimizer. *J. Hydrol.* **2023**, *624*, 129916. [CrossRef]
122. Lee, E.; Kam, J. Deciphering the black box of deep learning for multi-purpose dam operation modeling via explainable scenarios. *J. Hydrol.* **2023**, *626*, 130177. [CrossRef]
123. Yoshioka, H. Mathematical modeling and computation of a dam–reservoir system balancing environmental management and hydropower generation. *Energy Rep.* **2020**, *6*, 51–54. [CrossRef]
124. Saberian, M.; Mousavi, S.J.; Karray, F.; Ponnambalam, K. Cellular Automata-Based Optimization of Cascade Hydropower Systems Operations. In Proceedings of the 2019 IEEE 2nd International Conference on Renewable Energy and Power Engineering (REPE), Toronto, ON, Canada, 2–4 November 2019. [CrossRef]
125. Jahandideh-Tehrani, M.; Bozorg-Haddad, O.; Loáiciga, H.A. A review of applications of animal-inspired evolutionary algorithms in reservoir operation modelling. *Water Environ. J.* **2021**, *35*, 628–646. [CrossRef]
126. Shaw, A.R.; Sawyer, H.S.; LeBoeuf, E.J.; McDonald, M.P.; Hadjerioua, B. Hydropower optimization using artificial neural network surrogate models of a high-fidelity hydrodynamics and water quality Model. *Water Resour. Res.* **2017**, *53*, 9444–9461. [CrossRef]
127. Castillo-Botón, C.; Casillas-Pérez, D.; Casanova-Mateo, C.; Moreno-Saavedra, L.M.; Morales-Díaz, B.; Sanz-Justo, J.; Gutiérrez, P.A.; Salcedo-Sanz, S. Analysis and Prediction of Dammed Water Level in a Hydropower Reservoir Using Machine Learning and Persistence-Based Techniques. *Water* **2020**, *12*, 1528. [CrossRef]
128. Liu, X.; Zheng, X.; Wu, L.; Deng, S.; Pan, H.; Zou, J.; Zhang, X.; Luo, Y. Techno-ecological synergies of hydropower plants: Insights from GHG mitigation. *Sci. Total Environ.* **2022**, *853*, 158602. [CrossRef] [PubMed]
129. Zheng, T.; Qiang, M.; Chen, W.; Xia, B.; Wang, J. An externality evaluation model for hydropower projects: A case study of the Three Gorges Project. *Energy* **2016**, *108*, 74–85. [CrossRef]
130. Alexander, S.; Yang, G.; Addisu, G.; Block, P. Forecast-informed reservoir operations to guide hydropower and agriculture allocations in the Blue Nile basin, Ethiopia. *Int. J. Water Resour. Dev.* **2021**, *37*, 208–233. [CrossRef]
131. Séguin, S.; Audet, C.; Côté, P. Scenario-Tree Modeling for Stochastic Short-Term Hydropower Operations Planning. *J. Water Resour. Plan. Manag.* **2017**, *143*, 04017073. [CrossRef]
132. Xin-Yu, W.; Chun-Tian, C.; Jian-Jian, S.; Bin, l.; Sheng-Li, L.; Gang, L. A multi-objective short term hydropower scheduling model for peak shaving. *Int. J. Electr. Power Energy Syst.* **2015**, *68*, 278–293. [CrossRef]
133. Zanoli, S.M.; Cocchioni, F.; Pepe, C. Model Predictive Control with horizons online adaptation: A steel industry case study. In Proceedings of the 2018 European Control Conference (ECC), Limassol, Cyprus, 12–15 June 2018. [CrossRef]
134. Zanoli, S.M.; Pepe, C. A constraints softening decoupling strategy oriented to time delays handling with Model Predictive Control. In Proceedings of the 2016 American Control Conference (ACC), Boston, MA, USA, 6–8 July 2016. [CrossRef]
135. Zanoli, S.M.; Pepe, C.; Astolfi, G. Advanced Process Control of a cement plant grate cooler. In Proceedings of the 2022 26th International Conference on System Theory, Control and Computing (ICSTCC), Sinaia, Romania, 19–21 October 2022. [CrossRef]
136. Zanoli, S.M.; Pepe, C.; Rocchi, M.; Astolfi, G. Application of Advanced Process Control techniques for a cement rotary kiln. In Proceedings of the 2015 19th International Conference on System Theory, Control and Computing (ICSTCC), Cheile Gradistei, Romania, 14–16 October 2015. [CrossRef]
137. Ren, X.; Zhao, Y.; Hao, D.; Sun, Y.; Chen, S.; Gholinia, F. Predicting optimal hydropower generation with help optimal management of water resources by Developed Wildebeest Herd Optimization (DWHO). *Energy Rep.* **2021**, *7*, 968–980. [CrossRef]
138. Dehghani, M.; Riahi-Madvar, H.; Hooshyaripor, F.; Mosavi, A.; Shamshirband, S.; Zavadskas, E.K.; Chau, K.-w. Prediction of Hydropower Generation Using Grey Wolf Optimization Adaptive Neuro-Fuzzy Inference System. *Energies* **2019**, *12*, 289. [CrossRef]
139. Hammid, A.T.; Sulaiman, M.H.B.; Abdalla, A.N. Prediction of small hydropower plant power production in Himreen Lake dam (HLD) using artificial neural network. *Alex. Eng. J.* **2018**, *57*, 211–221. [CrossRef]
140. Pishgah Hadiyan, P.; Moeini, R.; Ehsanzadeh, E.; Karvanpour, M. Trend Analysis of Water Inflow Into the Dam Reservoirs Under Future Conditions Predicted By Dynamic NAR and NARX Models. *Water Resour. Manag.* **2022**, *36*, 2703–2723. [CrossRef]
141. Mainardi Fan, F.; Schwanenberg, D.; Collischonn, W.; Weerts, A. Verification of inflow into hydropower reservoirs using ensemble forecasts of the TIGGE database for large scale basins in Brazil. *J. Hydrol. Reg. Stud.* **2015**, *4*, 196–227. [CrossRef]

142. Anghileri, D.; Monhart, S.; Zhou, C.; Bogner, K.; Castelletti, A.; Burlando, P.; Zappa, M. The value of subseasonal hydrometeorological forecasts to hydropower operations: How much does preprocessing matter? *Water Resour. Res.* **2019**, *55*, 10159–10178. [CrossRef]
143. Fu, X.; Feng, Z.; Cao, H.; Feng, B.; Tan, Z.; Xu, Y.; Niu, W. Enhanced machine learning model via twin support vector regression for streamflow time series forecasting of hydropower reservoir. *Energy Rep.* **2023**, *10*, 2623–2639. [CrossRef]
144. Guo, Y.; Xu, Y.P.; Xie, J.; Chen, H.; Si, Y.; Liu, J. A weights combined model for middle and long-term streamflow forecasts and its value to hydropower maximization. *J. Hydrol.* **2021**, *602*, 126794. [CrossRef]
145. Boucher, M.A.; Ramos, M.H. Ensemble Streamflow Forecasts for Hydropower Systems. In *Handbook of Hydrometeorological Ensemble Forecasting*; Duan, Q., Pappenberger, F., Thielen, J., Wood, A., Cloke, H., Schaake, J., Eds.; Springer: Berlin/Heidelberg, Germany, 2018. [CrossRef]
146. Arsenault, R.; Côté, P. Analysis of the effects of biases in ensemble streamflow prediction (ESP) forecasts on electricity production in hydropower reservoir management. *Hydrol. Earth Syst. Sci.* **2019**, *23*, 2735–2750. [CrossRef]
147. Chen, J.; Brissette, F.P. Combining Stochastic Weather Generation and Ensemble Weather Forecasts for Short-Term Streamflow Prediction. *Water Resour. Manag.* **2015**, *29*, 3329–3342. [CrossRef]
148. Cassagnole, M.; Ramos, M.-H.; Zalachori, I.; Thirel, G.; Garçon, R.; Gailhard, J.; Ouillon, T. Impact of the quality of hydrological forecasts on the management and revenue of hydroelectric reservoirs—A conceptual approach. *Hydrol. Earth Syst. Sci.* **2021**, *25*, 1033–1052. [CrossRef]
149. Fan, F.M.; Schwanenberg, D.; Alvarado, R.; Assis dos Reis, A.; Collischonn, W.; Nauman, S. Performance of Deterministic and Probabilistic Hydrological Forecasts for the Short-Term Optimization of a Tropical Hydropower Reservoir. *Water Resour. Manag.* **2016**, *30*, 3609–3625. [CrossRef]
150. Wang, W.; Cheng, Q.; Chau, K.; Hu, H.; Zang, H.; Xu, D. An enhanced monthly runoff time series prediction using extreme learning machine optimized by salp swarm algorithm based on time varying filtering based empirical mode decomposition. *J. Hydrol.* **2023**, *620*, 129460. [CrossRef]
151. Wang, Y.; Liu, J.; Han, Y. Production capacity prediction of hydropower industries for energy optimization: Evidence based on novel extreme learning machine integrating Monte Carlo. *J. Clean. Prod.* **2020**, *272*, 122824. [CrossRef]
152. Feng, Z.K.; Niu, W.J.; Shi, P.F.; Yang, T. Adaptive Neural-Based Fuzzy Inference System and Cooperation Search Algorithm for Simulating and Predicting Discharge Time Series Under Hydropower Reservoir Operation. *Water Resour. Manag.* **2022**, *36*, 2795–2812. [CrossRef]
153. Peng, Y.; Xu, W.; Liu, B. Considering precipitation forecasts for real-time decision-making in hydropower operations. *Int. J. Water Resour. Dev.* **2017**, *33*, 987–1002. [CrossRef]
154. Zhang, X.; Peng, Y.; Xu, W.; Wang, B. An Optimal Operation Model for Hydropower Stations Considering Inflow Forecasts with Different Lead-Times. *Water Resour. Manag.* **2019**, *33*, 173–188. [CrossRef]
155. Wei, X.; Xun, Y. Evaluation of the effective forecast and decision horizon in optimal hydropower generation considering medium-range precipitation forecasts. *Water Supply* **2019**, *19*, 2147–2155. [CrossRef]
156. Kumar, A.; Yang, T.; Sharma, M.P. Long-term prediction of greenhouse gas risk to the Chinese hydropower reservoirs. *Sci. Total Environ.* **2019**, *646*, 300–308. [CrossRef] [PubMed]
157. Kumar, A.; Sharma, M.P. Assessment of risk of GHG emissions from Tehri hydropower reservoir, India. *Hum. Ecol. Risk Assess. Int. J.* **2016**, *22*, 71–85. [CrossRef]
158. Xu, X.; Lu, Y.; Vogel-Heuser, B.; Wang, L. Industry 4.0 and Industry 5.0—Inception, conception and perception. *J. Manuf. Syst.* **2021**, *61*, 530–535. [CrossRef]
159. Das, S.; Tanushree, P. A strategic outline of Industry 6.0: Exploring the Future (9 May 2022). Available online: https://ssrn.com/abstract=4104696 (accessed on 11 September 2023).

Disclaimer/Publisher's Note: The statements, opinions and data contained in all publications are solely those of the individual author(s) and contributor(s) and not of MDPI and/or the editor(s). MDPI and/or the editor(s) disclaim responsibility for any injury to people or property resulting from any ideas, methods, instructions or products referred to in the content.

MDPI
St. Alban-Anlage 66
4052 Basel
Switzerland
www.mdpi.com

Energies Editorial Office
E-mail: energies@mdpi.com
www.mdpi.com/journal/energies

Disclaimer/Publisher's Note: The statements, opinions and data contained in all publications are solely those of the individual author(s) and contributor(s) and not of MDPI and/or the editor(s). MDPI and/or the editor(s) disclaim responsibility for any injury to people or property resulting from any ideas, methods, instructions or products referred to in the content.

www.ingramcontent.com/pod-product-compliance
Lightning Source LLC
LaVergne TN
LVHW070502100526
838202LV00014B/1774